GSM, cdmaOne and 3G Systems

GSM, cdmaOne and 3G Systems

Raymond Steele
Chairman,
Multiple Access Communications Ltd,
Southampton, UK

Chin-Chun Lee
Professor of Commmunications,
Da-Yeh University, Chang-Hwa, Taiwan

Peter Gould
Director,
Multiple Access Communications Ltd,
Southampton, UK

JOHN WILEY & SONS, LTD
Chichester • Weinheim • New York • Brisbane • Singapore • Toronto

Other Wiley Editorial Offices

British Library Cataloguing in Publication Data

A catalogue record for this book is available from the British Library

ISBN 0 471 49185 3

Produced from PostScript files supplied by the authors
Printed and bound in Great Britain by Bookcraft (Bath) Ltd
This book is printed on acid-free paper responsibly manufactured from sustainable forestry,
in which at least two trees are planted for each one used for paper production.

Contents

Preface

This book is concerned with the description and analysis of the global second generation (2G) mobile radio systems: the Global System of Mobile Communications (GSM) and cdmaOne. A subsidiary goal is to examine how these two systems will evolve into third generation (3G) ones with their requirement to support multimedia mobile radio communications. The motivation for this book originated when we were asked to compare the capacities of GSM and, as cdmaOne was known then, IS-95. The multiple access method used by GSM is time division multiple access (TDMA), and this represented a significant change from the first generation (1G) analogue systems that operated with frequency division multiple access (FDMA). IS-95 had a more complex radio interface than GSM, employing code division multiple access (CDMA). Engineers at that time often held strong and somewhat uncompromising views regarding multiple access methods. We preferred CDMA from a spectral efficiency point of view, although that does not mean that CDMA should be deployed in preference to TDMA as there are many complex performance and economic factors to be considered when deciding on the type of system to select.

GSM was deployed before cdmaOne and is the market leader, entrenched in many parts of the world. Its success is due to numerous factors: its advanced backbone network, the introduction of subscriber identity modules (SIMs) that decoupled handsets from subscribers, its good security system, the low cost equipment due to open (i.e. public) interfaces, the relentless programme of evolution that has yielded substantial gains in spectral efficiency compared with the basic GSM system, and so on.

cdmaOne started as a radio interface. It was a bold step to use CDMA at a time when few thought CDMA could work in a cellular environment. But it did so, acquiring the necessary backbone network, and became a global standard offering tough competition to GSM. It is also worthy of note that Europe, which had designed and promoted GSM, has opted for wideband CDMA for its third generation (3G) networks.

Our cardinal objectives in this book are to present to the reader detailed descriptions of the basic GSM and cdmaOne systems, mainly from the radio interface point of view; as well as accompanying analyses. Our first chapter is designed to provide background material on TDMA, CDMA and cellular radio networks. The reader knowledgeable in mobile radio should omit reading this chapter and proceed directly to Chapter 2 which describes the basic GSM system. Chapter 3 provides an analysis of the performance of GSM networks. The same method of system description followed by a chapter dedicated to mathematical analysis is applied for cdmaOne in Chapters 4 and 5, respectively. The final chapter endeavours to describe how GSM is evolving to provide higher bit rate circuit-switched channels and packet transmissions that will have an ability to provide a range of multimedia services. The Universal Mobile Telecommunications System (UMTS) is then described, followed by a discussion of the evolution of cdmaOne to cdma2000. Both UMTS and cdma2000 are 3G systems.

The authors express their gratitude to those who have helped them in the gestation of this book. In particular they thank Dr Sheyam Lal Dhomeja for proof reading Chapters 2 and 5, Denise Harvey for her typing and helping to get the book to fruition, our colleagues at Multiple Access Communications Ltd for providing snipits of knowledge when required, and last, but not least, our loved ones for providing the support all authors need.

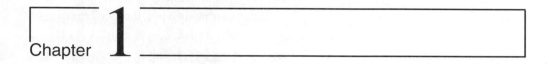

Chapter **1**

Introduction to Cellular Radio

This book is concerned with two digital mobile radio systems: the global system for mobile communications (GSM); and a code division multiple access (CDMA) system that was originally known as the American interim standard 95, or IS-95 and is now called cdmaOne [1–7]. While GSM was conceived and developed through the concerted efforts of regulators, operators and equipment manufacturers in Europe, cdmaOne owes its existence to one dynamic Californian company, Qualcomm Inc. The authors have been involved with both the pan-European mobile radio system, which became GSM, and the Qualcomm CDMA system for a number of years. The GSM system predates cdmaOne.

The two systems are very different. The radio interface of GSM relies on time division multiple access (TDMA), which means that its radio link is very different to that of cdmaOne. Also GSM is a complete network specification, from the subscriber unit through to the network gateway. Indeed its fixed network component is perhaps its most advanced feature [1,2]. cdmaOne, by contrast, has a more complex and advanced radio interface, and only later were fixed network issues addressed [3,7].

In the chapters to follow, the GSM and cdmaOne systems will be described and analysed while the final chapter deals with their evolution to third generation systems. This chapter is meant to provide background information on cellular radio [1–11]. *The reader who is well acquainted with the fundamentals of mobile radio communications should therefore bypass this chapter.*

For the reader who has elected to read this chapter we should state at the outset that our goal is to provide a clear exposition of the concepts of the subject rather than detailed analyses, which will follow in the later chapters. The first point to make is that a mobile radio network has a radio interface that enables a mobile station (MS) to communicate with the *fixed* part of the mobile network. Both components, the radio interface that facilitates user mobility, and the fixed network that enables the mobile to communicate with

1

other users via the public switch telephone network (PSTN) or the integrated services digital network (ISDN), are radically dissimilar and complex. This means that to have a good appreciation of mobile radio requires a wide knowledge that includes speech coding, channel coding, interleavers, radio modems, radio propagation, antennas, channel equalisation, RAKE receivers, diversity techniques, radio planning of cells, the significance of signal-to-interference ratios (*SIR*s), bit error rate (BER), teletraffic issues, protocol stacks, location databases, signalling systems, encryption, authentication procedures, switching, packetisation techniques, and so on. If some of these subjects are dealt with from a standing start in other chapters they will not be dealt with here. Neither will they be considered if they are outside the confines of this text. What we will consider here are topics that are needed when we come to our discussions of GSM and cdmaOne.

There are many ways of describing cellular radio, and the two most obvious are a bottom-up approach, or a top-down one. The former starts with the basic principles of radio propagation, to the concept of a cell, then clusters of cells to the radio links and multiple access methods, to setting-up, maintaining, and clearing-down of calls. The top-down approach is essentially the reverse process, starting with the big picture and ending up with radio propagation issues. We have opted for the bottom-up approach, building on concept after concept, until the overall concept of the network can be appreciated. Our starting point is the notion of a single cell.

1.1   A Single Cell

Consider a base station (BS) having an antenna located on a tower radiating an electromagnetic signal to a mobile station (MS). The received signal depends on many factors. The output port of the BS equipment delivers power at the appropriate radio frequency (RF) into the cable connected to the antenna. There are losses in the cable, e.g. a 40 W RF signal at the BS equipment may yield only 16 W of radiated power. The BS antenna is usually directional, which means that power is directed over a solid angle rather than over all angles. This means that compared with isotropic radiator there is a gain $G(\theta, \phi)$ of power in the θ and ϕ directions, where θ and ϕ are angles measured in the vertical and horizontal directions, respectively.

As the transmitted energy spreads out from the BS, the amount of power the MS antenna can receive diminishes [12, 13]. The mobile's antenna is usually located only one to two metres above the ground whereas the BS antenna may be at a height from several metres to in excess of a hundred metres. The heights of the antenna affect the path loss, i.e. the difference in the received signal power at the MS antenna compared with the BS transmitted power. The path loss (PL) is usually measured in decibels (dB). As an example, for the plane earth model there are two paths, a direct line-of-sight (LOS) path and a ground-reflected path. The expression for PL is

$$PL = \left(\frac{h_T h_R}{d^2}\right)^2 G_T(\theta,\phi) G_R(\theta,\phi), \tag{1.1}$$

where h_T and h_R are the heights of the transmitting and receiving antennas, respectively, d is the distance between the two antennas, and $G_T(\theta,\phi)$ and $G_R(\theta,\phi)$ are the gains of the transmitter and receiving antenna, respectively. When written in decibels, the path loss, L_p, becomes

$$\begin{aligned} L_p &= 10\log_{10} PL \\ &= 20\log_{10} h_T + 20\log_{10} h_R - 40\log_{10} d \\ &\quad + 10\log_{10} G_T(\theta,\phi) + 10\log_{10} G_R(\theta,\phi). \end{aligned} \tag{1.2}$$

This equation is only valid when

$$d > \frac{2\pi h_T h_R}{\lambda}, \tag{1.3}$$

where λ is the wavelength of the radiated wave.

The plane earth model is useful but may deviate significantly from reality. In the plane earth model, L_p decreases at 40 dB per decade increase in distance, i.e. if the distance increases by 10 times, the path loss will increase by 40 dB. This rate is often used in practical situations, although measurements show it may be closer to 35 dB per decade. If the transmitted power is sufficiently high a MS will often travel beyond the LOS of the BS antenna. When a mobile goes behind a large building the average received power will decrease and when it emerges from the building that casts the electromagnetic shadow, the average received power will rise. The fading due to large obstacles that produce electromagnetic shadows is called *shadow fading*. As a result of this fading effect, as the MS travels away from the BS the received power at the MS and the BS is subjected to considerable variations. These variations due to shadowing effects can be represented by a log-normal distribution of a shadow fading random variable ζ. Specifically we introduce this variable into Equation (1.2) to give

$$\begin{aligned} L_p &= 20\log_{10} h_T + 20\log_{10} h_R - 40\log_{10} d + \zeta \\ &\quad + 10\log_{10} G_T(\theta,\phi) + 10\log_{10} G_R(\theta,\phi), \end{aligned} \tag{1.4}$$

where ζ is measured in decibels and may be positive or negative.

In this book we will often use the expression for received signal power as

$$S(\text{dB}) = 10\log_{10} P - 10n\log_{10} d + \zeta \tag{1.5}$$

or, when not in decibels, the expression becomes

$$S = Pd^{-n}10^{\zeta/10}, \tag{1.6}$$

where P is the transmitted power from the BS and n is called the exponent of the PL. Observe that when we employ Equations (1.5) or (1.6), the terms relating to antenna heights and antenna gains are absent. This is because we often ignore the effects associated with the antennas on the path loss when we are concerned with signal-to-interference ratios (*SIRs*) since these parameters tend to cancel out on the signal and interference paths. Equation (1.6) is used extensively in Chapters 3 and 5.

The MS is not only subjected to shadow fading, but also to small scale fading, i.e. due to the received signal changing in amplitude and phase as a consequence of a small change in the spatial separation (e.g. fraction of a wavelength) between the MS and its BS [4]. This occurs because the MS is travelling through an electromagnetic field, receiving more than one version of the same transmitted signal that travelled via different paths. Each path results in a component of the received signal that has a specific attenuation and phase orientation. The received signal at the MS is therefore the vector sum of all these multipath signals. The vector sum may be large at one instant and a small movement of the MS may result in the multipath signal being very small. This variation often takes place over a distance of half a wavelength which is only $(3 \times 10^8)/(2 \times 10^9) = 15$ cm for a 1 GHz radio frequency carrier.

If the received paths are close together in time, we may represent the channel impulse function by a single delta function whose amplitude is Rayleigh distributed while its phase has a uniform distribution. The Fourier transform of a delta function is a flat spectrum. Since the weighting of the delta function varies due to the fading, the magnitude of the flat spectrum changes, and the condition is known as *flat fading*. This means all the frequencies in the received signal fade together and by the same amount.

Often we have a path arriving in the vicinity of the MS and subjected to local scattering producing a single delta function that is Rayleigh distributed. Then another ray arrives yielding another delta function that is also Rayleigh distributed. This process of each received ray causing a group of scattered rays that can be represented by a Rayleigh distributed delta function yields a channel impulse response that is itself made up of a number of impulses or delta functions at epochs $0, \tau_1, \tau_2, \ldots$, as shown in Figure 1.1. Since each delta function is fading independently the spectrum of the radio channel no longer fades uniformly for all frequencies. This type of fading is called *frequency selective fading*, which means that in the time domain the depths of the fades are, in general, much less than for flat fading. In the latter case the fading can be very deep, typically up to 40 dB, and this may cause bursts of symbol errors. As a consequence, having a wideband channel means that the signal is less likely to drop below the receiver sensitivity for a given transmitted power compared with a narrow band channel. However, the wideband channel has a wider impulse response, and since the received signal is the convolution of the transmitted signal with the impulse response of the radio channel, one data symbol is smeared into other symbols. This effect, called intersymbol interference (ISI), requires the receiver to un-smear the symbols. This is achieved using a channel equaliser in GSM and a RAKE receiver in cdmaOne. We will return to channel equalisation and RAKE receivers in some detail in later sections.

As a MS travels away from the BS, the received signal at the MS decreases as the path loss increases. The received signal will also exhibit large scale (shadowing) fading and small scale fading. Figure 1.2 shows an example of the variations in the received signal level (in dBs) as the MS travels. The dotted line represents the change in received signal level due to shadow fading. The rapid changes in the received signal level are the consequence of small scale fading, which for a particular carrier frequency depends on the MS speed. The faster the MS travels, the more rapid is the fading. A stationary MS may be in a deep fade. Fortunately the effect of small scale fading can be effectively combatted in modern digital mobile radio systems. Shadow fading and path loss is another matter.

Having passed through the radio channel, the RF signal transmitted by the BS will arrive at the MS antenna. This will usually connect directly into the receiver input but, unlike the BS, there are no cable losses. The antenna will be omni-directional whereby it is able to capture signal energy equally from all directions in the horizontal plane. In the case of a handheld MS, the signal may be attenuated by the user's body before arriving at the antenna, and network operators generally include a margin in their planning procedures to account for body loss.

As the MS travels there is a change in the frequency of the received carrier on each path due to the *Doppler effect*. For a MS travelling in a direction making an angle α_i with respect to a signal received on the ith path, the carrier frequency is changed from f_c to

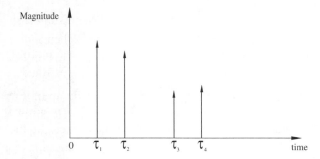

Figure 1.1: Magnitude of wideband channel impulse response, measured from the arrival of the first path.

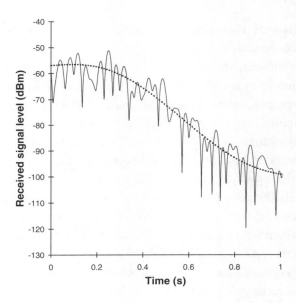

Figure 1.2: Combined shadow and fast fading.

$f_c + (v/\lambda)\cos\alpha_i$, where v is the speed of the MS and λ is the wavelength of the carrier $(= 3 \times 10^8/f_c$ m). Therefore, not only does each path in the received signal experience a different attenuation and phase shift, it is also subjected to a change in carrier frequency that can be positive or negative depending on α_i. The Doppler power spectral density (PSD) is parabolic about the carrier to frequencies $f_c \pm (v/\lambda)$ when the probability density function (PDF) of α_i is uniform, i.e. rays arrive at the MS from all directions with equal probability. In general there are only a few rays and their direction is often restricted, e.g. by local buildings. In this case the Doppler spectrum will be non-monotonic and rapidly changing. Fast changes in the Doppler spectrum manifest themselves as fast changes in the radio channel impulse response. Again, mobile radio equipment is well able to combat Doppler effects, unless the MS speed is excessive, e.g. in very high-speed trains.

So far we have considered a mobile travelling away from a BS and the received signal level decreasing with increasing BS to MS distance. The MS continues its travels with its receiver combating the fast fading, Doppler effects and channel dispersion due to its good design. The MS has a noise floor, which is reached when the mobile has travelled sufficiently far from the BS such that the receiver noise dominates the received signal level and the receiver behaves as if no signal is being received. Before this extreme condition is reached there is a received signal threshold known as the *receiver sensitivity*. When the received signal level is above this level, the bit error rate (BER) is acceptably low. Conversely, when the received signal level drops below the receiver sensitivity, the MS is no longer able to receive signals of an acceptable quality from the BS. The point in space at which this threshold occurs represents a boundary point for the down-link or forward link, i.e. the transmissions from the BS to the MS.

What about the up-link or reverse link, i.e. the transmission from the MS to the BS? The two links are never the same. They are similar in GSM and radically different in cdmaOne. The MS transmitter operates at significantly lower power levels than the BS and so the maximum radiated power levels are lower than those at the BS. The BS is able to compensate for the MS deficiencies by being able to operate at a lower receiver sensitivity and by employing techniques such as space diversity to enhance the received signal from the MS. It is important to note that the signal characteristics that we have already discussed in relation to the down-link (i.e. path loss, fast and slow fading, Doppler shift and ISI) will also be present in the received up-link signal. To simplify our discussion, we will assume that our boundary point is the same for either link, unless specifically stated.

If the MS takes a number of different routes away from the BS and on each route notes the location where the received signal goes below the receiver sensitivity, then by joining up these location points on a map we will form a contour around the BS. A stylised arbitrary irregularly shaped contour is shown in Figure 1.3. The area enclosed within the boundary is called a *cell*.

1.2 Multiple Cells

The dimensions of a cell are limited by the transmitter and receiver performances, the path loss, shadow fading and other factors described in the previous section. If we are going to cover wide areas we will need to tessellate cells, and switch a MS between BSs as it roams throughout the network. If hundreds or thousands of cells are required, then some cells must operate with the same carrier frequencies. This phenomenon is called *frequency reuse*.

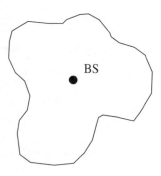

Figure 1.3: A single cell.

Let us consider the situation where each radio carrier supports N traffic channels, and the spacing between adjacent carriers is B_c Hz. Consequently, a traffic channel occupies an equivalent bandwidth of $B = B_c/N$ Hz. Suppose the spectrum regulator assigns W Hz for the up-link transmissions and W Hz for the down-link transmissions. The number of carriers for each down-link is approximately W/B_c. If we are going to reuse carriers in other cells we must ensure that each receiver can operate with an SIR that will give a sufficiently low BER. Let us for the moment consider that the only form of interference is from either users or BSs in other cells that are using the same traffic channel as a particular mobile in the zeroth cell, say. This interference is called *co-channel interference* or *intercellular interference*. To ensure that the interference is sufficiently low compared with the required signal power S (i.e. the *SIR* is sufficiently high) the interfering cells must be spaced sufficiently far apart. This may mean that other cells must be spaced between the zeroth cell and the interfering cells. Each cell is given a different channel set until all the bandwidth W is used. If this means M cells consume the bandwidth W, then we have M contiguous cells that form a *cluster* of cells. We now form another cluster of M cells and tessellate it with the first cluster. In each cluster all the channels are used, and the clusters are arranged such that two cells that use the same channel set are spaced as far apart as possible. Figure 1.4 shows two four-cell clusters where the cells marked A, B, C and D in each cluster, respectively, use the same channel sets.

The number of cells in the cluster, M, is called the *reuse factor*. The value of M depends on the *SIR*. If, for an acceptable BER, the SIR is required to be high, then we must have many cells in the cluster in order to space the 'reuse cells' sufficiently far apart such that the interference is low enough to satisfy the minimum *SIR* requirement. We will see that GSM requires $M \geq 3$, while cdmaOne can operate with $M = 1$.

Why is a low cluster size good? By operating with a smaller number of cells in a cluster the number of channels per cell, equal to $(N/M)(W/B_c)$, is high, since M is low. The carried

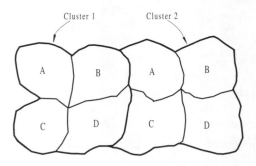

Figure 1.4: Two tesselated four-cell clusters.

traffic in Erlangs for a given blocking probability has a non-linear relationship with the number of channels per cell such that for more channels there is a disproportionate increase in the traffic that may be supported. We now observe an important aspect of *cellular* radio, i.e. for a mobile radio system that employs clusters of cells. If the radio link equipment is capable of operating with a low SIR, the cluster size becomes small and the carried traffic high. Another important point to note is that a cell becomes smaller in the presence of cochannel interference. By this we mean that the area around the cell site where the SIR is high enough to yield a sufficiently low bit error rate (BER) is decreased due to the presence of the interfering cells. This is illustrated in Figure 1.5. We also note that as the levels of interference power alter, so does the SIR, and so does the effective cell boundary for an acceptable BER. The cell boundaries shown in Figure 1.5 relate to a specific BER. For a higher BER, the cell size increases and vice versa. It is important to avoid the simple notion that a cell has a fixed area. It is better to think of it as *breathing*, i.e. changing its size as the traffic conditions within the network vary. Cell breathing is a feature of both GSM and cdmaOne, although it is more acute in the CDMA system. For analysis reasons we generally consider fixed cells and often worse case conditions.

Newcomers to cellular radio often consider spectral efficiency in terms of the number of channels, N, a carrier can support in a given bandwidth. This notion is related to modulation efficiency in terms of bits per second per Hertz of RF bandwidth. Since cellular radio must operate in an interference-limited environment, the crucial factor is not the modulation efficiency. For example, employing quadrature amplitude modulation (QAM), where each symbol carries multiple bits, gives a high modulation efficiency [10, 14]. However, QAM requires a high SIR value and hence large cluster sizes, resulting in low values of carried traffic per cell site, for a given bandwidth allocation. The choice of modulation and multiple access scheme is complex and will be addressed at a later stage. What we must note is that, given a modulation and multiple access scheme resulting in a cluster size of M, the number of users on the network is greatly increased if the cells, and thereby the clusters, are small. This is because each cluster carries a traffic of MA_c Erlangs, where A_c is the carried traffic at each BS, and if a cluster occupies an area S_c then the traffic carried per km^2 is MA_c/S_c Erlangs/km^2 for a bandwidth W. Using small cells, often called *microcells*, means S_c is small and the traffic density that may be supported is high.

1.2.1 Hexagonal cells

These types of cells are conceptual. The cell site is located at the centre of each hexagon, and the hexagonal cells are tessellated to form clusters [15]. Although these cells are fictitious, they are often used for comparing the performances of different cellular systems. Figure 1.6 shows clusters of tessellated hexagonal cells. Observe that for hexagonal cells there are always six near cochannel cells, irrespective of the cluster size. This is because

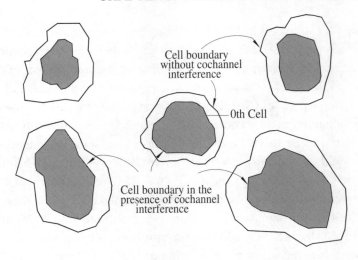

Figure 1.5: The zeroth cell and nearest cells using the same channel sets. The shaded areas are the cells after co-channel interference is introduced.

a hexagon has six sides. Figure 1.7 shows two co-channel cells shaded. From the figure,
$h = j \sin 60° = j(\sqrt{3}/2)$,

$$\sin \psi = \frac{j(\sqrt{3}/2)}{D}, \tag{1.7}$$

$$\cos \psi = \frac{i + (j/2)}{D}, \tag{1.8}$$

and as

$$\sin^2 \psi + \cos^2 \psi = 1, \tag{1.9}$$

$$D = (i^2 + ij + j^2)^{\frac{1}{2}}, \tag{1.10}$$

where D is the distance between cochannel BSs. From Figure 1.8 the distance between two cell sites is

$$2\mu = \sqrt{3}R, \tag{1.11}$$

where R is the distance from the centre of a cell to its apex, and from Figure 1.7,

$$i = l\, 2\mu \tag{1.12}$$

and

$$j = m\, 2\mu \tag{1.13}$$

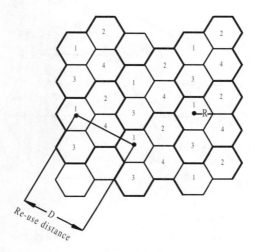

Figure 1.6: Hexagonal cells arranged into four-cell clusters.

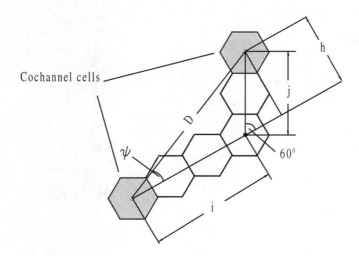

Figure 1.7: Part of a pattern of tesselated hexagonal cells showing two co-channel cells spaced by a distance D.

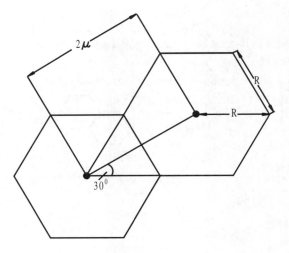

Figure 1.8: Distance between two cell sites.

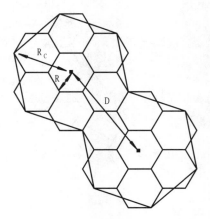

Figure 1.9: Approximation of a cluster of hexagonal cells by a single large hexagon.

where l and m are integers. Consequently, the distance between cochannel sites is

$$D = 2\mu \left(l^2 + lm + m^2 \right)^{\frac{1}{2}}. \tag{1.14}$$

A cluster of hexagonal cells can be approximated by a large hexagon of dimension R_c as shown in Figure 1.9. The number of cells, M, in this cluster is the ratio of the area of the cluster to the area of the cell, which is equal to the ratio of the distance squared between the centres of the clusters to the distance squared between the centres of adjacent hexagonal cells, i.e.

$$M = \frac{D^2}{(2\mu)^2} = l^2 + lm + m^2 \tag{1.15}$$

or, with the aid of Equation (1.11), we get the useful ratio used in later chapters:

$$\frac{D}{R} = \sqrt{3M}. \tag{1.16}$$

1.2.2 Sectorisation

The *SIR* can be increased by replacing the omnidirectional BS antennas with directional ones. Typically, directional antenna radiation patterns span $120°$ in the horizontal plane with the direction of the maximum radiation of each antenna spaced by $120°$ as shown in Figure 1.10. Each antenna radiation pattern is slightly in excess of $120°$, creating overlapping regions where each antenna is able to communicate with an MS. These regions are important as they facilitate an MS travelling between sectors to be switched from one cell site sectorised antenna to another, a process called *intrasector handover* or *hand off*. In Figure 1.10 we also show small back lobes in the antenna pattern. These cannot be avoided with practical antennas, and in general they do not create significant interference between sectors.

To simplify analysis we consider idealised antenna patterns that span a fixed number of degrees exactly, and ignore overlapping areas. Figure 1.11 shows a three-cell cluster arrangement with three sectors per cell, whereas Figure 1.12 is a four-cell cluster with again three sectors per cell but with different shaped sectors. As seen from the zeroth cell site there are only two sectors that cause significant interference. From Equation (1.14), $D = 2\sqrt{3}\mu$ and 4μ for the three- and four-cell per cluster arrangements, respectively. For an unsectorised arrangement, there are six significant interferers located at the same distances of $2\sqrt{3}\mu$ and 4μ. Consequently the interference is decreased by a factor of three.

While sectorisation does significantly increase the *SIR*s, it often decreases the carried traffic in time division multiple access (TDMA) and frequency division multiple access (FDMA) systems. Dividing the number of channels, N, at an omnidirectional cell site into three groups while maintaining the same probability of a cell being blocked means that the

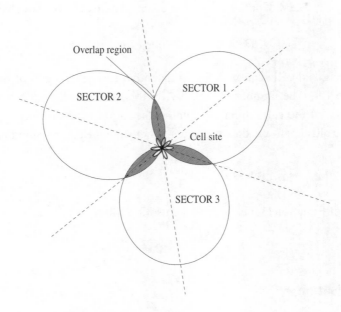

Figure 1.10: Antenna patterns for a cell site having three 120° sectors.

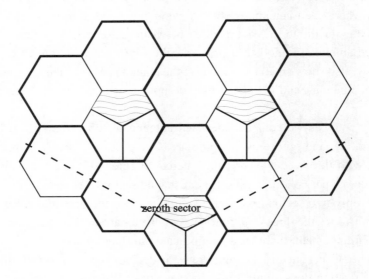

Figure 1.11: Up-link: three-cell cluster with three sectors per cell. The two most significant sectors are shown shaded.

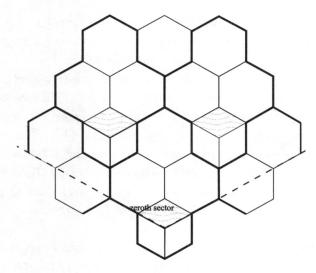

Figure 1.12: Up-link: four-cell clusters with three sectors per cell. The two most significant interfering sectors are shown shaded.

traffic is $3A_s$, where A_s is the traffic carried by a sector and $A_{omni} > 3A_s$, where A_{omni} is
the traffic carried by the omnidirectional site. The reason is that the traffic in Erlangs is
non-linearly related to the number of channels, and as each sector only has $N/3$ channels,
then each sector carries less than a third of A_{omni}. This effect is termed *trunking efficiency*.
However, a high SIR means good link quality. In some cases the use of sectorisation allows
the overall cluster size to be reduced without reducing the SIR beyond acceptable levels.
This will mean that more channels will be available at each BS and this will go some way
to offsetting the capacity reduction resulting from the loss in trunking efficiency.

In CDMA systems the situation is very different. The same channels may be reused in
each sector and there will be no trunking efficiency loss. In a system with perfect sectorisation the increase in capacity at a cell site will be equal to the number of sectors, i.e. a
three-fold increase for three sectors. In practice, interference caused by overlapping antenna
patterns and side and back lobes reduces this gain to around 80% of the ideal case.

1.3   The TDMA Radio Interface

1.3.1   Multiple access procedure for TDMA

In mobile radio communications, multiple users access the allotted radio spectrum in order
to communicate, via the fixed component of the mobile network, with another user in the
PSTN/ISDN or in its own or other mobile networks. There are different multiple access

methods but the basic one is FDMA in which each user is assigned a sub-band of the spectrum for the duration of the call. The sub-bands that support a traffic channel are arranged to be contiguous, as shown in Figure 1.13. Note that user k transmits on frequency f_{uk} on the up-link, so-called because it is from a mobile at a low elevation to a BS antenna at a higher elevation. The up-link is also referred to as the *reverse link*. The forward or down-link is used for transmissions from the BS to the MS, and therefore the MS receives on frequency f_{dk}. It is usual that each user has a pair of up-link and down-link channels that are always spaced apart by a frequency f_{dup}. Transmitting and receiving at the same time but on different frequencies is called frequency division duplexing (FDD). To assist the duplexer in protecting the strong transmitted signal from affecting the weak received signal, f_{dup} is sufficiently large to ensure that the transmitted energy at f_{uk} is very low at the received carrier frequency f_{dk}.

GSM uses FDMA and FDD, but instead of having one channel per FDMA carrier it has eight channels. The eight channels are arranged in a TDMA time frame. This frame has eight equal duration time slots, each slot carrying one channel. We will explain in Chapter 2 that these channels may be carrying traffic or signalling information in a complex manner. Suffice to say here that the frame endlessly repeats and if an MS has been assigned Slot-3, say, then it only transmits for the duration of Slot-3. This means that its traffic data, e.g. coded speech, which is being generated continuously and acquired over a complete frame, must be speeded up for transmission. For this reason the term *traffic burst* is used when it is the turn of the MS to transmit. Figure 1.14 shows the frame structure of an eight-slot TDMA carrier. The MS transmits nothing for seven slots and then, when it is its turn to transmit, its traffic burst modulates the carrier. Because the bit rate is much higher than in the equivalent FDMA system, the bandwidth of the transmitted signal is wider. Figure 1.15 shows the channel occupancy of single FDMA and TDMA channels.

There is another TDMA frame for the down-link. The MS will receive on Slot-3 because it transmits on Slot-3. However, in the case of GSM, the two TDMA frames (i.e. on the up-link and down-link) are offset by three time slots. This means that, after receiving its down-link burst, the MS has two time slots in which to retune to the corresponding up-link frequency in time to transmit its up-link burst. This offset means that the MS is not required to transmit and receive simultaneously and, as a consequence, it does not need a duplexer. The BS does require a duplexer. It has to transmit to up to eight MSs on its down-link while receiving up to eight signals from MSs on its up-link.

Suppose an operator is allotted 5 MHz of spectrum for the up-link and 5 MHz for the down-link. One carrier supports one eight-slot time frame and we will see later that, in the GSM system, this requires a channel occupancy of 200 kHz. By using five carriers contiguously spaced in frequency by 200 kHz, we can have five TDMA carriers per MHz, corresponding to 40 channels. For the 5 MHz band we have 24 carriers (allowing for a

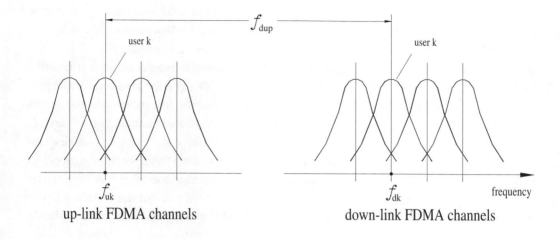

up-link FDMA channels   down-link FDMA channels

Figure 1.13: FDMA radio channels for FDD operation.

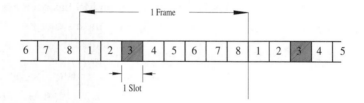

Figure 1.14: Transmitter TDMA framing structure.

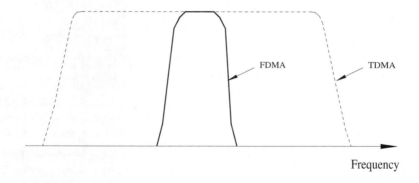

Figure 1.15: Channel occupancy of a single FDMA and TDMA channel (note that eight users can be accommodated per TDMA carrier).

guard band), and it is these carriers that are deployed in our cellular structure. Thus, if there are four cells per cluster and three sectors per cell, we may have two carriers or 16 channels per sector. Adjacent carrier frequencies must not be used at the same site as they will interfere with each other. This is known as *adjacent channel interference*. Planning which carriers to use in each sector is complex, and is known as *frequency planning*. Care must be exercised to maximise both the signal-to-cochannel interference ratios and the signal-to-adjacent channel interference ratios. Let us go one step farther. Given that we have 16 channels per sector, and that one is used for signalling, then with 15 traffic channels and a blocking probability P_n for new calls of 2% the carried traffic will be approximately 9 Erlangs (found from Erlang-B traffic capacity tables). Note that for a P_n of 2% the carried traffic is 98% of the offered traffic. If each user speaks on average for a total of 2 minutes in each hour, then the offered traffic is $2/60 = 33$ milliErlangs. As 9 Erlangs are available, the number of users that can be supported per sector is 270 users, or 3240 users per cluster. Thus we see that by using TDMA signals to modulate carriers in an FDMA mode, many users can be accommodated. This example can be easily modified to suit many different scenarios.

1.3.2   The TDMA radio link

In GSM, the up-link and down-link are essentially the same (this is not so in IS−95). While an MS only transmits on one slot in a frame, the BS may be transmitting on all of its slots. Similarly a BS receiver may need to receive on all of its slots, while an MS only receives on one slot on the down-link frame. The MS uses some of its slots to monitor the signal strength from nearby BSs, as described in detail in Chapter 2. For the purposes of exposition, this introductory chapter considers the basic TDMA link between a transmitter and a receiver. We will not consider the multiplexing or demultiplexing of other users. Figure 1.16 shows a simplified transmitter and receiver mobile link. We note that this could represent either the up-link (i.e. the mobile transmitter and base station receiver) or the down-link (i.e. the base station transmitter and mobile receiver). The transmitter consists of a speech encoder that digitises the speech signal from the microphone. Both GSM and cdmaOne use the analysis-by-synthesis type of codecs. The GSM codec is a regular pulse excited linear predictive codec (RPE-LTP) that generates bits at a fixed rate. A variable rate code excited linear predictive codec (CELP) is used in cdmaOne. Since our analysis of both GSM and cdmaOne presupposes that the speech is in a coded format, we advise the interested reader to read Chapters 3 and 8 in Reference [2] which deal with analysis-by-synthesis speech coding.

Both GSM and cdmaOne employ convolutional coding which accepts the digitised speech and essentially adds redundancy bits prior to transmission in order that the convolutional decoder can correct some of the bit errors that occurred as a result of transmission over

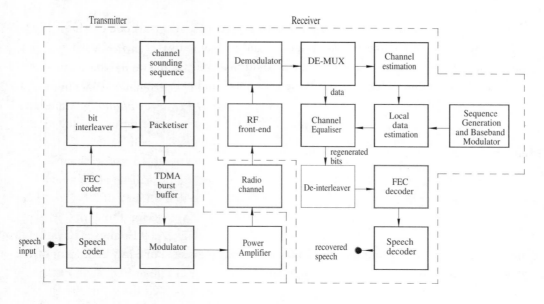

Figure 1.16: The basic TDMA mobile radio link.

the mobile radio channel. It is therefore appropriate that we say a few words concerning convolutional coding, but for a more comprehensive discourse consult Reference [2].

Let us start by saying that mobile radio channels are not benign like optical channels or even copper wire links in the PSTN. Mobile radio channels often cause high error rates unless considerable counter measures are deployed. We have mentioned that as the mobile travels away from the BS there is an increasing path loss, there is shadow fading, fast fading, dispersion effects, receiver noise, co-channel interference and adjacent channel interference. All of these factors may impair the received signal, and cause bits to be erroneously regenerated at the receiver. A host of measures are therefore employed to decrease the probability of bit errors, and if bit errors do occur, the role of forward error correction (FEC) codes is to correct as many of them as the power of the code will allow. In FEC coding the coder generally takes k input message bits at a time and maps them into n-bit code words. The amount of redundancy introduced by the coder is measured by the ratio n/k, and the inverse of it, namely k/n, is defined as the *coding rate*. The redundancy bits, $n - k$, are used to increase the relative Hamming distance, which is the number of different symbols between two code words or coded symbol sequences. An FEC decoder is able to provide error correction, although this is limited by the Hamming distance provided.

We are interested in convolutional codes as they are used by both cdmaOne and GSM. A convolutional coder accepts the latest k-bit and the previous $(K - 1)k$-bit inputs to generate an n-bit code word, where K indicates the number of k-bit inputs required to produce a code word, and is referred to as the *constraint length* [2]. A convolutional code can basically be defined by the three parameters, n, k, and K, and is denoted by $cc(n,k,K)$, where k/n is called the coding rate. An example of a convolutional coder is the $cc(2,1,3)$ coder shown in Figure 1.17. In this coder, the code word generator can be described by two vectors

$$g_1 = [101] \qquad \text{and} \qquad g_2 = [111] \tag{1.17}$$

and the corresponding code word $[c_2, c_1]$, where

$$c_1 = \begin{bmatrix} b_2 \\ b_1 \\ b_0 \end{bmatrix} g_1 \qquad \text{and} \qquad c_2 = \begin{bmatrix} b_2 \\ b_1 \\ b_0 \end{bmatrix} g_2, \tag{1.18}$$

and where b_0, b_1, and b_2 are the three bits in the shift register. An equivalent form of the generator vectors is two generator polynomials

$$g_1(z) = 1 + z^2 \qquad \text{and} \qquad g_2(z) = 1 + z + z^2, \tag{1.19}$$

where 1 denotes the present input bit, and z and z^2 represent the previous input bits having one and two clock period delays, respectively, and $+$ is modulo 2 addition. The convolutional coder is a finite-state machine which can be described by its state diagram. Figure 1.18 shows the state diagram of the $cc(2,1,3)$ convolutional code, where the states are the content of the previous $(K-1)k = 2$ bits. The state transitions in response to an input bit of 1 or 0 are shown as dashed and solid lines, respectively, in the diagram. As the message bits shift into the register $k = 1$ bit at a time, new coded symbols are formed in response to these state transitions. On the basis of the state diagram, the trellis diagram is generated to represent the coding process, and is formed by concatenating the consecutive instants of the state transition diagram. The trellis diagram for $cc(2,1,3)$ is shown in Figure 1.19, where the dashed and solid lines correspond to the latest input bit of 1 and 0, respectively. In the trellis diagram, each node, represented by a dot, corresponds to the state shown on the left-hand side of the figure. Similar to the state diagram of Figure 1.18, the dashed and solid lines indicate the state transitions due to an input bit of 1 or 0, respectively. For a input data sequence of 01101, the corresponding paths in the trellis diagram are displayed by thick dashed and solid lines, and the coded symbol sequence is shown at the bottom of the figure. Since the error correction performance of convolutional codes is related to their Hamming distance, we redraw the state diagram as shown in Figure 1.20 to examine the Hamming distance of $cc(2,1,3)$. Instead of labelling each branch with its corresponding output code word, we label it with $D^0 = 1$, $D^1 = D$ or D^2, where the exponent of D represents the Hamming distance corresponding to that branch compared with the all-zero branch. We also label each branch with H or 1 to indicate a transition corresponding to an

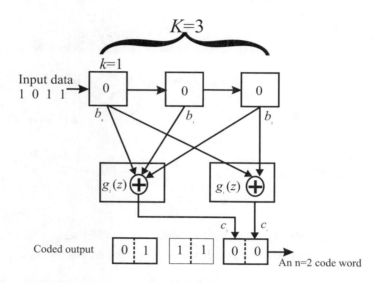

Figure 1.17: The schematic arrangement of a cc$(2, 1, 3)$ convolutional coder.

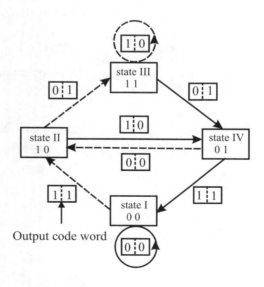

Figure 1.18: State diagram for cc$(2, 1, 3)$.

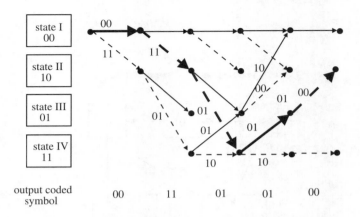

Figure 1.19: The trellis diagram for cc(2,1,3).

input of 1 or 0, respectively. In the figure, X_I is the input of the state, while the other four states in the diagram have outputs shown by the following state equations:

$$
\begin{aligned}
X_{II} &= D^2 H X_I + H X_{IV}, \\
X_{III} &= D H X_{II} + D H X_{III}, \\
X_{IV} &= D X_{II} + D X_{III}, \\
X_V &= D^2 X_{IV}.
\end{aligned}
\tag{1.20}
$$

For an infinite-length coded symbol sequence, the transfer function of the cc(2,1,3) code is defined as

$$
T(D,H) = \frac{X_V}{X_I}.
\tag{1.21}
$$

From Equation (1.20) and after some manipulations, the transfer function can be shown to be

$$
\begin{aligned}
T(D,H) &= \frac{D^5 H}{1 - 2HD} \\
&= D^5 H + 2D^6 H^2 + 4D^7 H^3 + \ldots + 2^k D^{k+5} H^{k+1} + \ldots \\
&= \sum_{d=5}^{\infty} 2^{d-5} H^{d-4} D^d \\
&= \sum_{d=5}^{\infty} \beta_d D^d,
\end{aligned}
\tag{1.22}
$$

where

$$
\beta_d = 2^{d-5} H^{d-4}
\tag{1.23}
$$

is the coefficient of the transfer function which indicates the number of paths having a Hamming distance of d. The minimal value of $d = 5$ is referred to as the *minimal free distance*, d_f, of the code.

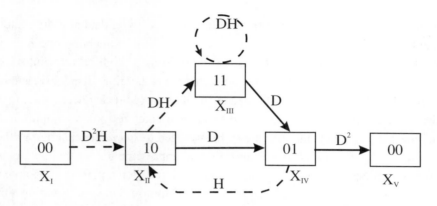

Figure 1.20: An alternative state diagram for $cc(2,1,3)$.

We will complete this discussion on the convolutional codec by briefly describing the convolutional decoding process. If a hard-decision circuit is used to regenerate the received coded symbol sequence from the demodulated signal, it is referred to as *hard-decision decoding*. If the demodulated signal is sampled and quantised, de-interleaved, and then applied to the convolutional decoder, the decoding process is referred to as *soft-decision decoding*. In conventional convolutional decoding, the decoding process operates on the received coded symbols by estimating the most likely path of state transitions in the trellis. The coder input sequence corresponding to this path is then considered to be the most likely message sequence. In the maximal likelihood decoding algorithm, or the Viterbi algorithm, all the possible paths in the trellis are searched. The path having the smallest deviation or 'distance' compared with the received path is selected, and the message sequence is regenerated accordingly.

Errors occur in bursts in cellular radio, and for the convolutional decoder to work efficiently, these errors must be randomised. To achieve this, the coded bits at the transmitter are interleaved. This means that if an error burst occurs, the receiver, on de-interleaving will transform the burst errors into random ones. Bit interleaving can be as simple as reading into a *square buffer* in rows and reading out in columns, and vice versa for de-interleaving. It can also be more complex, as described in Reference [2]. Observe that bit interleaving is a form of time-diversity that lowers the BER at the expense of an added delay in recovering the information.

Having digressed into channel coding, we now return to the basic TDMA link shown in Figure 1.16. The interleaved FEC coded speech is loaded into a buffer or packetiser, to form a packet. The input to the buffer is at the coded rate r_{code} and will be removed from the buffer as a burst in its assigned time slot and at the much higher rate r_{burst}. The burst data modulate an FDMA carrier and are transmitted. By opting for TDMA, the RF equipment is simplified, but the signal processing at baseband in the receiver is increased. This is a consequence of the TDMA structure transmitting and receiving data at a high rate due to the method of bursty transmissions. As a crude approximation the burst rate r_{burst} is r_{code} increased by the number of slots per frame. Transmitting at high bit rates introduces ISI where one bit is smeared over many. The radio channel is therefore dispersive, and in order to regenerate the bits at the receiver we need to equalise the radio channel in order to remove these dispersive effects. This in turn requires us to estimate the complex impulse response of the mobile radio channel, and to achieve this we need to *sound* the channel. Rather than apply an impulse to the radio channel, or rather an approximation to one, we opt to send a pseudo-random sequence that has an autocorrelation function (ACF) that is impulse-like. By this means we can estimate the channel impulse response. However, sending a pseudo-random sequence in each TDMA packet represents a costly overhead in that, if it were not required, the bits could be assigned to the speech or FEC coding.

Referring to Figure 1.16, we have now described that speech is encoded, followed by channel coding and bit interleaving to combat bit error bursts. We also see that the packetiser accepts both the interleaved coded speech and a sounding bit sequence. This sounding sequence is placed in the centre of the packet with the coded data on either side. If the sounding sequence were placed at the front of the packet, then the estimation of the channel impulse response at the receiver would be considerably inaccurate for the last data bits in the packet. By placing the sounding sequence in the centre of the packet the channel impulse response will not significantly change over half a packet length.

The TDMA burst buffer is responsible for packetising the coded speech at a slow rate and applying it to the modulator at the TDMA rate. Modulators used for TDMA usually allow Class C amplifiers to be used. Forms of frequency shift keying or phase shift keying are popular. GSM uses Gaussian minimum shift keying (GMSK) because its bandwidth occupancy is relatively small, although it introduces ISI prior to transmission.

From Figure 1.16 we can see that once the signal leaves the power amplifier at the transmitter, it then passes through a radio channel, via the transmitting antenna, before arriving at the receiver. As we have already seen in this chapter, the radio channel will cause a number of deleterious effects to the transmitted signal. Its amplitude will vary with the path loss and fast and slow fading, and ISI may be introduced as a result of multiple, different length paths between the transmitter and receiver. In the case of GSM, any ISI introduced by the radio channel will be in addition to the ISI already introduced by the modulation scheme.

The RF front-end of the receiver will consist of a receiving antenna followed by a low noise amplifier to boost the received signal whilst adding very little thermal noise. Following this amplifier stage the signal will be passed through a band-pass filter that will pass the frequencies within the operating band of the TDMA system, but reject the signals from systems operating in adjacent bands. After the RF front-end, the signal is down-converted to an intermediate frequency (IF), before being narrowly filtered at the bandwidth of the

TDMA carrier. The signal is next converted into its in-phase (I) and quadrature (Q) baseband components by the quadrature demodulator. This is achieved by mixing the IF signal with two quadrature carriers also at the IF. The resulting signals are then low-pass filtered to leave only the baseband components.

The I and Q signals will be sampled in an analogue-to-digital converter and the remainder of the receiver is implemented in digital form, i.e. by a digital signal processing (DSP) chip. The received digitised I and Q signals are demultiplexed into the traffic data and the sounding sequence. The latter is applied to a matched filter to provide an estimate of the channel impulse response. We have previously mentioned that both the modulator and the mobile radio channel cause each received bit to have a duration that no longer spans the original bit period, but a number of bit periods. This means that each modulation symbol will interfere with both its preceding and successive bits, an effect known as intersymbol interference (ISI). We know the amount of ISI that was deliberately introduced by the modulator to restrain the bandwidth occupancy of the modulated signal. We now have to decide how much additional bit spreading, or *dispersion*, caused by the channel, we wish to accommodate. There is a trade-off between the amount of channel dispersion that can be accommodated and the receiver complexity, i.e. the larger the dispersion we wish to accommodate, the higher the complexity of the channel equaliser at the receiver. Having decided how much spreading will be accommodated, say three bits for the modulator and two for the radio channel, then spreading in excess of five-bits may cause regenerated bits to be in error. We know that five bits can be arranged in 32 different ways, and we apply each five bit sequence to a baseband digital modulator identical to the one used at the transmitter. Now we see why we needed to sound the radio channel and get an estimate of its impulse response; because, armed with this response, we convolve each of the 32 local modulated signals to get 32 estimates of the received signal. These 32 estimates, or templates, apply to receiving a logical 1 or 0 with all the combinations that the two adjacent bits on either side of the bit being processed could have.

When the traffic data are processed, each traffic bit is compared with all 32 templates, and the mean square error between the actual data bit and each of the 32 template waveforms yields 32 incremental metrics that are used in the Viterbi processor [16]. Note that if the channel estimate were perfect, then one of these incremental metrics would be zero. A description of the Viterbi processor is beyond the scope of this book and the reader is referred to Reference [2] for a comprehensive description. Suffice to say here that the Viterbi processor used for channel equalisation is similar to the workings of the convolutional decoder. The equalisation process requires these incremental metrics, and hence channel sounding is essential. The data in the burst are not regenerated until the last bit is finally processed.

This entire process has effectively equalised the effects of the radio channel (and the ISI introduced in the modulation process) thereby allowing the transmitted data bits to be recovered. Also, since the Viterbi algorithm is employed, the entire process is termed *Viterbi equalisation*. We note that the equalisation process does not have to occur at the burst data rate. Instead, a burst will be captured at the burst data rate, but it may be processed at a slower rate, provided the processing has been completed by the time the next burst arrives in the next frame.

Following equalisation, the recovered data bits are de-interleaved using the exact inverse of the interleaving process employed at the transmitter. This has the effect of taking the error bursts, which are a characteristic of the mobile radio channel, and distributing them throughout the data so that the errors appear to be random. The de-interleaved data are then passed through a convolutional decoder to correct any errors in the recovered data. This type of error correction is far more effective when errors occur randomly rather than in bursts, and hence the requirement for interleaving. Observe that, although GSM employs Viterbi equalisation, other forms of equalisation can be used which require less processing for an exchange in channel signal-to-noise ratio (*SNR*) for a given BER [17].

Finally, the decoded data are sent to the speech decoder where they are converted into an analogue electrical signal and finally into an acoustic signal using a speaker. Should multimedia signals, other than speech, be processed, then the speech codec and speaker are replaced by the appropriate source codec and transducer. Other elements of the radio link may also be modified to suit the data being transmitted, e.g. the interleaving depth may be increased, or the power of the FEC coding may be increased.

1.4   The CDMA Radio Interface

Direct sequence spread spectrum communications involve sending a symbol in the form of a *code* that itself is composed of many symbols called *chips*. For example, a bit of logical value 1 having a duration of T seconds may be replaced by a code having chips, i.e. narrow pulses of value ± 1, each of duration T_c. If the bit is a logical 0, then all the chips in the code are inverted. As $T \gg T_c$, the coded signal requires a much higher bandwidth than the message composed of bits of duration T. This expansion of the message signal bandwidth by the factor T/T_c is called *spectrum spreading*, and communicating by the use of codes is called *spread spectrum communications*. When multiple users are given their own codes to access the radio medium, the multiple access method is called code division multiple access (CDMA) [3, 7] .

This preamble is to introduce the words *spread spectrum* and *CDMA*, to whet your appetite before we open Pandora's box and see what is inside. Suffice to say at this juncture that spread spectrum communication systems have their origins in military communications because of their ability to withstand high levels of jamming interference. They also have the virtue that the spread signals can be transmitted at low power making them useful for covert operations. We will describe CDMA using the bottom-up approach adopted in the previous section. This means we will start by discussing a simple radio link before spreading is even introduced. Then we will examine the spreading and despreading processes, and then move to the use of CDMA in cellular radio.

1.4.1 Binary phase shift keying

We will later describe a CDMA radio link employing binary phase shift keying (BPSK), and so it is appropriate initially to describe the basic BPSK radio link. In BPSK, when the binary data is a logical 1, the phase of the transmitted radio carrier is $0°$, and it changes to $180°$ when the data are a logical 0. The data signal $b(t)$, which is composed of bits of amplitude ± 1 and duration T, arrives at the BPSK modulator and switches the phase of the radio carrier between $0°$ and $180°$ for durations of T or multiples of T. The demodulator at the receiver makes a decision as to the logical value of the bit transmitted by deciding if the value of the demodulated phasor is closer to $0°$ or $180°$, and regenerating the appropriate bit.

The subject of BPSK is treated extensively in many textbooks and Reference [11] provides a good description. Within this discussion we highlight some of the key features of BPSK modulation. The transmitted signal, $s(t)$, may be represented by the expression

$$s(t) = \begin{cases} g_T(t)\cos\omega_c t, & b = \text{logical 1}, \\ -g_T(t)\cos\omega_c t, & b = \text{logical 0}, \end{cases} \tag{1.24}$$

where $g_T(t)$ is the impulse response of the filter that shapes the bit b for transmission, defined over a period T, and ω_c is the angular carrier frequency. We may also write this equation as

$$s(t) = bg_T(t)\cos\omega_c t \tag{1.25}$$

since

$$b = \begin{cases} 1, & \text{logical 1}, \\ -1, & \text{logical 0}. \end{cases} \tag{1.26}$$

The energy of $s(t)$ over a bit period is

$$\mathcal{E}_s = \int_0^T s^2(t)\mathrm{d}t = \frac{1}{2}\int_0^T g_T^2(t)\mathrm{d}t = \mathcal{E}/2, \tag{1.27}$$

where $\mathcal{E}$ is the energy of a filtered baseband bit.

The signal $s(t)$ is transmitted over the channel, and for an additive white Gaussian noise channel it can be shown that the probability of a bit error is

$$P_e = Q\left(\sqrt{2\mathcal{E}/N_0}\right) = \frac{1}{\sqrt{2\pi}}\int_{\sqrt{2\mathcal{E}/N_0}}^{\infty} e^{-x^2/2}\mathrm{d}x \tag{1.28}$$

and the *SNR* is

$$SNR = \gamma = \mathcal{E}/N_0 \cdot \tag{1.29}$$

A BPSK link having a Gaussian channel is shown in Figure 1.21. For reasons of exposition, let the current bit b in the input sequence $\{b_k\}$ applied to the transmitter be a logical 1. After being shaped by the input filter, up-conversion to the carrier angular frequency ω_c ensues to give the transmitted signal $\tilde{s}(t)$. The received signal,

$$r(t) = g_T(t)\cos\omega_c t + n(t), \quad \text{logical 1 transmitted,} \tag{1.30}$$

is coherently demodulated by multiplying it by $g_T(t)\cos\omega_c t$, followed by integration over a bit period, namely

$$
\begin{aligned}
Z(T) &= \frac{1}{2}\int_0^T g_T^2(t)(1+\cos 2\omega_c t)dt + \int_0^T n(t)g_T(t)\cos\omega_c t\,dt \\
&= \frac{1}{2}\int_0^T g_T^2(t)dt + N_1 \\
&= \frac{\mathcal{E}}{2} + N_1,
\end{aligned}
\tag{1.31}
$$

where $n(t)$ is the additive white Gaussian noise at the receiver input, and N_1 is the noise component in $Z(T)$. The output of the integrator is sampled at the end of the bit period, and if

$$
\begin{aligned}
Z(T) &\geq 0, \text{then} \quad \text{logical 1 regenerated (no error),} \\
Z(T) &< 0, \text{then} \quad \text{logical 0 is regenerated (error).}
\end{aligned}
\tag{1.32}
$$

Having regenerated the bit, the integrator is reset to zero. Observe that the receiver requires carrier synchronisation for the coherent carrier reference, and bit synchronisation to establish the clock signal for the integrate and dump process. Figure 1.22 shows a more general BPSK system.

Another way of representing a BPSK signal is

$$s(t) = \sqrt{2Pb}(t)\cos\omega_c t, \tag{1.33}$$

where P is the power in $s(t)$. Note that we are considering power and not energy as $s(t)$ is not now confined to one bit duration and $b(t)$ is a sequence of data bits. We have, for simplicity, omitted the transmitter and receiver filters. If $b(t)$ is a random signal, then its PSD is

$$S_b(f) = T\operatorname{sinc}^2(fT) \tag{1.34}$$

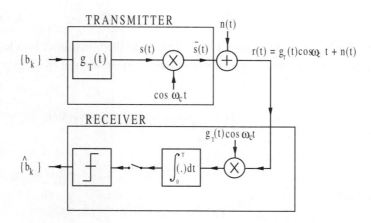

Figure 1.21: A BPSK system having coherent demodulation and an integrator and dump circuit.

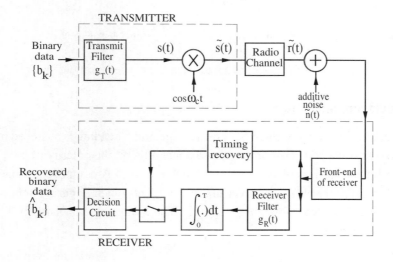

Figure 1.22: General BPSK system.

and its ACF is

$$R_b(\tau) = \begin{cases} 1 - \frac{|\tau|}{T}, & |\tau| < T, \\ 0, & \text{elsewhere.} \end{cases} \tag{1.35}$$

The carrier $\sqrt{2P}\cos\omega_c t$ has a Fourier transform given by

$$\sqrt{2P}\left[\frac{1}{2}\delta(f - f_c) + \frac{1}{2}\delta(f + f_c)\right], \tag{1.36}$$

where $\delta(\cdot)$ is a delta function, f is the frequency and f_c is the carrier frequency. The PSD of the carrier is therefore,

$$S_c(f) = (P/2)\left[\delta(f - f_c) + \delta(f + f_c)\right] \cdot \tag{1.37}$$

To determine the PSD of $s(t)$ we use the frequency convolution theorem; namely, given two functions $x_1(t)$ and $x_2(t)$, the spectrum of their product is

$$\int_{-\infty}^{\infty} X_1(\theta)X_2(f - \theta)d\theta, \tag{1.38}$$

which becomes for $s(t)$:

$$S_s(f) = \int_{-\infty}^{\infty} T\mathrm{sinc}^2\left[(f - f_1)T\right] \cdot \frac{P}{2}\left[\delta(f_1 - f_c) + \delta(f_1 + f_c)\right]df_1 \cdot \tag{1.39}$$

Since $S(f_1 \pm f_c)$ only exists at $\mp f_c$, then

$$S_s(f) = \frac{PT}{2}\mathrm{sinc}^2\left[(f - f_c)T\right] + \frac{PT}{2}\mathrm{sinc}^2\left[(f + f_c)T\right]. \tag{1.40}$$

The PSD of $s(t)$ consists of two sinc^2 functions at the carrier frequencies $\pm f_c$.

1.4.2   Spectrum spreading

As we remarked at the beginning of Section 1.4, spectral spreading is achieved by replacing a data bit in $b(t)$ by a code $c(t)$ or its inverse depending on the polarity of a bit b. In other words each bit in $b(t)$ is replaced by $bc(t)$, and since $c(t)$ is composed of narrow binary pulses or chips of duration T_c, the spectrum of $b(t)c(t)$ is much larger than the spectrum of $b(t)$. This effect is called *spectrum spreading*. The frequency expansion is T/T_c and the transmitted signal is

$$s(t) = c(t)\left\{b(t)\sqrt{2P}\cos\omega_c t\right\} \cdot \tag{1.41}$$

Although the multiplication of $b(t)(=\pm 1)$ by $c(t)(=\pm 1)$ is done at baseband, we may view the process of multiplying the BPSK signal by $c(t)$, as shown in Figure 1.23. We will refer to Equation (1.41) as the transmitted CDMA signal, it being appreciated that at the moment we are only considering one user.

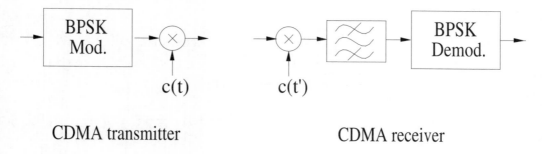

CDMA transmitter   CDMA receiver

Figure 1.23: CDMA transmitter and receiver.

At the receiver, the received signal $r(t)$ is multiplied by a delayed version of the code used at the transmitter, $c(t')$. If the receiver code, $c(t')$, is synchronised to the code used at the transmitter, allowing for the propagation delay, then the product of the two codes will be $+1$ if the codes have the same polarity, and -1 if one of them is inverted. The product of the two codes, when aligned or synchronised, is always 1, so the product of $bc(t) \cdot c(t)$ is always b, equal to ± 1 over the duration T. In Figure 1.23, $c(t')$ is $c(t)$ after allowing for propagation delays. In Figure 1.24 we display the data symbols for a logical 1 and a logical 0, and underneath, the corresponding product $bc(t)$. This product is $c(t)$, our code for the logical 1 symbol. The product $b(t) \cdot c(t)$ are the spread signals. The despread symbols are of course the original ones (in the absence of transmission errors.)

The CDMA receiver multiplies the received signal $r(t)$ by $c(t')$ to return the signal to the BPSK format, whence BPSK demodulation ensues. The $c(t')$ multiplication can be done at baseband, rather than at RF.

1.4.2.1   Why do we use these codes?

In CDMA, all users occupy the same transmission bandwidth or the same receiver bandwidth at the same time. Each user has a different code, and extraction of a wanted signal with code c_k is achieved by multiplying the received signal from N users by the code c_k. This act causes despreading of the user's data associated with code c_k, while

$$c_k c_j = 0, \quad \text{for } j \neq k, \tag{1.42}$$

in the ideal case of all user codes being orthogonal and the channel ideal. In multipath channels, or with non-synchronised reception of the codes, the orthogonality between codes does not prevent multiple access interference (MAI).

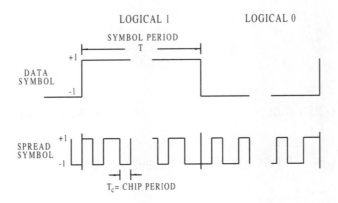

Figure 1.24: The data symbols and the spread symbols.

1.4.2.2   Appropriate codes

Sequences known as m-sequences have a length of $n = 2^m - 1$ chips, consisting of 2^{m-1} logical ones and $2^{m-1} - 1$ zeros. Their ACF is periodic in n, having a peak of n for zero shift and -1 for all other shifts, as shown in Figure 1.25. The ACF is therefore spike-like with triangular peaks every n chips whose duration at the base is two chips. The PSD is a sinc^2 function. The cross-correlation function (CCF) of a pair of m-sequences is shown in Figure 1.26.

Some pairs of m-sequences, known as *preferred sequences*, have CCFs with three values which are significantly lower than the peak of the ACF. A graphical illustration of the CCF of two preferred sequences is shown in Figure 1.27. For a pair of preferred sequences a set of $n + 2$ sequences can be generated, called Gold codes. Although constructed from maximal sequences, i.e. the longest codes that can be generated by a particular shift register, Gold codes are not maximal. They can be produced at relatively high rates as they are formed by modulo-2 addition of a pair of m-sequences of the same length. There are numerous Gold codes, and the CCFs between the codes are -1, $t(m)$, and $t(m) - 2$, where

$$t(m) = \begin{cases} 2^{(m+1)/2} + 1, & m \text{ odd,} \\ 2^{(m+2)/2} + 1, & m \text{ even,} \end{cases} \tag{1.43}$$

while the ACF away from the peak is bounded below $t(m)$. These characteristics make them attractive for use in CDMA.

cdmaOne employs three types of codes on its down-link, as we will discuss in Chapter 4. There is a user code which is a $2^{42} - 1$ m-sequence, two 2^{15} length short codes, which are derived from $2^{15} - 1$ length m-sequences, and a set of 64 Walsh codes that identify down-

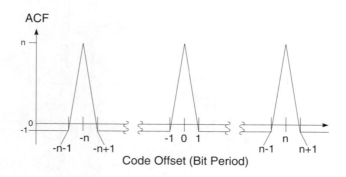

Figure 1.25: The autocorrelation function of an *m*-sequence.

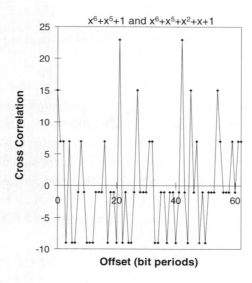

Figure 1.26: The cross-correlation function of two *m*-sequences.

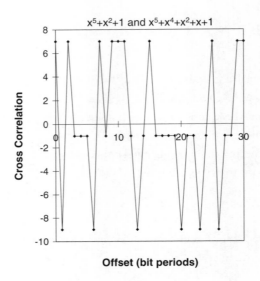

Figure 1.27: The cross-correlation function of two preferred sequences.

link channel numbers. Since the Walsh codes have a high profile role in cdmaOne we will describe them here [18]. Walsh codes are binary orthogonal codes. When used on their own they are not effective in CDMA. For example, in some textbooks on CDMA they are not even mentioned. If a single code is to be used, unlike the situation in cdmaOne, then a user would do better not to use Walsh codes [19]. However, as used in cdmaOne they are effective.

Walsh codes are easy to generate using a Hadamard matrix. The matrix of order two contains the zero-order and first-order Walsh codes, namely

$$H_2 = \begin{bmatrix} 1 & 1 \\ 1 & -1 \end{bmatrix}. \tag{1.44}$$

The matrix is square and the zero-order Walsh code is found in the first row and in the first column. The first-order Walsh code is the second row and the second column. The

Hadamard matrix of order four is

$$H_4 = \begin{bmatrix} H_2 & H_2 \\ H_2 & -H_2 \end{bmatrix} = \begin{bmatrix} 1 & 1 & 1 & 1 \\ 1 & -1 & 1 & -1 \\ 1 & 1 & -1 & -1 \\ 1 & -1 & -1 & 1 \end{bmatrix}, \tag{1.45}$$

where row 1 and column 1 are Walsh code 0, or w_0; row 3 and column 3 are w_1; row 4 and column 4 are w_2; and row 2 and column 2 are w_3. The Walsh code number w_j is the average number of zero crossings between $+1$ and -1. If the numbers are identified according to the row number, we will call these W_j, then $W_1 = w_0$, $W_2 = w_3$, $W_3 = w_1$ and $W_4 = w_2$. In cdmaOne the row numbering in the Hadamard matrix is used and *not* the Walsh number.

Higher order Hadamard matrices are built up using the same procedure, i.e.

$$H_8 = \begin{bmatrix} H_4 & H_4 \\ H_4 & -H_4 \end{bmatrix} \tag{1.46}$$

and the codes used in cdmaOne are

$$H_{64} = \begin{bmatrix} H_{32} & H_{32} \\ H_{32} & -H_{32} \end{bmatrix}. \tag{1.47}$$

Figure 4.1 shows the set of 64 Walsh codes used in cdmaOne, where the numbering relates to W_j, $j = 0, 1, \ldots, 63$.

1.4.3   The spread signal

The spectrum of the BPSK signal is given by Equation (1.39). We now consider the spreading of this spectrum. Let us assume for simplicity that the spreading code $c(t)$ is sufficiently long that its ACF is triangular,

$$\mathcal{R}_c(\tau) = \begin{cases} 1 - \frac{|\tau|}{T_c}, & |\tau| < T_c, \\ 0, & \text{elsewhere,} \end{cases} \tag{1.48}$$

where T_c is the chip period. We see that $\mathcal{R}_c(\tau)$ has the same form as $\mathcal{R}_b(\tau)$ of Equation (1.35), and therefore the PSD of $c(t)$ is

$$S_c(f) = T_c \, \text{sinc}^2(fT_c). \tag{1.49}$$

From Equation (1.41) the PSD of $s(t)$ is the frequency convolution of $S_c(f)$ and $S_s(f)$, namely

$$S_{sp}(f) \simeq \frac{1}{2} P T T_c \left[\text{sinc}^2 \left\{ (f - f_c)T_c \right\} + \text{sinc}^2 \left\{ (f + f_c)T_c \right\} \right]. \tag{1.50}$$

The wideband PSD, $S_c(f)$, has been convolved with the PSD, $S_s(f)$, of the narrow band BPSK signal. The PSD of $c(t)$ has been shifted to $\pm f_c$ and only marginally affected by the shape of $S_s(f)$, as can be seen from Equation (1.50).

1.4.3.1 Single user transmission over a Gaussian channel with a single frequency jammer

The transmitted CDMA signal is given by Equation (1.41) and the jammer signal is

$$j(t) = \sqrt{2J}\cos(\omega_c t + \theta), \tag{1.51}$$

where J is the jammer power and θ is an arbitrary phase angle. The additive noise is

$$n(t) = A_R(t)\cos(\omega_c t + \psi), \tag{1.52}$$

where $A_R(t)$ is Rayleigh distributed and ψ is uniformly distributed. The received signal is

$$r(t) = s(t - T_P) + j(t) + n(t), \tag{1.53}$$

where T_p is the propagation delay between the transmitter and receiver. When the receiver multiplies $r(t)$ by $c(t)$, the wanted narrow band BPSK signal becomes

$$\hat{s}(t) = \sqrt{2Pb}(t - T_p)\ \cos(\omega_c(t - T_p)). \tag{1.54}$$

The jammer signal has been spread by the receiver code and is now wideband,

$$\hat{j}(t) = c(t - T_p)\sqrt{2J}\ \cos(\omega_c t + \theta), \tag{1.55}$$

having a PSD of

$$S_j(f) = \frac{1}{2}JT_c\left[\text{sinc}^2\left\{(f - f_c)T_c\right\} + \text{sinc}^2\left\{(f + f_c)T_c\right\}\right]. \tag{1.56}$$

The noise also remains wideband since

$$\hat{n}(t) = c(t - T_p)A_R(t)\cos(\omega_c t + \psi) \tag{1.57}$$

with a PSD of

$$S_n(f) = \sigma_n^2 \int_{-\infty}^{\infty} T_c\ \text{sinc}^2\left\{(f - f_1)T_c\right\}\mathrm{d}f_1. \tag{1.58}$$

The signal $r(t)c(t - T_p)$ is filtered from $f_c - (1/2T)$ to $f_c + (1/2T)$. The wanted signal is passed as is the jammer noise that is within the filter bandwidth yielding a noise power of

$$J_f = J(T_c/T) \tag{1.59}$$

since $\text{sinc}^2\{\cdot\} = 1$ over the filter bandwidth. Similarly the Gaussian noise passed is

$$\int_{f_c - \frac{1}{2T}}^{f_c + \frac{1}{2T}} S_n(f)\mathrm{d}f = \sigma_n^2(T_c/T). \tag{1.60}$$

Observe that the act of despreading has decreased the unwanted interference power and noise power by the same factor, T_c/T. We call this the *processing gain*:

$$G = T/T_c. \tag{1.61}$$

The signal supplied to the BPSK demodulator is

$$r_{BPSK}(t) = \sqrt{2P}\, b(t - T_p)\cos(\omega_c(t - T_p)) + N(t), \tag{1.62}$$

where $N(t)$ is a Gaussian noise signal having power $(J + \sigma_n^2)/G$. The output of the BPSK demodulator is

$$\int_0^T r_{BPSK}(t) 2\cos(\omega_c(t - T_p))dt = \sqrt{2P}b + \hat{N}, \tag{1.63}$$

where $\hat{N}$ is the noise component after integration. An error occurs if

$$\begin{aligned}
\text{logical 1 transmitted;} &\quad \sqrt{2P} + \hat{N} < 0, \\
\text{logical 0 transmitted;} &\quad \sqrt{2P} + \hat{N} \geq 0.
\end{aligned} \tag{1.64}$$

1.4.3.2 Summary

We have discussed BPSK, and then introduced the notion of spreading a BPSK signal by replacing each bit to be transmitted by a code $c(t)$ or its inverse depending on the logical value of the data bit. Then we considered the case of a single user spreading the binary signal $b(t)$ followed by BPSK modulation. The channel was a simple Gaussian one, except that a narrow band jammer was present. We then found that the receiver could recover the wanted signal in the presence of receiver noise and a jamming signal by multiplying by the code of the wanted signal. This had the effect of suppressing both the receiver noise and the jammer interference, relative to the wanted signal, by a factor equal to the processing gain, G.

1.4.4 Multiple CDMA users

Consider the case of N mobile users transmitting over wideband fading channels in the presence of additive white Gaussian noise (AWGN) [19–21]. The transmitted CDMA signal from the kth user is

$$s_k(t) = \sqrt{2P}c_k(t)b_k(t)\cos(\omega_c t + \theta_k), \tag{1.65}$$

where θ_k is the phase of the kth carrier. The kth MS transmits over the kth radio channel whose low pass complex impulse response is

$$h_k(t) = a_k \sum_{i=1}^{L} \beta_i \delta(t - \tau_i)e^{j\phi_i}, \tag{1.66}$$

where

$$a_k = d_k^{-\alpha} 10^{\frac{\zeta}{10}} \tag{1.67}$$

allows for path loss and shadow fading, as described in connection with Equation (1.6), where α is the path loss component and ζ is the shadow fading random variable. In Equation (1.66), β_i is the magnitude and ϕ_i is the phase of the ith received path at time τ_i, $\delta(t - \tau_i)$ is a delta function at time τ_i and L is the number of paths. The received signal for N users is

$$r(t) = \sum_{k=1}^{N} [s_k(t) * h_k(t)] + n(t), \tag{1.68}$$

where $n(t)$ is the AWGN having a two-side PSD of $N_0/2$. On taking the convolution,

$$\begin{aligned} r(t) &= \sqrt{2P} \sum_{k=1}^{N} a_k \sum_{i=1}^{L} \beta_i e^{j\phi_i} c_k(t - \tau_i) b_k(t - \tau_i) \\ &\times \cos(\omega_c t - \omega_c \tau_i + \theta_k) + n(t), \end{aligned} \tag{1.69}$$

where τ_k is the random delay of the kth user and is uniformly distributed over $[0, T]$.

Suppose the receiver recovers the phase in order to achieve coherent demodulation, and is locked to, say, the jth path. On multiplying by the coherent carrier and code for the mth user, followed by integration over a bit period T yields

$$Z_m^{[j]}(T) = \sqrt{P/2}\,T\left(a_m \beta_j^{(m)} b_0^{(m)} + Z_s + Z_{MAI}\right) + Z_n(T), \tag{1.70}$$

where the first term in this sum of terms is the wanted one, i.e. the scaled current bit of the mth user, $b_0^{(m)}$. An error term arises because we are locked onto the jth path, whereas there are received components from L paths, and these other $(L - 1)$ versions of the signal that occur at τ_i, $i \neq j$, cause the interference term Z_s. Another form of interference Z_{MAI} is due to the MAI from other user signals, while $Z_n(T)$ is the Gaussian noise term. In Reference [19], Z_s and Z_{MAI} are calculated in terms of partial cross-correlation functions between the spreading codes. The bit error probability at the decision circuit is

$$P_e = \frac{1}{2}\mathrm{erfc}\left[\frac{\sqrt{P/2}\,T(a_m \beta_j^m + Z_s + Z_{MAI})}{\sqrt{2}\sqrt{\mathrm{var}[Z_n(T)]}}\right]. \tag{1.71}$$

The variance of the AWGN component at the output of the matched filter can be shown to be

$$\mathrm{var}[Z_n(T)] = \frac{N_0 T}{4}, \tag{1.72}$$

where N_0 is the double-sided PSD of the AGWN. Substituting $\mathrm{var}[Z_n(T)]$ into Equation (1.71) gives

$$P_e = \frac{1}{2}\mathrm{erfc}\left[\sqrt{\frac{E_b}{N_0}}(a_m \beta_j^m + Z_s + Z_{MAI})\right]. \tag{1.73}$$

In the above equation, the bit error probability relies on the knowledge of the partial cross-correlation properties between the spreading codes. For conventional CDMA systems, each bit is spread by a unique code having a code length of an integer number of spreading chips per bit. The partial cross-correlation functions can be derived for a given spreading code. However, in cellular CDMA systems such as cdmaOne, the user code is a segment of a PN code which has a length that is much greater than the number of spreading chips per bit. Consequently, the property of the partial cross-correlation functions varies from segment to segment, and it becomes difficult to use the above approach to analyse the performance of these CDMA systems.

In cellular environments, the interference not only comes from the users in the MS's cell, but also from all the users in other cells. The total multiple access interference tends to be Gaussian distributed due to the Central Limit Theorem. Therefore, an alternative approach to calculating P_e is to employ the Gaussian approximation method. This models the summation of the total interference and AWGN at the receiver input as a Gaussian distribution. For the Gaussian approximation model, Equation (1.70) can be rewritten as

$$Z_m^{[j]}(T) = \sqrt{P/2} T a_m \beta_j^m b_0^m + \sqrt{\frac{P}{2}} T [Z_s + Z_{MAI}] + Z_n(T) , \tag{1.74}$$

where the first term is the wanted signal and the others are AWGN. The corresponding bit error probability is

$$P_e = \frac{1}{2} \text{erfc} \left[\frac{\sqrt{P/2} T a_m \beta_j^m}{\sqrt{2} \left\{ \sqrt{\text{var}[\sqrt{P/2} T (Z_s + Z_{MAI})] + \text{var}[Z_n(T)]} \right\}} \right] . \tag{1.75}$$

Substituting $\text{var}[Z_n(T)]$ into Equation (1.71) with (1.72), we have

$$P_e = \frac{1}{2} \text{erfc} \left[\frac{a_m \beta_j^m}{\sqrt{2\text{var}[(Z_s + Z_{MAI})] + N_0/E_b}} \right] . \tag{1.76}$$

1.4.5 Simple capacity equation

This equation should be applied with caution because it is based on rather simplistic assumptions. Nevertheless it is often used to give an estimate of how many users a cell site can support. The number of users a system can support is highest for the cdmaOne type of CDMA systems if each MS adjusts its transmitted power so that the received power at the BS is S. The interference to any MS from the other $(N-1)$ MSs is $(N-1)S$, and for a receiver noise power of N_0W, the SIR for a single cell is

$$SIR = \frac{S}{(N-1)S + N_0W}, \tag{1.77}$$

where N_0 is the PSD of the receiver noise and W is the bandwidth of the radio channel. Another representation of the *SIR* is

$$SIR = \frac{E_b \text{ (energy/bit)} \cdot R_b \text{ (bits/sec)}}{I_0 \text{ (PSD of the interference)} \cdot W} . \qquad (1.78)$$

Equating these two expressions for the *SIR* gives

$$\frac{E_b}{I_0} = \frac{G}{(N-1) + \frac{N_0 W}{S}} \qquad (1.79)$$

with G, the processing gain, given by

$$G = \frac{T}{T_c} \simeq \frac{W}{R_b} . \qquad (1.80)$$

Hence the number of users is

$$N = 1 - \frac{N_0 W}{S} + \frac{G}{E_b/I_0} \qquad (1.81)$$

and the signal-to-receiver noise ratio $(S/N_0 W)$ is reasonably high so

$$N \simeq \frac{G}{E_b/I_0} . \qquad (1.82)$$

Note that although the *SNR* $\ll 1$ because it is approximately $1/N$, the noise power at the input of the receiver, namely $(N-1)S + N_0 W$, is decreased by the processing gain, G, to give a positive E_b/I_0 at the output of the matched filter.

Of course, in practice a power control system that always ensures every MS delivers an exact power, S, at the cell site cannot be realised. A factor h can be introduced to account for imperfect power control. Then there is discontinuous transmission whereby an MS essentially ceases transmissions when the user is not speaking. We represent this by a factor μ. Users in neighbouring cells will also produce interference at the BS which we call *intercellular interference*. An up-link reuse efficiency j is included to allow for this interference. If there are g sectors per cell site we end up with an approximate expression for the number of users per site as

$$N \simeq \left(\frac{G}{E_b/I_0} \right) \left(\frac{hjg}{\mu} \right) . \qquad (1.83)$$

Other multiplicative factors can be introduced to allow for the effects of handovers between cells and between sectors.

In Chapter 5 we will deal in more detail with the calculation of N. The above discussion is meant to introduce the reader to a simple method of calculating how many users can be accommodated at a cell site.

1.4.6 Cellular CDMA

A key attribute of CDMA is that it is able to operate in single cell clusters, and when sectorisation is introduced, each sector can also use the same carrier frequencies. Thus we have a reuse of unity, not just at the cell level, but at the sector level. It is this ability to deploy all the carriers at each sector that makes CDMA spectrally efficient in a cellular environment. There is intercellular and intersector interference, but the limiting interference is from other users within a mobile's sector. This interference is called *intracellular* interference. It is not surprising that CDMA can cope with high levels of interference; indeed, it was conceived to function in a jamming environment.

1.4.6.1 Forward and reverse links

In general, cellular systems employ frequency division duplex (FDD) in which signals to and from the mobiles are sent in different frequency bands. Transmissions from the BS to the mobiles are sent over a link called the forward link in the United States and the down-link in Europe. The transmissions from the mobiles to the BS travel over the reverse link or the up-link. On the forward link, BSs interfere with mobiles in adjacent cells, whereas on the reverse link, MSs interfere with adjacent BSs. Figures 1.28 and 1.29 illustrate the down-link and up-link situations, respectively. The two links are radically different in CDMA. On the forward link all the users' signals are CDMA coded, along with CDMA coding of a synchronisation channel, paging channels and a pilot channel. All these channels are summed, applied to a modulator and linearly amplified. The reverse link is the weaker link because the problem of controlling both the received bit energy and the interference at the BS is difficult, as we will subsequently describe.

As we will see, the pilot channel performs many tasks in cdmaOne, such as BS identification, open-loop power control, synchronisation assistance, and coherent demodulation. The pilot CDMA signal does not carry any information data. It is a pseudo-random code that is known to all mobiles. The mobiles autocorrelate this code to get an estimate of the complex baseband impulse response $h_e^*(t)$ of the radio channel. In this way the mobiles obtain updated estimates of their channel impulse responses $\hat{h}_e^*(t)$ as they roam. Knowing $\hat{h}_e^*(t)$, the mobiles are able to perform coherent demodulation on the other forward channels using their RAKE receivers. The RAKE receiver is discussed in detail in Section 4.2.2.6. Suffice to say here that it gets its name from a garden rake, as it has prongs that are essentially receivers, each servicing one of the incoming paths in the mobile radio environment. The signals from each path are combined in a process known as *maximal ratio combining* to provide path diversity reception.

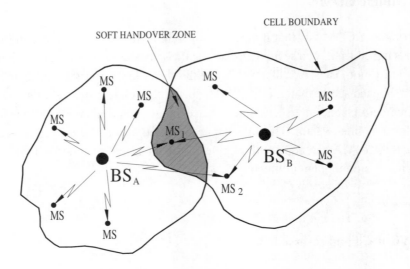

Figure 1.28: Forward or down-links. BS_A causes intercellular interference to MS_2 and BS_A and BS_B provide down-link macrodiversity to MS_1 in the soft handover zone.

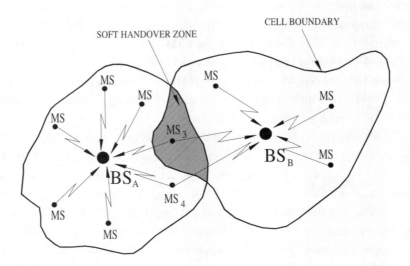

Figure 1.29: Reverse or up-links. MS_4 causes intercellular interference to BS_B, and BS_A and BS_B provide up-link macrodiversity to MS_3 in the soft handover zone.

1.4.6.2 Power control

The transmitted power of the mobile is controlled using both open-loop and closed-loop systems. The combined control systems must provide rapid changes to the mobile transmitted power over a wide dynamic power range. All mobiles adjust their transmitted power in an attempt to make their received power at the BS close to a target E_b/I_0. Errors in controlling the received power, and therefore errors in E_b/I_0 for each user, can significantly decrease the number of users that a BS can handle for a given BER.

The fast fading is different on each of the two FDD links and the closed-loop system attempts to track this fading. This is only successfully accomplished at low mobile speeds. At fast vehicular speeds link quality is maintained by the FEC codec and its interleaver. The shadow fading is the same on both FDD links and therefore it can be combatted by the open-loop system. The mobiles filter the output of the pilot correlator so that the variations in the received pilot level can be tracked at the slow fading rates. Remember that the pilot signal is transmitted at a constant power level.

Neighbouring cells also exercise up-link power control on their mobiles, but not on the interference that these mobiles can cause to neighbouring BSs. This interference is called *intercellular interference*.

On the forward link, a mobile receives the complete CDMA signal from its BS. The signals for the other mobiles in its cell constitute intracellular interference, but we may view the power control as perfect since they are all transported on the same carrier from the same place (the cell site) and with the same relative power. Because of neighbouring cells this idyllic situation is spoilt because some mobiles near a cell boundary are subjected to intercellular interference from adjacent BSs. In this case, the transmitted forward link power for these mobiles is increased slightly to overcome this additional interference. However, the dynamic range of the transmitted power variations on the forward link is small and the updating of the power level is slow compared with the control exercised on the reverse link, i.e. on the transmitted power of each mobile.

1.4.6.3 Importance of E_b/I_0

A key parameter is the expected value of the ratio of the energy per bit, E_b, to the PSD of the interference plus receiver noise, I_0, referred to the input of the decision circuit. We deal with the expected values of this ratio, i.e. $E\{E_b/I_0\}$, as E_b, an energy, relates to the transmitted power, the path loss, the shadow fading and the number of paths in the RAKE receiver; while I_0 is dependent on the intracellular and intercellular interferences, which in turn depend on the number of users making calls, the voice activity, mobile speed, errors in the power control and so forth.

We consider I_0 to be AWGN, like the receiver noise. This is because the MAI is the sum of many user signals, which is Gaussian by the Central Limit Theorem. If the RAKE, FEC coding with interleaving and space diversity are able to remove the effects of fading, then demodulation of the data is done in a Gaussian channel.

The capacity of a CDMA system is determined by the $E\{E_b/I_0\}$ on the reverse link. The BER depends on the Q-function of E_b/I_0, and for a maximum acceptable BER there is an $(E_b/I_0)_{acc}$. This $(E_b/I_0)_{acc}$ is not a constant, but varies with mobile speed. For a Gaussian channel, $(E_b/I_0)_{acc}$ is small. For poor channels we have a maximum $(E_b/I_0)_{acc}$, i.e. the interference plus noise PSD, I_0, becomes large. Controlling I_0 is more difficult on the up-link. Power control errors have a significant effect on capacity.

1.4.6.4   Intercellular interference

This interference results from users in neighbouring cells and is therefore dependent on the loading of these cells, i.e. on the number of users making calls, the path loss exponent α and the shadow fading. The lower the α the greater the interference, and α is not homogeneous, particularly in microcells. Typically, the number of users that a cell can support may decrease due to intercellular interference by some 50% in terrestrial cellular systems, and by more in mobile satellite systems.

1.4.6.5   Processing gain

The processing gain G is defined as the symbol duration T to the chip duration T_c. Usually T_c is fixed, but T is a function of the service. When a low bit rate service is being used, G is large, and the interference is suppressed giving a high $E\{E_b/I_0\}$ and low BER. The situation is reversed for high symbol rate services.

1.4.6.6   Cell size

The dimensions of a cell are dependent on the number of users and the number of users in adjacent cells. If we consider an isolated cell with one user the dimension of the cell is very large, as I_0 is merely the noise due to the receiver. Suppose the mobile is at the cell boundary where the mobile transmits its maximum power to overcome the path loss. If more users become active in this single cell, the interference to our mobile on the boundary increases and because the mobile is already transmitting at maximum power, E_b/I_0 is unacceptably low. Consequently the mobile on the boundary must move towards the BS, increasing the received E_b, until the acceptable E_b/I_0 is reached. The distance from the BS where this occurs marks the new boundary. Thus the cell boundary is not fixed.

Suppose now a neighbouring cell site is activated. Intercellular interference occurs and our mobile on the boundary has a value of E_b/I_0 at the receiver that is too low. To increase E_b at the receiver, it again has to move closer to its BS and now we have a new boundary surrounding a smaller size cell. Therefore, as the number of users increases in a set of contiguous cells, the cell sizes decrease and vice versa. It is therefore essential to provide generously proportioned soft handover zones in which the cell sizes can fluctuate. The biggest factor affecting cell size is the intracellular interference.

1.4.6.7 Effect of demodulation on the reverse link capacity

Another reason for the reverse link having the poorer performance is that non-coherent demodulation is used in narrow-band CDMA, because the BS cannot cope from an interference point of view with receiving N pilots from the N mobiles, in addition to their traffic channels. It is possible to employ coherent demodulation, but at the expense of capacity. This is done in the CODIT system [22]. Wideband DS-CDMA can transmit a pilot from each mobile, provided the pilot level is at a much lower power than its traffic channel.

1.4.6.8 FEC coding, interleaving and modulation

On both links it is customary to FEC code the data. The designer is not inhibited in his use of FEC coding from a bandwidth expansion point of view as the FEC coding gain more than compensates for the loss of processing gain resulting from spectral spreading by the CDMA code. Indeed, if complexity were not a limitation the FEC code could expand the data rate to the chip rate bandwidth. It would still be necessary to use a code for the purpose of multiple access.

Although wideband channels do not have the fading depths of narrow-band channels, errors will still occur in bursts. Accordingly the bits after FEC coding are interleaved in order that the receiver on de-interleaving will disperse the errors to facilitate efficient FEC decoding.

Basic types of modulation tend to be used for CDMA. BPSK is appropriate. Quadrature phase shift keying (QPSK), where the same data are applied to each quadrature channel, but different spreading codes are used, is better. Figure 1.30 shows a balanced QPSK CDMA modem where the same BPSK signal is applied to two quadrature arms and spread by different codes. M-ary modulation can also be employed, but the gains that are achieved can be more easily realised by other methods, such as decreasing the FEC code rate.

1.4.6.9 Narrow-band versus wideband CDMA

Narrow-band direct sequence CDMA (DS-CDMA), typified by cdmaOne, has the advantage that is does not require a large spectrum allocation for its introduction. Furthermore, it is easier to introduce into existing analogue networks, like AMPS or TACS, as only a limited number of analogue channels need to be released for narrow-band DS-CDMA compared with wideband DS-CDMA.

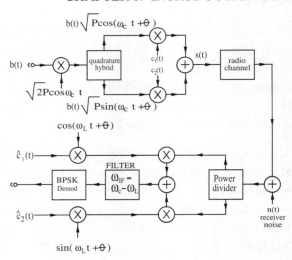

Figure 1.30: Balanced QPSK CDMA modem.

Wideband CDMA has a much greater processing gain and is therefore able to cope with greater levels of interference caused by more users. It is also easier to carry a wide range of multimedia services. Unlike narrow-band CDMA that needs its own spectrum, wideband CDMA may be able to accommodate sufficient numbers of users while allowing interfering narrow-band signals within its spectrum. The wideband CDMA receivers use notch filters to remove the interfering narrow-band signals. It is as if a frequency fade occurs in these bands and wideband CDMA is well able to cope with these types of fades. The narrow-band signals, e.g. point-to-point radio links, do not experience much interference from the wideband CDMA transmissions because the PSD of this signal is very low.

1.4.6.10   Exploiting source activity

When the source is momentarily inactive, for example during conversational silences in the case of speech data, it is better to cease transmissions. This decreases the MAI. Known as discontinuous transmission (DTX) it is widely used, but because of the larger numbers of interferers in the CDMA system compared with TDMA and FDMA systems, the DTX gains are reliably realised in CDMA. The interference at a BS is decreased by the voice activity factor, which is typically around 40% in a normal conversation.

1.4.6.11   Macrodiversity

When each cell uses the same radio carrier a mobile near a cell boundary can conveniently communicate with two cell sites at the same time. This is called *macrodiversity*. The mobile examines the pilots from each BS and chooses the K strongest paths, irrespective of which BS they are associated with, where K is the maximum number of paths that may be tracked by the RAKE receiver. The mobile then exercises maximum ratio combining (MRC) on the traffic channel data. Although MRC can be performed by combining the received signals at each BS, it is more convenient to employ switch diversity reception at the base station controller (BSC).

1.4.6.12 Handover

As the mobile approaches the boundary of its cell and its link performance deteriorates, its BS is asked for assistance. The result is that the BS that services the cell on the other side of the boundary is directed to also communicate with the mobile. Macrodiversity occurs and the process of soft handover, also called soft hand off (SHO), is said to begin. When the mobile travels sufficiently far into the new cell, the old BS ceases to communicate with the mobile leaving the communications between the new BS and the mobile. When this occurs the SHO process is completed. SHO is, therefore a *make-before-break* handover. Softer handovers (SSHOs) are when soft handovers occur in the boundary zones of adjacent sectors of the same cell site.

When an operator has been allocated many carriers he may arrange for each cell site to use all of these carriers. When a mobile is near a boundary operating on carrier f_1, the adjacent BS may have no capacity on its carrier f_1 and so an SHO cannot be performed. The mobile will then experience a hard handover (HHO) to a different CDMA carrier when the quality of the link associated with carrier f_1 becomes unacceptable. The HHO is characterised by a *break before make* protocol. The HHO means that because macrodiversity cannot be accommodated, the mobile must transmit and receive at higher power levels compared with when an SHO is implemented. The consequence is higher interference to other users and a decrease in system capacity. Consequently the network should avoid HHOs whenever possible.

Note that in an SHO the mobile transmits as if it were transmitting to its original BS, using the synchronisation codes, etc. that were assigned to it on call set-up. It receives, however, by observing the same data transmitted by both BSs. (SHO can involve more than two BSs.) The new BS would already be receiving interference from the mobile, but now starts to decode its signals as wanted ones. It does have to transmit to the mobile, but at low power levels because the mobile is using MRC. The network operator must have channels available for SHO transmissions. The percentage of channels for this purpose may be high, e.g. 30%, since the soft handover decision zones near the boundary can be substantial. The continuity of traffic flow during SHO is important for data services as it avoids delays due to buffering in systems that employ HHO, like GSM.

1.4.6.13   Diversity techniques in DS-CDMA

1. Spaced diversity on the reverse link where the BS uses two or more antennas. RAKE receivers utilise the best set of paths from the pool of paths received from one or more antennas.

2. Deep fades associated with narrow-band TDMA and FDMA are avoided by the act of frequency spreading. Instead, fading may occur at some frequencies, but the fading is not flat across the band. In the time domain the fading depth is therefore less than in narrow-band systems.

3. By using RAKE reception we are able to demodulate versions of the transmitted signal that are received over different paths. The effect of multipaths is partially removed leading to a more Gaussian-like channel. The RAKE receiver is said to provide path diversity.

4. The macrodiversity associated with SHO, where more than one BS communicates with a mobile, can also be employed in indoor cells where the RAKE may be ineffective if there are insufficient numbers of resolvable paths. It can also be applied in mobile satellite communications where the mobile utilises signals from more than one spot-beam to compensate for being in high path loss situations.

5. Time diversity is achieved by powerful channel coding followed by time interleaving. The time diversity achieved by the interleaving of symbols over a window randomises the error bursts and thereby assists channel decoding, but results in an interleaving delay. This delay is made a function of the offered service, being relatively small for speech and long for computer file transfers. When long delays are permissible, concatenated coding is deployed to decrease the BER.

1.4.6.14   No frequency planning

CDMA networks can use all the allocated carriers at each cell site. By using single cell clusters frequency planning is avoided. This is an enormous advantage compared with networks employing TDMA and FDMA, where frequency planning is time consuming and expensive. The problem of frequency planning in three-dimensional microcells is formidable. The problem of frequency planning shifts in CDMA to one of coverage planning. Cells have irregular shapes and fitting them together with suitable soft handover zones is non-trivial. Hence careful planning of transmitter power and BS siting is important.

1.4.6.15 Capacity in DS-CDMA

This is high compared with TDMA and FDMA using fixed channel allocation (FCA) mainly because CDMA can operate in single cell clusters and get large gains from sectorisation. In accessing the network users are assigned codes which may be orthogonal at the mobile transmitters, but have non-zero cross-correlation at the BS receiver due to multipath fading. This manifests itself as intracellular interference from users within the same cell and intercellular interference from users in adjacent cells. The interference limits the number of users.

To combat interference we use good radio link design so that the link can function in high levels of interference, e.g. by using FEC coding, and we employ techniques to decrease the interference experienced by each user, e.g. sectorisation. Another approach is to remove the MAI due to the non-zero cross-correlation of the users' codes at the receiver. This approach is known as *interference cancellation* [23, 24, 28–31] and is not used in the current cellular CDMA systems.

1.4.6.16 Frequency hopping CDMA

All our descriptions of CDMA are of DS-CDMA. There is another generic form of CDMA known as frequency hopped CDMA (FH-CDMA). Before completing our discussion of CDMA we will say a few words regarding the hopping variety. There are two types of FH-CDMA, one is called slow frequency hopped CDMA (SFH-CDMA) and the other fast frequency hopped CDMA (FFH-CDMA). As the names imply, FH-CDMA transmits its information at any instant on a carrier drawn from a bank of carriers. The channel occupancy of each carrier must be sufficient to accommodate the modulated signal.

The length of time the transmitted carrier is unchanged is called the dwell time (T_D). After this time has elapsed the transmitted carrier may change, i.e. hop, to another carrier. There is a sequence of frequency hops that is given to a user. This sequence is a frequency hopping code, and in general this code sequence continuously repeats while a user is transmitting data. Figure 1.31 shows a code sequence for the situation of eight carrier frequencies. Each shaded block represents a data modulated signal positioned at a specific frequency. Suppose the modulated data are binary frequency shift keying (BFSK) where a logical 1 is represented by frequency f_A and a logical 0 by a frequency f_B. If the duration of a data bit is T_b, then the BFSK output is either f_A or f_B and lasts for T_b seconds. Should $T_D < T_b$, then it follows that for one bit there will be a number of frequency hops. This is the case of FFH-CDMA. On the other hand, if $T_D \geq T_b$, each hop may last for a packet of data and this is referred to as SFH-CDMA. The military often use FFH-CDMA, where the hopping rate may be very fast making it difficult to effectively jam the transmitted signal. However, the technology is complex and expensive for the commercial market where SFH-CDMA is preferred.

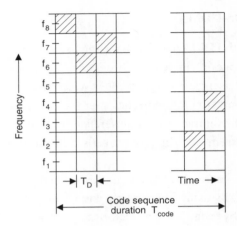

Figure 1.31: FH-CDMA hopping sequence.

Consider the case of SFH-CDMA using BFSK. The basic arrangements of the transmitter and receiver are shown in Figure 1.32. The input binary data are applied to the BFSK modulator and a sequence of frequencies containing f_A and f_B that represents the logical bits is formed. A $\log_2 n$-bit code generator selects f_i, one of n possible carrier frequencies. The BFSK signal is translated in frequency to f_i, and remains at this frequency while the BFSK frequency sequence corresponding to a packet of input data bits occurs. After time T_D, a new data packet occurs and is transmitted on a new value of f_i according to the dictates of the code. At the receiver, the same carrier f_i used in the transmitter is generated, but shifted by the IF. As a consequence the output of the mixer is always the BFSK signal at the IF. The BFSK signal is then demodulated by, say, a discriminator.

When other mobiles are present, the receiver will receive the wanted signal in the presence of signals that are hopping according to different codes. We can make these codes orthogonal. For example, we may give users the same code, but with a time offset. If the code is of length n, then n different codes may be used without any two users occupying the same hopping frequency at any one time. At the receiver, the unwanted signals will not be down-converted to IF, but to other frequencies outside the bandwidth of the IF amplifier. In other words, the wanted signal is despread to the IF, while the unwanted signals remain wideband and are filtered out.

Owing to the need to reuse the set of hopping frequencies over a wide geographical area, it is inevitable that sometimes another user will use the same hopped frequency at the same time. Demodulation of the data packet will now have to be done in the presence of an interferer, and bit errors are inevitable, unless suitable error correcting procedures are employed. It is an appropriate juncture to observe that if SFH-CDMA were not used, and only BFSK

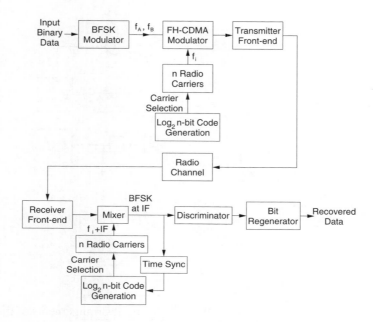

Figure 1.32: FH-CDMA using BFSK modulation of the data.

employed, then for the same interference power spread over n hopping frequencies, the demodulation at any hop is performed with $1/n$ of the interference power. The processing gain G is therefore equal to the number of hopping frequencies employed, $G = n$.

BFSK with non-coherent detection has a probability of bit error of

$$P_e = \frac{1}{2}e^{-E_b/2I_0},\tag{1.84}$$

where I_0 is the PSD of the interference plus receiver noise. When FEC coding with hard decision decoding is employed,

$$P_e = \frac{1}{2}e^{-r_c E_b/2I_0},\tag{1.85}$$

where r_c is the code rate of a linear block code.

A frequency hopped differential phase shift keying (DPSK) system is shown in Figure 1.33 [25, 26]. The user's address is housed in the m-ary FSK generator which may produce a change of frequencies every dwell period, i.e. chip period, T_c. The address is repeated every symbol period, T_s, giving $n = T_s/T_c$ chips in the address. All users share the same frequency band W, and have their own unique address. The input bit stream may be formed into $\log_2(n)$ symbols and each symbol is applied to a code generator which produces a code word w_k having n chips. There are n different code words, w_k, $k = 1, 2, \ldots, n$, and

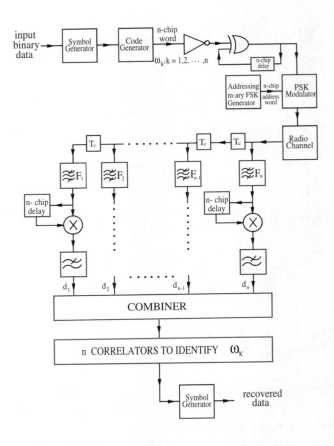

Figure 1.33: Frequency hopped DPSK.

they are usually orthogonal, such as a set of n Walsh functions. The bipolar Walsh codes are inverted and applied to an X-OR gate, along with the output of the X-OR gate that occurred during the previous code word period. This arrangement provides differential coding. The output of the X-OR gate now produces DPSK as the phase of a chip in the m-ary FSK signal is changed by π if the differential encoded chip is a logical 0 (binary -1). If a logical 1 occurs, then there is no phase change in the m-ary FSK signal. Each transmitted symbol has the frequency–time profile of the address and the phase of the frequency in any chip period depends on the logical value of the differential encoded data [25].

The receiver consists of $(n-1)$ delays, each of T_c. The output of each delay is connected to a filter F_k, $k = 1, 2, \ldots, n$, that is matched to the chip of period T_c and has a noise bandwidth $1/T_c$. By the time the first chip has got to the end of the delays, the last chip in the code has arrived. Consequently all n chips proceed at the same time through F_k, $k = 1, 2, \ldots, n$.

Their phase relative to the previous word is detected with the aid of the n-chip delays and the product detectors. The low pass filters remove out-of-band frequencies to yield the detected outputs d_j, $j = 1, 2, \ldots, n$.

The pass-bands of filters F_k are selected according to the user's address, i.e. to the m-ary FSK code signal. Hence the output of the phase detectors gives the logical value of the n-chip code word ω_k representing the data symbol. The receiver must now identify ω_k and hence the data symbol associated with ω_k. This identification is achieved by correlating the sequence d_j, with all possible words ω_k,

$$R_k = \sum_{j=1}^{N} \omega_{kj} d_j, \ k = 1, 2, \ldots, n, \tag{1.86}$$

where ω_{kj} is the jth bit of the kth word, and the value of k that gives the maximum R_k is deemed to be the transmitted data code word from which the data symbol is immediately known.

1.4.6.17 Some features of FH-CDMA

FH-CDMA and DS-CDMA both spread the spectrum of the data symbols using codes and thereby achieve frequency diversity on a user's link, while the interference is averaged from all users and decreased by the processing gain. The frequency hopping enables the effects of fading to be mitigated, where the fading on each hop should be uncorrelated. FEC coding and interleaving may be employed. Users in interfering cells use different frequency hopping sequences. The timing of each mobile can be adjusted under BS control to ensure that the signals from all mobiles in a cell are time aligned at the BS. Because the mobiles use orthogonal codes, they remain so at the BS. If quasi-orthogonal hopping sequences are used, then time alignment is not necessary due to the low cross-correlation properties between any two hopping sequences.

Dispersion due to multipaths is, in general, small in FH-CDMA. Guard times are also used with each burst. Consequently power control is not stringent and is similar to that used in TDMA systems (like GSM) and for the same reasons, namely to decrease intercellular interference. A dynamic range of 30 dB is sufficient with step sizes of ± 3 dB, compared with 85 dB with step sizes of ± 1 dB and an update rate of about a 1 kHz in DS-CDMA. FH-CDMA does not need a contiguous band of frequencies, like DS-CDMA, and, as each hop represents a narrow-band channel, it does not require the large guard bands associated with band edges of a contiguous set of FDMA DS-CDMA signals. Discontinuous transmission (DTX) is used.

1.5   Cellular Network Architecture

1.5.1   Physical and logical channels

In the previous sections we have discussed various methods that may be used to share a common radio resource between a number of users, i.e. TDMA, FDMA and CDMA. These techniques are used to divide the radio resources into a number of *physical* channels. Taking TDMA as an example, a single timeslot on a single carrier frequency is a physical channel. The user-generated and control information is mapped onto the physical channels using *logical* channels which carry information of a similar type. A single physical channel may be used to carry a number of different logical channels. In the following section we will examine the different types of logical channel.

1.5.2   Traffic and signalling channels

In any cellular radio system there are two main categories of logical channel that may exist on the radio link between the mobile unit and the BS [1, 2].

The *traffic* channels carry the user-generated information between the MS and the network. This information may take a number of different forms, including encoded speech data or fax data. The maximum bit rate supported by the traffic channels defines the types of service the system may support, e.g. speech, fax, video.

Both during and outside of a call, a degree of control information must flow between the MS and the network, and this is carried on *signalling* (or control) channels. This signalling information is used to control the MS during such processes as call set-up and handover. The signalling channels may be classified into a number of categories depending on the manner in which they are allocated and used.

The first category of signalling channels are those that are present whilst a MS is engaged in a connection with a BS. In GSM terminology, these channels are termed *associated* control channels, since they are always associated with a traffic channel. These channels are used by the network to control the MS during a call and by the MS to report measurement information back to the network. GSM and cdmaOne both use different strategies to support these associated control channels and these will be discussed in the chapters dealing with each system.

In cellular systems there are times when an MS must exchange signalling information with the network outside of a normal call. One example of this is location updating, whereby an MS will autonomously, i.e. without any user interaction, register its location with the network as it moves between cells. In GSM terminology, these channels are termed *dedicated* control channels, since they are allocated to a particular MS in a similar way to traffic channels. In general, dedicated control channels support a significantly lower data rate than traffic channels (in GSM it is eight times smaller).

Cellular systems must provide an effective means by which an MS can access the network, either in response to an incoming call or in response to the user making a call. MSs will be alerted to the presence of an incoming call, or mobile terminated call, by means of a *paging* channel. Every MS in a cell is required to continuously monitor this channel whilst it is not engaged in a call. When an MS needs to set up a connection with the network, either in response to a paging call or in response to the user making a call, i.e. a mobile originated call, it will do so using an *access* channel. The paging and access channels are generally termed *common* control channels, since they may be used by any MS within a cell and they are not allocated to individual MSs.

The final category of signalling channel used in cellular systems is the *broadcast* channels. As the name would suggest, these channels are used by the network to broadcast general synchronisation and system information to all the MSs within a cell. This information is used by the MS initially to *find* and synchronise with a BS at switch-on. The broadcast channels will also carry general system parameters relating to the BS configuration.

Each BS must support channels from each of the different categories. The broadcast channel configuration will generally be the same at each BS, since the amount of information conveyed by these channels is unlikely to vary significantly across the entire network. The number of paging channels required at a particular cell is determined by the rate at which paging calls must be transmitted. To examine the factors that affect the paging rate, we must first introduce the important concept of the *location area*.

As an MS moves around a cellular network it must always ensure that the system is aware of its location so that it can be paged when an incoming call arrives. This is achieved by a process known as *location updating*, whereby an MS will access the network and supply some brief details about the subscriber. Each MS will be registered within a location area which may consist of a single cell or a number of cells. In the event of an incoming call, the MS will be paged in every cell within the location area in which it is registered. This means that larger location areas will result in each BS having to transmit more paging messages, and the number of paging channels must be increased to support this additional load.

When an MS is paged, it will only reply to one BS, regardless of the size of the location area. This means that the number of access requests will be determined by the number of users located within an individual cell.

The number of dedicated control channels required at each cell is governed by a number of factors. An MS must register its location with the network each time it moves between different location areas. This means that smaller location areas will lead to a larger number of registrations, and since these will be supported using dedicated control channels, this will increase the number of these channels required at a cell. The number of dedicated control channels will also be governed by the types of services supported within a cell, e.g. short message service (SMS), which make use of these dedicated channels.

Finally, the number of traffic channels required at each BS will be governed by the number of users within each cell and the amount of time they spend engaged in calls. The number of traffic channels required is usually determined using the Erlang-B formula [27].

1.5.3   Network topology

In this section we take a brief look at some of the main components of a mobile radio system. Chapter 2 contains a more detailed description of these components as they apply to GSM. cdmaOne is essentially an air-interface specification and we do not describe any of the supporting network in Chapter 4, which deals with this system. The main component that the general public would associate with a cellular radio system is an MS that is only a mobile phone. In many cases this is the only part of the cellular network that the subscriber ever encounters. MSs may take a number of different forms, from the car phone, where the unit is installed in a vehicle and makes use of an exterior antenna and the vehicle's power supply, to the hand-held portable, which has its own battery power supply and is small enough to be carried in a shirt pocket. As a minimum, an MS will contain a microphone and loud speaker, and a keypad for entering telephone numbers and accessing the MS's features. In the case of a hand-held portable, the antenna may be entirely internal, or it may be external with the facility to be extended to improve link quality.

The MS communicates with the cellular network, and any other networks, by means of a radio link to its nearest BS. BSs are the other part of a cellular network that is clearly visible to the general public – on the tops of tall buildings or on tall masts by the sides of major roads. Current BSs generally consist of a large weatherproof and vandal-proof cabinet which houses all the electronics and power supplies. A cable runs from the cabinet to the antenna system, which may take a number of different forms depending on the cell configuration. Simple, non-sectorised cells may use two omnidirectional antennas in a 'rugby goal post', or 'American football goal post' arrangement. Although only one antenna is required for transmission, the second antenna is used to provide space diversity for the reception of signals from the MS. Sectorised cells require one antenna, or two antennas for diversity reception, in each sector, and these are positioned on and around a building or tower according to the sector plan.

As the demand for cellular services continues to expand at a phenomenal rate, the network operating companies are continually being forced to reduce the size of their cells to support the additional traffic. This means that much smaller BSs are required where antennae can be mounted below the urban sky-line on the sides of buildings or on lamp posts. These BSs use small, unobtrusive antennas, similar to those currently mounted on vehicles.

The BSs are connected to a switching and control centre via dedicated links, usually non-radio ones. We will refer to this centre as the mobile switching centre (MSC) (although we will see later how GSM interposes a base station controller (BSC) between its MSC and its BSs). Generically, there are a number of MSCs that connect to each other and to the BSs they are responsible for. Like the PSTN switches, the MSCs are concerned with routing traffic. However, they must also support MS mobility. An MSC will allocate and supervise the radio channels at each of its BSs. Generally, the BSs are relatively simple, with the intelligence in the MSC. Some MSCs have a gateway role, interfacing with other fixed and cellular networks.

In addition to the MS, BS and MSC, which are essentially all that is required to provide a simple connection between two users on the same, or different network, there are a number of other components of a cellular system which are used to support the mobility of MSs and control the services that are provided. *Home location registers* (HLRs) are used to store information particular to each individual subscriber on a cellular network. An HLR contains information relating to the types of services to be supported, e.g. international or premium rate calls, and the subscriber's current location within the network. Every subscriber to a cellular network has an entry in its HLR and this may be accessed by the subscriber's telephone number or their electronic serial number (ESN).

A second database, known as the *visitor location register* (VLR), is used to hold information relating to those subscribers within a particular part of the network. Each VLR may be associated with one, or a number of location areas. When an MS registers in a new location area, the subscriber's details are copied from the HLR into the VLR for that area. If the subscriber subsequently moves into a location area controlled by a different VLR, their details will be deleted from the old VLR and added to the new VLR. Using the VLR to hold a local copy of the subscriber's details means that they can be accessed quickly and efficiently when required, e.g. during call set-up, rather than have the network access the HLR. This is particularly important in systems that support international roaming. In this case the HLR and the MS may be located in different countries.

Although not essential to the operation of the network, a further database may be provided to track stolen or malfunctioning MSs. In GSM, this is termed the *equipment identity register* (EIR). Where a cellular network supports security functions, e.g. subscriber authentication or radio path encryption, a further database is required to hold any security related information. In GSM this database is termed the *authentication centre* (AUC).

A cellular network must also provide some means by which the operator may monitor and control its operation. This may take the form of monitoring the number of call attempts that are blocked at a particular BS, or changing the manner in which the radio resources are allocated to a particular cell. The interface between the network operator and the cellular network is termed the *operations and maintenance centre* (OMC). This may be concentrated within a central location or distributed throughout a number of different locations.

A cellular network must support features which assist the network operator in administering and billing subscribers. This is provided by the *administration centre* (ADC).

In this section we have described some of the generic functional blocks which make up a cellular system. Different systems may use different names for these blocks or divide their functionality in different ways; however, a cellular system must support all the basic functions described above in some form or another.

1.5.4   Making a call to a mobile subscriber

Armed with the discussions in the previous sections, we now examine the processes that occur when a person uses a telephone connected to a PSTN or ISDN, to make a call to a mobile subscriber who, for argument's sake, is travelling along the M3 motorway between Southampton and London at 70 mph. One prerequisite for this entire process is that the mobile subscriber actually has her mobile phone, or MS, switched on and that she is currently within the coverage area of the cellular network. At switch-on the MS searches for an appropriate cellular network by scanning the relevant frequency band for some form of recognisable control channel transmitted by a nearby BS. The MS then informs the network of its current location by performing a process known as *location updating* whereby it accesses the network and provides it with its unique serial number. In some cases this location up-dating process may not be necessary, for example where an MS is switched off and then on again in the same area. When a location update is performed, the network enters the subscriber's details in the VLR associated with the location area in which the MS has registered. Also, details of the MS's location are stored within the subscriber's entry in the HLR. Once an MS has successfully registered its location with the network it enters the *idle mode* whereby it listens to the paging channels from the selected BS.

Returning to our example of the mobile subscriber travelling along the M3 motorway, we will assume that the MS was switched on at the start of the journey in Southampton's City Centre. In this case the MS will have identified a BS in this area, registered its location with the network and then commenced monitoring one of the BS's paging channels. As the subscriber drives out of the City Centre and towards the outskirts of the city to join the M3 motorway, the MS will notice that the signal received from the City Centre BS begins to fall. If this was allowed to continue unchecked the subscriber would eventually drive out of range of the City Centre BS and the MS would no longer be in a position to make or receive calls. However, in the case of a cellular system, as the signal begins to deteriorate from one BS, the MS will start to look for a more appropriate BS to take over.

When the MS can identify a more appropriate BS it will examine its control channels to determine the location area to which it belongs. If it belongs to the same location area as the previous BS, the MS simply retunes to a paging channel on the new BS and continues to monitor this new channel for incoming paging calls. Where the MS has moved between BSs in different locations areas, it must perform a location update and inform the network of its new position. In any case this transition between BSs whilst in the idle mode is termed an *idle mode handover*.

Returning to the person making the call from the fixed network, the entire process will be initiated by the person lifting the handset and dialling the telephone number of the mobile subscriber. This telephone number will include an area code, which will indicate that the call is destined for a mobile subscriber and it will also indicate the home network of the subscriber. Where the fixed network subscriber and the mobile subscriber are in different countries, the telephone number will also include a country code to indicate the country of the subscriber's home network.

On receiving a number with this area code, the PSTN/ISDN network will route the call to the *gateway switch* of the mobile network. The PSTN/ISDN network will also supply the mobile network with the telephone number of the mobile subscriber. The gateway switch represents the interface between the mobile network and other networks. Using the subscriber's telephone number the gateway switch will interrogate the mobile network's HLR to recover the subscriber's records. These will show that the subscriber is registered in a location area in Southampton and the gateway will route the call to the MSC that serves this location area.

With the introduction of second generation networks, the subscriber has access to a number of services that may be used to manage her incoming calls. For example, the subscriber may choose to route all calls to an answering service or a different telephone number if she is otherwise engaged. This information may be stored in the HLR and it will be used to route incoming calls appropriately. In our example we will assume that the mobile subscriber has not redirected her calls in any way.

Once the call arrives at the MSC, the MS must be paged to alert it to the presence of an incoming call. A paging call will be issued from each BS in the location area in which the subscriber is registered. On receiving a paging call containing its address, the MS responds by initiating the *access procedure*. This process is different for different types of cellular system; however, we will base this discussion on the processes that occur in the GSM system. The access procedure commences with the MS sending a message to the BS requesting a channel. The BS will reply by sending the MS details of a dedicated channel and the MS will retune to this channel. A certain degree of handshaking will occur to ensure that the identity of the subscriber is correct and that both the BS and the MS are correctly decoding the information sent on the dedicated channel. This channel will carry signalling information used for the initial call set-up and, in the case of GSM, it will be a dedicated signalling channel. Assuming the incoming call is a voice call, the speech channel will be assigned at a later stage.

Once the dedicated signalling channel has been established, any security procedures, e.g. subscriber authentication, will take place over this channel. When this process is completed, a few more messages will pass between the MS and BS to ensure that the MS has the ability to receive the call. Following this the network will allocate a dedicated speech channel and both the BS and MS will retune to this channel and establish a connection. It is important to note that all of these processes have been performed autonomously by the MS and no interaction is required from the subscriber. In fact, to this point the mobile subscriber is completely unaware of the presence of an incoming call.

It is only once all these processes have been completed that the MS begins to ring and the subscriber is made aware that someone is calling her. The subscriber will indicate that she wishes to receive the call by pressing the 'off-hook' button, or in some cases the MS may automatically answer the call. The affirmation is signalled back to the network by the MS and the incoming call is finally routed to the subscriber's MS, and the two parties may start conversing.

We recall that our mobile subscriber is travelling along the M3 motorway and shortly after the call commences the subscriber moves out of the coverage area of the current BS and into the coverage area of a new BS. It is important, if the call is to be maintained, that it is routed to the MS through the new BS. The process by which an MS moves between BSs while a call is in progress is called a *handover*. There are a number of different ways in which the cellular system can determine which new BS it should use as it moves out of the coverage area of its current BS. In the case of both the GSM and cdmaOne systems, the handover process is mobile assisted in that the MS continually measures the received signal strength from its own BS and also from the neighbouring BSs. These measurements are reported back to the network and they are used to determine the most appropriate cell for handover purposes. Once the network has decided that an MS is to be handed over between two BSs, it issues a handover command from the current BS on the dedicated speech channel and this will contain details of the new dedicated speech channel on the new BS. The MS will then retune to this new channel and a small amount of handshaking will occur to ensure that the new connection has been established correctly. Once this has occurred the call is routed to the MS via the new BS and the call continues normally. The handover process will result in a short break in the speech path; however, this will normally be imperceptible to the people involved in the conversation. We have essentially described a hard handover. cdmaOne employs soft handovers where the link to the new BS is established before the old link is relinquished. This means that there will be no break in the signal path, which is very attractive from the point of view of data transmission where breaks can result in data loss.

Once the call has ended and one party has put down the telephone (or pressed the on-hook button in the case of the mobile subscriber) the *call clear down* process will be initiated. This consists of a small exchange of signalling information which ensures that both the network and the MS know that the call has been ended by one of the parties. Once this process is complete the MS will return to its idle mode and it will monitor the paging channel of its current cell.

Calls initiated by the mobile subscriber follow a very similar process to that outlined above; however, the process will commence with the subscriber dialling a number into the MS and pressing the 'send' button. Following this, the MS will access the network using the access procedure and the process will continue as described above.

◇ ◇ ◇

We trust that this chapter has given you some insight into cellular radio. We now get into specifics by describing the GSM system.

Bibliography

[1] Mouly M. and M.-B. Pautet, *The GSM System for Mobile Communications*, Published by the authors, 1992.

[2] Steele R. [Ed.], *Mobile Radio Communications*, John Wiley and IEEE Press, 1992.

[3] Viterbi A.J., *Principles of Spread Spectrum Communications*, Addison–Wesley, 1995.

[4] Rappaport T.S., *Wireless Communications*, Prentice-Hall, 1996.

[5] Macario R.C.V. [Ed.], Modern personal radio systems, *IEE Telecommun. Ser.* **33**, 1996.

[6] Gibson J.D. [Ed.], *The Mobile Communications Handbook*, CRC Press and IEE Press, 1966.

[7] Prasad R., *CDMA for Wireless Personal Communications*, Artech House, 1996.

[8] Jabbari B., P. Godlewski and X. Lagrange [Eds], *Multiaccess, Mobility and Teletraffic for Personal Communications*, Kluwer Academic, 1966.

[9] Nanda S. and D.J. Goodman, *Third Generation Wireless Information Networks*, Kluwer Academic, 1992.

[10] Wong W.C., R. Steele and C.-E. Sundberg, *Source-matched Mobile Communications*, Pentech Press and IEEE Press, 1995.

[11] Proakis J.G. *Digital Communications*, McGraw-Hill, 1995.

[12] Parsons J.D. and J.G. Gardiner, *Mobile Communications Systems*, Blackie, 1989.

[13] Griffiths J., *Radio Wave Propagation and Antennas – An Introduction*, Prentice-Hall International, 1987.

[14] Webb W.T. and L. Hanzo, *Quadrature Amplitude Modulation*, John Wiley and IEEE Press, 1994.

[15] MacDonald V.H., The cellular concept, *Bell Syst. Tech. J.*, **58**, 1979, 15–41.

[16] Forney G.D. (Jr), The Viterbi algorithm, *Proc. IEEE*, **61**, Mar 1973, 268–278.

[17] Cheung J.C.S. and R. Steele, Soft-decision feedback equaliser for continuous phase modulated signals in wideband mobile radio channels, *IEEE Trans. Commun.*, **42**(2/3/4), Feb/Mar/April 1994, 1628–1638.

[18] Harmuth H.F., *Transmission of Information by Orthogonal Functions*, Springer-Verlag, New York, 1969.

[19] Lam W.H. and R. Steele, Performance of direct-sequence, spread-spectrum multiple-access systems in mobile radio, *IEE Proc.-I*, **138**(1), Feb 1991, 1–14.

[20] Pursley M.B. and D.V. Sarwate, Performance for evaluation of phase-coded spread spectrum multiple access communication. Part I: System analysis, *IEEE Trans. Commun.* **25**, 1977, 795–799.

[21] Pursley M.B. and D.V. Sarwate, Performance for evaluation of phase-coded spread spectrum multiple access communication. Part II: Code sequence analysis, *IEEE Trans. Commun.* **25**, 1977, 800–803.

[22] Baier A., U.-C. Fiebig, W. Granzow, W. Koch, P. Teder and J. Thielecke, Design study for a CDMA-based third generation mobile radio system, *IEEE J. Sel. Area. Commun.*, **12**(4), May 1994, 733–743.

[23] Li Y. and R. Steele, Serial interference cancellation method for CDMA, *Electron. Lett.*, **30**(19), 15 Sept 1994, 1581–83.

[24] Mowbray R.S., R.D. Pringle and P.M. Grant, Increased capacity CDMA system capacity through adaptive cochannel interference regeneration and cancellation, *IEE Proc.-I*, **139**(5), Oct 1992, 515–524.

[25] Henry P.S., Spectrum efficiency of a frequency-hopped-DSPK spread spectrum mobile radio system, *IEEE Trans. Veh. Technol.*, **VT-28**(4), Nov 1979, 327–332.

[26] Cooper G.R. and R.W. Nettleton, A spread-spectrum technique for high capacity mobile communications, *IEEE Trans. Veh. Technol.*, **VT-27**, Nov 1978, 264–275.

[27] Boucher J.R., *Voice Teletraffic System Engineering*, Artech House, 1988.

[28] Oon T.-B., R. Steele and Y. Lee, Cancellation frame size for a quasi-single-bit detector in asynchronous CDMA channel, *Electron. Lett.*, **33**, Feb 1997, 258–259.

[29] Oon T.-B., R. Steele and Y. Lee, Performance of an adaptive successive serial-parallel CDMA cancellation in flat Rayleigh fading channels, *IEEE Veh. Tech. Conf.*, **1**, 1997, 193–197.

[30] Oon T.-B. and R. Steele, Maximum likelihood channel estimation of flat Rayleigh multi-user CDMA channels, *IEE Colloq. (Digest)*, **129**, May 1997, 5/1–5/7.

[31] Oon T.-B. and R. Steele, Successive serial parallel cancellation and joint detection CDMA in flat fading asynchronous channels, *Electron. Lett.*, **34**, Jan 1998, 42–43.

Chapter 2

The GSM System

2.1 Introduction

In 1982, the main governing body of the European telecommunication operators, known as CEPT (Conférence Européene des Postes et Télécommunications), created the Groupe Spécial Mobile (GSM) committee and tasked it with specifying a pan-European cellular radio system to operate in the 900 MHz band. The system was conceived to overcome the perceived capacity limitations of the successful analogue systems already deployed in several European countries (e.g. the Nordic Mobile Telephone system, NMT, in the Nordic countries). The pan-European cellular standard would support international roaming and provide a boost for the European telecommunications industry. The power centres behind the proposed system were the 12 countries of the European Economic Community (EEC), the 26 countries involved in CEPT and the French and German PTTs. There was also strong support from the Nordic countries and the UK Government and industry. The French and German alliance, formed in 1983, was joined by Italy in 1985, and in 1986 the UK joined to form the Quadripartite [1].

After initial discussions, three working parties (WPs) were created to deal with specific aspects of the system definition, and later on a fourth WP was added. In 1986, a permanent nucleus was set up in Paris to co-ordinate the efforts of the working parties and also manage the generation of the system recommendations. The WPs were required to define the system interfaces that would allow a mobile, in the form of either a hand-held or vehicular mounted unit, to roam throughout the countries where the new system had been deployed and have access to the full range of services. Compared with the existing analogue systems, the new system was required to have a higher capacity, comparable or lower operating costs and a comparable or better speech quality. The system was also required to co-exist with the analogue systems. A common pan-European bandwidth allocation for the new system of 890–915 MHz and 935–960 MHz was agreed; however, by the time the system was to be deployed, parts of this band would be occupied by analogue cellular systems in some countries (e.g. the Total Access Communications System, TACS, in the United Kingdom). In these countries only a portion of the band would be used initially for GSM.

Although studies in various European countries had concluded that digital systems were to be preferred over analogue systems, the choice of the multiple access scheme was not as clear-cut. It was decided that a number of different system proposals, put forward by companies and consortia from a number of different European countries, should be evaluated in prototype form. There were eight different system proposals. The MATS-D system proposed by TEKADE incorporated three different multiple access schemes, namely code division multiple access (CDMA), frequency division multiple access (FDMA) and time division multiple access (TDMA). The CD900 system proposed by SEL was a wideband TDMA system in conjunction with spectral spreading [2, 3]. The remaining six proposals were all based on narrow-band TDMA. The SFH900 system proposed by LCT used frequency hopping in combination with Gaussian minimum shift keying (GMSK) modulation, Viterbi equalisation and Reed–Solomon channel coding. Bosch proposed the S900-D system, which used four-level frequency shift keying (FSK) modulation, and Ericsson proposed the DMS90 system which used frequency hopping, GMSK modulation and an adaptive decision feedback equaliser (DFE). The Mobira system and the MAX II system proposed by Televerket were similar to the DMS90 system. Finally, the system proposed by ELAB of Norway employed adaptive digital phase modulation (ADPM) and a Viterbi equaliser to combat the effects of intersymbol interference (ISI). Some of the basic features of the eight different systems are given in Table 2.1 [4].

The different systems were trialled in Paris at the end of 1986 and the most spectrally efficient (and 'unofficial winner') was the system proposed by ELAB. During 1987 the results of the trial were discussed and eventually agreement was reached on the main characteristics of the new system. The wideband solutions advocated by the French and Germans were not adopted for a number of reasons, including the probability that the 1 μm VLSI technology, needed to support the complex baseband signal processing required by these systems, might not be available within the proposed time-scales. By June 1987 there was complete agreement that the system should employ narrow-band TDMA and that it would have many of the features of the ELAB system. The system would initially support eight channels per carrier with eventual evolution to 16 channels per carrier.

The speech codec was chosen based on a subjective comparison of six different codecs at 16 kb/s. The two codecs which performed significantly better than the others were a residual excited linear prediction (RELP) codec and a multipulse excitation linear prediction codec (MPE-LPC). These two designs were merged to produce a regular pulse excitation LPC (RPE-LPC) with a net bit rate of 13 kb/s.

Table 2.1: Some basic features of the GSM prototype systems.

	Multiple access method	Transmission bit rate (kb/s)	Carrier spacing (kHz)	Modulation scheme	Channels per carrier
CD-900	CDMA/TDMA	7980	4500	4-PSK	63
MATS-D	CDMA/TDMA	2496	1250	QAM	32
	FDMA	19.5	25	GTFM	1
ELAB	TDMA	512	600	ADPM	12
DMS90	TDMA	340	300	GMSK	10
MOBIRA	TDMA	252	250	GMSK	9
SFH-900	TDMA	200	150	GMSK	3
S900-D	TDMA	128	250	4-FSK	10
MAX II	TDMA	104.7	50	8-PSK	4

The success of the ELAB system in the Paris trials focused attention on the Viterbi equaliser, which out-performed the DFE used in other systems. Although Reed–Solomon channel coding was heavily favoured amongst the prototype systems, the high levels of synergism between the Viterbi equaliser and the convolutional decoder, which also employs the Viterbi algorithm, meant that convolutional channel coding was chosen for the new system. The adaptive digital phase modulation (ADPM) scheme employed by the ELAB system was initially chosen as the main candidate for the new system. However, GMSK was later preferred because of its improved spectral efficiency.

The initial drafts of the GSM specifications became available around the middle of 1988 and by the end of that year the GSM working parties and the associated expert groups had completed a substantial part of the specifications of the pan-European system. Around this time it became clear that it would not be possible to fully specify every feature of the proposed system in time for the launch in 1991. For this reason, the system specification was divided into two phases. The most common services (e.g. call forwarding and call barring) were included in the Phase 1 specifications which were frozen in 1990. The remaining services (e.g. supplementary services and facsimile) were delayed until the Phase 2 release. The second phase was also used to rectify faults in, and improve the performance of, the Phase 1 system.

At the request of the United Kingdom, a version of GSM, operating in the 1800 MHz band, was included in the specification process, tailored to the requirements of the emerging Personal Communications Networks (PCN). This system became known as the Digital Cellular System at 1800 MHz (DCS1800). From this point onwards we shall use the term GSM900 to describe the system operating in the 900 MHz band to distinguish it from the 1800 MHz system. The term GSM will be used to refer to both systems. It is important to note that the term GSM1800 is also commonly used to refer to the DCS1800 system. The DCS1800 system adaptation began in 1990 and, in 1991, Phase 1 of the DCS1800 system specifications were frozen and were subsequently released as a set of amendments to the GSM900 Phase 1 specifications. The amendments were termed delta recommendations or Δ-recs. In Phase 2 of the specifications, which were frozen in June 1993, the GSM900 system and the DCS1800 system were combined into the same set of documents.

During the development of the Phase 2 specifications it became clear that the task of revising the specifications for a third time would be huge. For this reason it was decided that beyond Phase 2 the GSM system should gradually evolve as new features arrive and this continual evolution has become known as Phase 2+. Some of the more significant features proposed for Phase 2+ included the half-rate speech coder, an increase in the maximum mobile speed for reliable communications and a higher power 4 W mobile for the DCS1800 system.

In 1988, GSM became a Technical Committee of the newly created European Telecommunications Standards Institute (ETSI). Each of the four working parties became Sub-Technical Committees (STCs). At the end of 1991 the scope of the GSM Technical Committee was widened to include the specification of a successor to GSM and, for this reason, the technical committee was renamed the Special Mobile Group (SMG) with the STCs becoming SMG1 to 4. SMG5 was added with the task of specifying the Universal Mobile Telecommunication System (UMTS), GSM's successor [5]. Several other STCs have been added and their responsibilities are summarised in Table 2.2. The term GSM is still used to describe the system, but it has been renamed 'The Global System for Mobile Communications'. SMG5 has since been discontinued and the task of developing the specifications for UMTS has been distributed among the other committees.

In this chapter we give an overview of GSM. We have concentrated our description on the GSM radio interface, since this has a direct impact on the capacity of a cellular system. The reader wishing to learn more about GSM is either referred to books that are solely dedicated to describing the system [6, 7], or the complete GSM specifications themselves which run to some 5000 pages and describe all the complexities of GSM.

2.2   An Overview of the GSM Network Architecture

In this section we briefly examine the different components that together make up a GSM network. Many of these components are common to any cellular network; however, a few are peculiar to GSM. We also note that GSM sometimes uses its own terminology to describe familiar components. A block diagram showing the simplified hierarchical structure of the GSM public land mobile network (PLMN) is given in Figure 2.1.

Table 2.2: Responsibilities of SMG committees within ETSI.

SMG	Responsibilities
1	Definition of the services and facilities of the systems within the scope of SMG (i.e. GSM900, DCS1800, UMTS)
2	Specification of the physical layer of the radio interface of GSM900, DCS1800 and UMTS
3	Specification of the network aspects of GSM900, DCS1800 and UMTS
4	Specification of the data and telematics services for GSM900, DCS1800 and UMTS
5	Co-ordination of the specification of UMTS (discontinued)
6	Specification of the network management functions of GSM900, DCS1800 and UMTS
7	Specification of the mobile station conformity requirements for GSM900 and DCS1800
8	Development of base station system testing procedures for GSM900, DCS1800 and UMTS
9	Specification of the subscriber identity module and mobile equipment interface for GSM900, DCS1800 and UMTS
10	Specification of the security aspects of GSM900, DCS1800 and UMTS
11	Specification of the speech coding aspects of GSM900, DCS1800 and UMTS
12	Specification of the system architecture aspects of GSM900, DCS1800 and UMTS

2.2.1 The mobile station

A subscriber will use a mobile station (MS) to make and receive calls via the GSM network. The MS is composed of two distinct functional entities, the subscriber identity module (SIM), which is a removable smart card containing information that is specific to a particular subscriber, and the mobile equipment (ME), which is essentially the mobile phone itself without the SIM.

The ME may be sub-divided into three functional blocks. The first is the terminal equipment (TE) and this performs functions that are specific to a particular service, for example a fax machine. The TE does not handle any functions that are specific to the GSM system. The second functional block is the mobile termination (MT) and this carries out all the func-

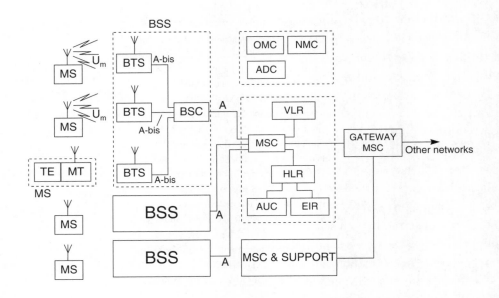

Figure 2.1: GSM network architecture.

tions relating to the transmission of information over the GSM radio interface. Finally, the third functional block is the terminal adapter (TA), which is used to ensure compatibility between the MT and the TA. For example, a TA would be required to interface between an ISDN-compatible MT and a TA with a modem interface.

The SIM is a 'credit-card' size (or smaller, in the case of some handheld units) smart card which can be used by a subscriber to 'personalise' an ME. We emphasize that, in GSM terminology, the term MS refers to the combination of a SIM and an ME. The SIM has an area of non-volatile memory which is used to store information specific to a particular subscriber [8] and this includes the subscriber's unique international mobile subscriber identity (IMSI) number. This number is used to identify each individual subscriber within the GSM network and it consists of not more than 15 decimal digits. The first three digits of the IMSI form the mobile country code (MCC) and this is used to identify the country of the particular subscriber's home network, i.e. the network with which the subscriber is registered. The subscriber will always be billed through her home network, even when she incurs call charges on other networks.

The next two digits of the IMSI form the mobile network code (MNC) and this identifies the subscriber's home PLMN within the country indicated by the MCC. The MNCs are allocated by a relevant authority within each country. The remaining digits of the IMSI are the mobile subscriber identification number (MSIN) which is used to uniquely identify each subscriber within the context of their home PLMN. From this discussion it is clear that the IMSI is unique to each individual subscriber and it may also be used to determine the subscriber's home network.

The SIM will also contain the subscriber's secret authentication key, K_i, the authentication algorithm, A3, and the cipher key generation algorithm, A8. The functions of each of these items will be examined in detail in Section 2.5, suffice to say at this point that they are used to implement the security features of GSM and they are stored in the SIM under heavy protection. The language preference indicator is also located in the SIM and this is used to indicate the language to be used on the MS screen.

The items described above are mandatory and must be present in any SIM that conforms to the GSM specifications. The SIM may also contain a number of optional items which will include the subscriber's abbreviated dialling numbers. This is a list of the subscriber's commonly used telephone numbers that may be accessed using short numeric codes or using a system of menus. The SIM may also contain a list of the last number(s) that the subscriber has dialled and an area of storage for the subscriber's short messages. GSM provides the facility for a subscriber to send and receive short alpha-numeric text messages from their MS and this facility has been termed the *short message service* (SMS).

Inserting an SIM card into an ME effectively personalises the equipment to the particular subscriber. Any incoming calls for the subscriber will be routed to the ME and any charges incurred using the ME will be billed to the subscriber's account. This feature allows subscribers easily to switch between different MEs when, for example, their ME has been returned for repair and a different ME must be used temporarily.

One of the major motives behind the development of the GSM system was to allow subscribers the freedom to roam throughout Europe whilst maintaining the ability to make and receive calls using the same MS. This is only possible where compatible networks exist in each country and it is not possible to roam between GSM900 and DCS1800 networks using the same MS, unless the MS has a dual-mode capability allowing it to operate with the 900 MHz and 1800 MHz systems. However, the SIM card has also introduced the concept of 'SIM roaming' whereby a subscriber may roam between different, incompatible networks by renting an appropriate ME and personalising it with her SIM card. This facility has become particularly attractive with the introduction in the United States of the PCS1900 system (or GSM1900 as it is commonly called), which is a derivative of the GSM system operating in the US 1900 MHz PCS band. Although GSM900 and DCS1800 equipment will be incompatible with the PCS1900 system, a GSM900 subscriber from Europe could rent a PCS1900 ME whilst visiting the United States and, by inserting her SIM card into the rented equipment, would be able to receive calls via her normal telephone number and have the resulting charges billed to her home account. The ability to perform normal roaming or SIM card roaming will be subject to the existence of an appropriate roaming agreement between the subscriber's home network operator and a network operator in the country where the subscriber would like to roam.

In first generation analogue cellular systems, a user's unique electronic serial number (ESN) is programmed directly into the subscriber unit (i.e. the mobile phone). In the event that a subscriber decides to switch to a different network, the mobile phone must be exchanged or reprogrammed. The introduction of the SIM card allows the subscriber complete freedom to switch between different networks without the need to exchange or reprogramme the ME itself. For example, if a GSM900 subscriber wishes to switch from Network 1 to Network 2, the old SIM card can be replaced with a new one obtained from the operator of Network 2 and the same GSM900 mobile phone may be used on the new network.

The situation is slightly more complicated in some countries (e.g. the United Kingdom) where the handsets are subsidised by the network operator and this subsidy must be returned if a subscriber decides to change networks. In this case a locking mechanism will be included such that a handset cannot be used with an SIM from a different network operator until an unlocking code has been entered. The subscriber must make a payment to the existing network operator to receive this unlocking code and this represents the repayment of the handset subsidy. The interface between the SIM and the ME is fully defined in the specifications and is referred to as the SIM–ME interface. This ensures compatibility between the SIMs and MEs of different manufacturers.

2.2.2   The base station subsystem

An MS communicates with a base transceiver station (BTS) via the radio interface, U_m. A BTS performs all the transmission and reception functions relating to the GSM radio interface along with a degree of signal processing. In some ways a BTS can be considered to be a complex radio modem that takes the up-link radio signal from an MS and converts it into data for transmission to other machines within the GSM network, and accepts data from the GSM network and converts it into a radio signal that can be transmitted to the MS. The BTSs are used to form the coverage cells in GSM and it is their position that determines the network's coverage and capacity.

Although a BTS is concerned with transmission and reception over the radio interface, it plays only a minor role in the way the radio resources are allocated to the different MSs. Instead, the management of the radio interface is performed by a base station controller (BSC). The management functions include the allocation of radio channels to MSs on call set-up, determining when a handover is required and identifying a suitable target BTS and controlling the transmitted power of an MS to ensure that it is just sufficient to reach its serving BTS. BSCs vary from manufacturer to manufacturer, but a BSC might typically control up to 40 BTSs. In addition to its processing capacity, a BSC will also have a limited switching capability, enabling it to route calls between the different BTSs under its control. The interface between a BSC and an associated BTS is known as the *A-bis interface* and it is fully defined by an open, or public, specification. In theory this allows a network operator the freedom to procure their BSCs and BTSs from different equipment manufacturers. The BTS and BSC are collectively known as the base station subsystem (BSS).

2.2.3   The mobile services switching centre

Referring to Figure 2.1, we see that each BSS is connected to a mobile services switching centre (MSC). The MSC is concerned with the routing of calls to and from the mobile users. It possesses a large switching capability that varies between manufacturers, but a typical MSC will control a few tens of BSCs and it will have a capacity of several tens of thousands of subscribers. The MSC is similar to the switching exchange in a fixed network. However, it must include additional functions to cope with the mobility of the subscribers, e.g. functions to cope with location registration and handover. The GSM specifications use the term *MSC area* to describe the part of a network that is covered by a particular MSC and its associated BSCs and BTSs. The interface between the MSC and BSS is known as the A interface and it is fully defined in the specifications, giving the network operator the freedom to choose their MSCs and BSCs from different manufacturers. The interface between different MSCs is called the E interface.

   The network operator may also select one or a number of MSCs to act as gateway MSCs (GMSC). As its name would suggest, the GMSC provides the interface between the PLMN and external networks. In the event of an incoming call from another network, the GMSC communicates with the relevant network databases to ensure that the call is routed to the appropriate MS.

2.2.4   The GSM network databases

In the previous sections we have examined the various components within the GSM network that are used to form the communication path between an MS and another MS or a user on another network, e.g. the public switched telephone network (PSTN). Equally important in a commercial network are the means of charging and billing subscribers, maintaining accurate subscription records and preventing fraudulent network access. In a cellular network where subscribers are free to roam throughout the coverage area, the network must also possess some way to track MSs so that it is able successfully to route incoming calls to them. All of these functions are supported using a combination of databases or registers.

The home location register (HLR) is used to store information that is specific to each subscriber. It will contain details of a particular user's subscription, e.g. the services to which they have access, and some information relating to the location of each subscriber, e.g. the details of the MSC area within which the subscriber is currently registered. The information contained within the HLR may be accessed using either the subscriber's IMSI or mobile station international ISDN (MSISDN) number, which is essentially the subscriber's telephone number. Every GSM subscriber will have an entry in the HLR of their home network. The interface between an HLR and an MSC is called the C interface.

Another GSM database that is very closely associated with the HLR is the authentication centre (AuC). The AuC is solely used to store information that is concerned with GSM's security features, i.e. user authentication and radio path encryption. It will contain the subscriber's secret K_i key and the $A3$ and $A8$ security algorithms. The functions of the K_i key and the security algorithms are described in detail in Section 2.5. The AuC will only ever communicate with the HLR and it does this using the H interface.

Another important database used in the GSM system is the visitor location register (VLR). A VLR is associated with one or a number of MSCs and it contains information relating to those subscribers that are currently registered within the MSC area(s) of its associated MSC(s). The area that is served by a particular VLR is termed the *VLR area*. It is termed the *visitor* location register because it holds information on those subscribers that are visiting its VLR area. The main function of the VLR is to provide a local copy of the subscriber's information for the purposes of call handling and it removes the need to continually access the HLR to retrieve information about a particular subscriber. This becomes important in a system such as GSM where subscribers may use networks in countries other than the country of their home network. The VLR also contains information that enables the network to 'find' a particular subscriber in the event of an incoming call.

The process of locating a subscriber is facilitated by subdividing the network's coverage area into a number of *location areas*, each consisting of one or a number of cells or sectors. The VLR will contain the details of the location area in which each subscriber is registered. In the event of an incoming call, an MS will be paged in each of the cells within its location area and this means that the MS may move freely between the cells of a location area without having to inform the network. However, when an MS moves between cells belonging to different location areas, it must register in the new area using the *location updating* procedure. Where a subscriber moves between location areas controlled by different VLRs, its details are copied from the HLR to the new VLR. The HLR will also ensure that the subscriber's details are removed from the old VLR. The interface between the HLR and the VLR is called the D interface and the interface between an MSC and its associated VLR is called the B interface. An interface also exists between different VLRs and this is termed the G interface.

The introduction of the SIM card in GSM means that tracking a subscriber no longer implies the tracking of a piece of equipment, and vice versa. For this reason the equipment identity register (EIR) has been introduced to allow the network operator to track stolen and malfunctioning MEs. Each ME is assigned a unique 15-digit international mobile equipment identity (IMEI) at the point of manufacture. Each model of ME must undergo a process known as type approval, wherein a number of its features are tested using a GSM system simulator. The type approval testing is carried out by accredited laboratories that are independent of any manufacturing or operating companies and it is used to ensure that all GSM ME models meet a minimum standard, regardless of the manufacturer. Once an ME model has been type approved it will be assigned a six-digit type approval code (TAC) and this forms the first six digits of an ME's IMEI. The next two digits of the IMEI represent the final assembly code (FAC) and this is assigned by the manufacturer to identify the place where the ME was finally assembled or manufactured. The next six digits of the IMEI represent the ME's serial number (SNR) and this will be unique to every MS for a given combination of TAC and FAC. The remaining digit of the 15-digit IMEI is defined as 'spare'.

The EIR is used to store three different lists of IMEIs. The *white* list contains the series of IMEIs that have been allocated to MEs that may be used on the GSM network. The *black* list contains the IMEIs of all MEs that must be barred from using the GSM network. This will contain the IMEIs of stolen and malfunctioning MEs. Finally, the network operator may also use a *grey* list to hold the IMEIs of MEs that must be tracked by the network for evaluation purposes.

During an access attempt or during a call, the network has the ability to command an MS to supply its IMEI at any time. If the IMEI is on the black list or it is not on the white list, the network will terminate the call or access attempt and the subscriber will be sent an 'illegal ME' message. Once an MS has failed an IMEI check it will be prevented from making any further access attempts, location updates or paging call responses. However, this MS may still be used to make emergency calls. The IMEI check is performed within the EIR and the IMEI is passed to the EIR by the MSC that is currently serving the MS. The results of the IMEI check are then returned by the EIR to the relevant MSC. The interface between the EIR and the MSC is termed the F interface.

2.2.5 The management of GSM networks

From an operator's viewpoint, an effective network management system is an important part of any telecommunications network. It is essential for the network operator to be able to identify problems in the network at an early stage and correct them quickly and efficiently. It is also important for the network operator to be able to make changes to the network configuration with a minimum of effort and without affecting the service provided to its subscribers. The functional blocks associated with the management of the GSM network, as shown in Figure 2.1, are the operations and maintenance centre (OMC), the network management centre (NMC) and the administration centre (ADC).

The OMC provides the means by which the operator controls the network. Each OMC will typically be in charge of a subsystem, e.g. the BSS or the Network Switching Subsystem, NSS (i.e. the MSC, HLR, VLR, etc.) The NMC is concerned with the management of the entire network and it generally has a wider operational role than an OMC. The ADC is concerned, as its name would suggest, with the administrative functions required within the network.

2.3   The GSM Radio Interface

The radio interface provides the means by which an MS communicates with the BTSs of a GSM network whilst it moves within the coverage area. The performance of the radio interface, and particularly its ability to provide acceptable speech links in the face of co-channel interference from other users within the system, acutely affects the overall capacity of a cellular system. In this section we examine the features of the GSM radio interface. Figure 2.2 shows a simplified block diagram of the GSM radio link. In the following sections we will examine the function and operation of each of these blocks in some detail. It is difficult to keep to a strict top-down or bottom-up description of the GSM radio interface because, in some cases, in order to appreciate the reason for a particular feature of the radio interface, it is necessary to understand some features associated with a higher or lower layer protocol. For this reason we have adopted a somewhat unconventional approach to the description of the radio interface. We will begin by examining the modulation scheme and the carrier frequencies used in GSM. Then we will discuss the construction of the TDMA bursts, or packets, and the way in which these may be demodulated in the presence of intersymbol interference (ISI) caused by the radio channel and the modulation process itself. Following this we will discuss the different channels that are available in GSM and the manner in which the radio resources are allocated to each of the channels. At this point we will have effectively built up a picture of the radio interface as a 'bit pipe' where data are applied to the transmitter and the same data, possibly with a number of errors, are recovered at the receiver.

We will then turn our attention to the coding, interleaving and ciphering processes that occur on the GSM radio interface. These processes are different for speech information, user data (e.g. fax transmissions) and signalling information and, therefore, we will deal with each of these different types of information separately. Finally, we will bring the two halves of our radio interface description together by describing the manner in which the coded, interleaved and ciphered, or encrypted, data are mapped onto the TDMA bursts.

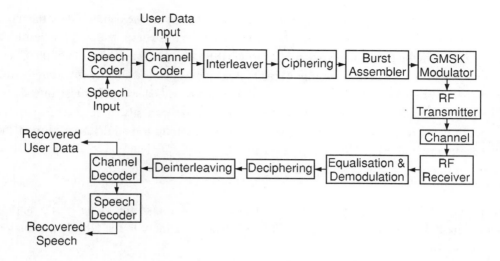

Figure 2.2: Block diagram of a GSM transmitter and receiver.

2.3.1 The GSM modulation scheme

The modulation scheme used in GSM is Gaussian minimum shift keying (GMSK) with a normalised bandwidth product, *BT*, of 0.3 and the modulation symbol rate is around 271 kb/s. For the reader not familiar with GMSK modulation we will include a brief description of its fundamentals. GMSK is based on a simpler modulation scheme known as minimum shift keying (MSK) in which the carrier amplitude remains constant and the information is carried in the form of phase variations. A logical '1' will cause the carrier phase to increase by 90° over a bit period and a logical '0' will cause the carrier phase to decrease by the same amount. This phase change is produced by instantaneously switching the carrier frequency between two different values, f_1 and f_2, according to the input data and, therefore, MSK is a special case of FSK modulation. The frequencies f_1 and f_2 are given by

$$
\begin{aligned}
f_1 &= f_c + R_b/4, \\
f_2 &= f_c - R_b/4,
\end{aligned} \tag{2.1}
$$

where R_b is the modulation symbol rate ($\approx$ 271 kb/s in GSM) and f_c is the nominal carrier frequency. It is interesting to note that, in MSK, the carrier frequency, f_c, is never transmitted.

This shows that MSK requires instanteous changes in the carrier frequency and, consequently, the modulated spectrum is, in theory, infinitely wide. The spectrum of an MSK modulated signal may be compressed by filtering the modulating baseband pulses to produce much smoother changes in frequency, thereby compressing the bandwidth of the modulated signal. The type of filter used has a Gaussian impulse response and the resulting modulation scheme is called Gaussian MSK or GMSK. The relative bandwidth of the Gaussian filter defines the spectrum compression that is achieved, i.e. a smaller filter bandwidth results in a narrower modulated spectrum. Unfortunately, the Gaussian filter also introduces ISI whereby each modulation symbol spreads into adjacent symbols.

The ith data bit, d_i, is differentially encoded by performing a modulo-2 addition of the current and previous bits. This is expressed as

$$\hat{d}_i = d_i \oplus d_{i-1}, \tag{2.2}$$

where $\hat{d}_i$ is the differentially encoded ith data bit, d_i may take the value 0 or 1 and $\oplus$ denotes modulo-2 addition [9]. The modulating data at the input to the GMSK modulator, α_i, is given by

$$\alpha_i = 1 - 2\hat{d}_i, \quad \hat{d}_i = 0, 1, \tag{2.3}$$

where α_i may take the values ± 1. The process detailed in Equation(2.3) has the effect of mapping the differentially encoded data bits, $\hat{d}_i$, onto the logical levels ± 1 such that

$$\hat{d}_i = 0 \rightarrow \alpha_i = +1,$$
$$\hat{d}_i = 1 \rightarrow \alpha_i = -1. \tag{2.4}$$

The modulating data, α_i, are then passed through a linear filter with a Gaussian-shaped impulse response, $h(t)$, given by

$$h(t) = \frac{1}{\sqrt{2\pi}\sigma T} \, e^{-\frac{t^2}{2\sigma^2 T^2}}. \tag{2.5}$$

where

$$\sigma = \sqrt{\ln(2)}/2\pi BT, \tag{2.6}$$

T is the bit period and B is the 3 dB filter bandwith. The BT product is the relative bandwidth of the baseband Gaussian filter and in GSM it is set to 0.3. This effectively means that each bit is spread over (or has an effect on) three modulation symbols. The resulting ISI must be removed at the receiver using an equaliser. The impulse response, $h(t)$, and the frequency response, $H(f)$, of this filter are shown in Figure 2.3(a) and (b), respectively. We note that in each figure the amplitude has been normalised to give a maximum value of 1, the time axis in Figure 2.3(a) has been normalised to T and the frequency axis in Figure 2.3(b) has been normalised to $1/T$.

The pulse response of this filter, $g(t)$, i.e. the signal that appears at the output of the filter when a pulse of width T is applied to the input, is given by

$$g(t) = h(t) * \text{rect}(t/T), \tag{2.7}$$

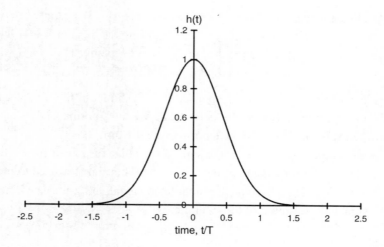

(a) Impulse response

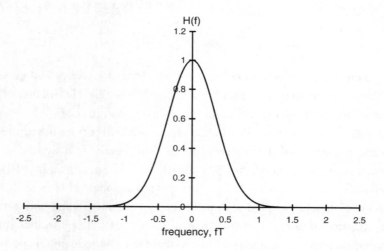

(b) Frequency response

Figure 2.3: The impulse and frequency responses of the Gaussian filter used on GMSK.

where rect(t/T) is defined as

$$\text{rect}(t/T) = \begin{cases} 1/T, & |t| < T/2, \\ 0, & \text{otherwise,} \end{cases} \tag{2.8}$$

and $*$ denotes convolution. The pulse response, $g(t)$, is shown in Figure 2.4 and we note that it extends for approximately three bit periods, T. We have normalised the amplitude of $g(t)$ to a maximum value of 1 and we have normalised the time axis to T.

The signal at the output of the filter is the sum of the pulse responses for each input data bit, as shown in Figure 2.5 for a data sequence of 0010. This signal is used to modulate the frequency of the carrier. The phase of the modulated signal, $\varphi(t)$, may be determined by integrating the signal at the output of the filter, i.e.

$$\varphi(t) = \sum_i \alpha_i \pi m \int_{-\infty}^{t-iT} g(u)du, \tag{2.9}$$

where the modulation index, m, is 1/2 (i.e. the maximum phase change over a data interval is $\pi/2$ radians). Given Equation (2.9), the modulated RF carrier signal may be expressed as

$$x(t) = \sqrt{\frac{2E_c}{T}} \cos(2\pi f_0 t + \varphi(t) + \varphi_0), \tag{2.10}$$

where E_c is the energy per modulating bit, f_0 is the carrier frequency and φ_0 is a random phase offset that will remain constant for the duration of a single TDMA burst.

An example of the spectrum of a GSM carrier is shown in Figure 2.6. We observe that the power has only decreased by some 35 dB at an offset of 200 kHz from the centre frequency, which represents the centre of the adjacent carrier. This results in a significant amount of adjacent channel interference between GSM carriers and the specifications [10] define that a receiver will only perform satisfactorily if the wanted channel is no more than 9 dB less than the adjacent channel. Coupled with the effects of shadow fading and power control, this precludes the use of adjacent channels in the same cell. The specifications define a number of transmitted spectrum masks to ensure that the radio transmitters do not generate unacceptable levels of adjacent channel interference. An example of one of these masks is given in Figure 2.7. The transmitted signal must remain below the mask (shown by a dark line) at each frequency offset from the carrier. For example, at a 400 kHz offset from the centre frequency the transmitted power must be 60 dB less than the power at the centre frequency.

2.3.2   The GSM radio carriers

GSM uses a combined time division multiple access (TDMA) and frequency division multiple access (FDMA) scheme. The available spectrum is partitioned into a number of bands,

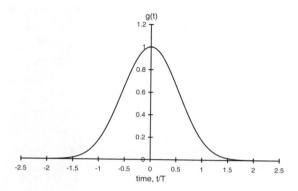

Figure 2.4: The pulse response of the GMSK filter.

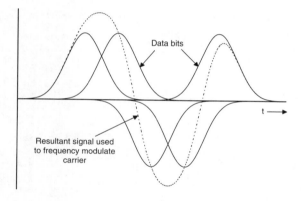

Figure 2.5: The output of the baseband filter.

each 200 kHz wide. Each of these bands may be occupied by a GMSK modulated RF carrier supporting a number of TDMA time slots. The RF carriers are paired to allow a simultaneous data flow in both directions, i.e. full duplex. The GSM900 frequency bands defined in Phase 1 of the specifications [10] are 890 MHz to 915 MHz for the up-link (i.e. MS to BTS) and 935 MHz to 960 MHz for the down-link (i.e. BTS to MS), respectively. In Phase 2 of the specifications an extension frequency band has been added to allow GSM900 operators to provide more capacity in urban areas. For this reason, the frequency bands described above are sometimes called the primary GSM900 bands (P-GSM900). The extended GSM900 bands (E-GSM900) are 880 MHz to 890 MHz and 925 MHz to 935 MHz for the up-link and down-link, respectively. In the case of the DCS1800 system, the Phase 2 specifications define the 1710 MHz to 1785 MHz frequency band for the up-link transmis-

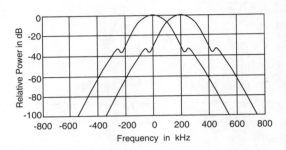

Figure 2.6: A typical GMSK modulated spectrum.

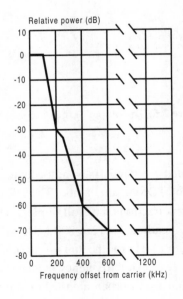

Figure 2.7: The modulated spectrum mask.

sions and the 1805 MHz to 1880 MHz frequency band for the down-link transmissions.

There is a guard band of 200 kHz at the lower end of each frequency band and it is likely that the RF channels at either end of the allocations will not be used. Each RF carrier frequency pair is assigned an absolute radio frequency channel number (ARFCN). In the specifications [10], $Fl(n)$ is used to describe the frequency of the carrier in the lower up-link frequency band with an ARFCN of n, and $Fu(n)$ is used for the upper down-link frequency band. Using this notation the relationship between frequency and ARFCN is given in Table 2.3, where all frequencies are in MHz.

In addition to the frequency separation between the duplex carriers, which is 45 MHz for GSM900 and 95 MHz for DCS1800, the down-link and up-link bursts of a duplex link are separated by three timeslots. This removes the necessity for the MS to transmit and receive simultaneously. Where the propagation delay between the MS and BTS is very small, the MS will receive a down-link burst from the BTS, retune to the up-link frequency and transmit an up-link burst three timeslots later. The timing schedule at the BTS is shown in Figure 2.8. This shows that each duplex carrier supports a number of timeslots that are 15/26 ms ($\approx$577 μs) in duration. These are arranged into TDMA frames consisting of eight time slots with a duration of 60/13 ms ($\approx$4.615 ms). Each timeslot within a TDMA frame is numbered from zero to seven and these numbers repeat for each consecutive frame. The time slot and frame durations are derived from the fact that 26 TDMA frames are transmitted in 120 ms. The reasons for choosing these particular numbers will become clear when we examine GSM's complex frame structure. Suffice to say at this point that the TDMA frame duration is

$$\frac{120}{26}\text{ms} = \frac{60}{13}\text{ms}$$

(2.11)

and the timeslot duration is

$$\frac{120}{26 \times 8}\text{ms} = \frac{15}{26}\text{ms}.$$

(2.12)

Table 2.3: Absolute radio frequency channel numbers.

Band	Frequency			Channel numbers
P-GSM900	$Fl(n) = 890 + 0.2n$	$Fu(n) =$	$Fl(n)$ $+45$	$1 \leq n \leq 124$
E-GSM900	$Fl(n) = 890 + 0.2n$ $Fl(n) = 890 + 0.2(n - 1024)$	$Fu(n) =$	$Fl(n)$ $+45$	$0 \leq n \leq 124$ $975 \leq n \leq 1023$
DCS1800	$Fl(n) = 1710.2 + 0.2(n - 512)$	$Fu(n) =$	$Fl(n)$ $+95$	$512 \leq n \leq 885$

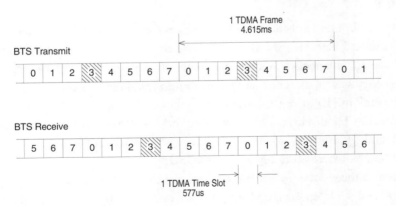

Figure 2.8: Burst schedule at the BTS.

2.3.3   The GSM power classes

Having examined the modulation scheme used in GSM and described the way in which the radio carriers are used by transmitting a burst within a particular time slot, we now look at the transmitted power of these bursts at the MS and the BTS.

The specifications define five classes of MS for GSM900 and two classes for DCS1800 based on their output power capabilities. These classes are shown in Table 2.4 [10]. The typical handheld units are Class 4 for GSM900 and Class 1 for DCS1800 and the typical GSM900 vehicular unit is Class 2. Each MS has the ability to reduce its output power in steps of 2 dB from its maximum down to a minimum of 5 dBm (3.2 mW) for a GSM900 MS and 0 dBm (1 mW) for a DCS1800 MS in response to commands from a BTS. This facility is used to implement up-link power control, whereby an MS's transmitted power is adjusted to ensure that it is just sufficient to provide a satisfactory up-link quality. This process is used to conserve MS battery power and also reduce the up-link interference throughout the system.

Table 2.4: Mobile station power classes.

Power Class	Maximum output power GSM900	Maximum output power DCS1800
1	20 W (43 dBm)	1 W (30 dBm)
2	8 W (39 dBm)	0.25 W (24 dBm)
3	5 W (37 dBm)	
4	2 W (33 dBm)	
5	0.8 W (29 dBm)	

2.3. THE GSM RADIO INTERFACE

In Phase 1 of the GSM specifications, eight classes of GSM900 maximum output powers ranging from 2.5 W up to 320 W, and fc BTS were defined with maximum output powers ranging from 2.5 W power BTS classes were included in Phase 2 for each system and t *BTSs*, as they are intended for use in smaller cells, e.g. microcells. Tl ＿＿classes defined in Phase 2 of the specifications are summarised in Table 2.5 [10]. The actual output power of the BTS may be adjusted in at least six steps of around 2 dB to allow a fine adjustment of the coverage by the network operator. The BTS output power may also be adjusted by up to 15 steps, each of 2 dB, to allow power control to be implemented on the down-link.

2.3.4 The GSM bursts

As we have already seen, each GSM RF carrier supports eight timeslots and the data are transmitted in the form of bursts that are designed to fit within these slots. In this section we will examine the content of these bursts. The GSM specifications define five different bursts, four of which are shown in Figure 2.9.

The normal burst (NB) is the most commonly used burst in GSM. It consists of a 26-bit training sequence surrounded by two 58-bit information blocks. Three tail bits are added at the beginning and the end of the burst. The total duration of the burst is 148 bits leaving a guard period equivalent in duration to 8.25 bits. The training sequence is used to 'sound' the radio channel and produce an estimate of its impulse response at the receiver. This estimate is used in the demodulation process to equalise the effects of multipath propagation. The channel estimate will only be accurate at the instant the sounding is taken and, for this reason, the training sequence is placed in the centre of each burst to minimise the error in the information bits farthest from the training sequence. Consequently the first section of the burst must be stored before demodulation can proceed. The training sequence consists of a 16-bit sequence extended in both directions by copying the first five bits at the end of the sequence and the last five bits at the beginning. The central 16 bits are chosen to have a highly-peaked autocorrelation function, following GMSK modulation, and the repeated bits at either end ensure that the resulting channel estimate may be up to five bits wide before being corrupted by the information bits. The specifications [11] define eight different training sequences for use in the normal burst, each with low cross-correlation properties following GMSK modulation. Each training sequence is described by a training sequence code (TSC). A list of these sequences is given in Table 2.6. Potential co-channel cells will use different training sequences to prevent the channel estimate being corrupted by an interfering signal. The tail bits in the normal burst are always set to zero to ensure that the Viterbi decoder begins and ends in a known state.

The frequency correction burst (FB) is used by the MS to detect a special carrier which is transmitted by every BTS in a GSM network. This carrier is called the broadcast control

Table 2.5: BTS power classes.

	Maximum output power	
BTS power class	GSM900	DCS1800
1	320–(<640) W	20–(<40) W
2	160–(<320) W	10–(<20) W
3	80–(<160) W	5–(<10) W
4	40–(<80) W	2.5–(<5) W
5	20–(<40) W	-
6	10–(<20) W	-
7	5–(<10) W	-
8	2.5–(<5) W	-
Micro-BTS 1	(>0.08)–0.25 W	(>0.5)–1.6 W
Micro-BTS 2	(>0.03)–0.08 W	(>0.16)–0.5 W
Micro-BTS 3	(>0.00)–0.03 W	(>0.05)–0.16 W

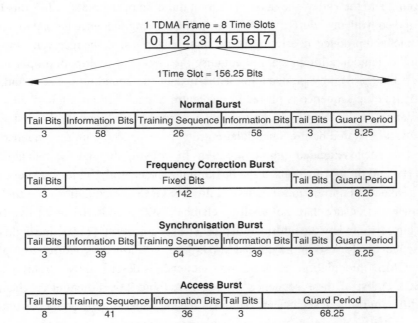

Figure 2.9: The GSM bursts.

Table 2.6: The GSM training sequences.

Training sequence code (TSC)	Training sequence bits (b61, b62, ..., b86)
0	(0,0,1,0,0,1,0,1,1,1,0,0,0,0,1,0,0,0,1,0,0,1,0,1,1,1)
1	(0,0,1,0,1,1,0,1,1,1,0,1,1,1,1,0,0,0,1,0,1,1,0,1,1,1)
2	(0,1,0,0,0,0,1,1,1,0,1,1,0,1,0,0,0,1,0,0,0,1,1,1,1,0)
3	(0,1,0,0,0,1,1,1,1,0,1,1,0,1,0,0,0,1,0,0,0,1,1,1,1,0)
4	(0,0,0,1,1,0,1,0,1,1,1,0,0,1,0,0,0,0,0,1,1,0,1,0,1,1)
5	(0,1,0,0,1,1,1,0,1,0,1,1,0,0,0,0,0,1,0,0,1,1,1,0,1,0)
6	(1,0,1,0,0,1,1,1,1,1,0,1,1,0,0,0,1,0,1,0,0,1,1,1,1,1)
7	(1,1,1,0,1,1,1,1,0,0,0,1,0,0,1,0,1,1,1,0,1,1,1,1,0,0)

channel (BCCH) carrier and we will examine its uses later in this chapter. Suffice to say at this point that it acts as a form of beacon and MSs will search for BCCH carriers to detect the presence of a GSM network. The frequency correction burst is also used by MSs as a frequency reference for their internal timebases. Every bit in the frequency correction burst (including the tail bits) is set to zero and, after GMSK modulation, this results in a pure sinewave at a frequency around 68 kHz (1625/24 kHz) higher than the RF carrier centre frequency.

The synchronisation burst (SB) carries 78 bits of coded data formed into two blocks of 39 bits on either side of a 64-bit training sequence. As its name suggests, this burst carries details of the GSM frame structure and allows an MS to fully synchronise with the BTS. The synchronisation burst is the first burst that the MS has to demodulate and, for this reason, the training sequence is extended to 64 bits. This extended sequence provides a larger autocorrelation peak than the 26-bit sequence of the normal burst. It also allows larger multipath delay spreads to be resolved. All synchronisation bursts use the same training sequence for bit number 42 to bit number 105 in the burst. The arrangement of the training sequence is as shown below

$$
\begin{aligned}
\text{b42, b43, ..., b105} = (&1,0,1,1,1,0,0,1,0,1,1,0,0,0,1,0, \\
&0,0,0,0,0,1,0,0,0,0,0,0,1,1,1,1, \\
&0,0,1,0,1,1,0,1,0,1,0,0,0,0,1,0,1, \\
&0,1,1,1,0,1,1,0,0,0,0,1,1,0,1,1).
\end{aligned}
$$

An MS can use this training sequence to synchronise to the BTS transmissions to within a quarter-bit accuracy.

The final GSM burst shown in Figure 2.9 is the access burst (AB). This consists of a 41-bit training sequence followed by 36 information bits. The access burst is used by the MS to access the network initially and it is the first up-link burst that a BTS will have to demodulate from a particular MS. As with the synchronisation burst, the training sequence is extended to ease the demodulation process. The number of tail bits at the beginning of the burst is increased to eight. The extended tail bits at the front of the burst are

$$b0, b1, b2, ..., b7 \; = \; (0,0,1,1,1,0,1,0).$$

This is followed by the training sequence,

$$
\begin{aligned}
b8, b9, ..., b48 \; = \; &(0,1,0,0,1,0,1,1,0,1, \\
&1,1,1,1,1,1,1,1,0,0,1, \\
&1,0,0,1,1,0,1,0,1,0, \\
&1,0,0,0,1,1,1,1,0,0,0).
\end{aligned}
$$

The tail bits at the end of the burst are all set to zero. We note that the AB is much shorter than the other bursts and this results in a large guard period of 68.25 bit periods. This guard period is included to compensate for the propagation delay between the MS and BTS. Once a duplex link has been established, a closed loop timing advance mechanism is activated to ensure that the MS up-link bursts arrive at the BTS within the correct time slots. However, this is not possible on the AB. Accordingly, a guard period of 68.25 bit periods, equivalent to 252 μs, allows the MS to be up to 38 km from the BTS before its up-link bursts will spill into the next time slot.

We will later describe how the maximum timing advance of the MS decides the maximum range of a BTS, although there are ways to increase the maximum range at the expense of BTS capacity. As a result of its small size, the AB carries relatively little information and this has an impact on the access procedure.

The fifth type of burst is not shown in Figure 2.9. It is the dummy burst (DB) and is similar to the NB in that it has the same structure and uses the same training sequences. The main difference between the DB and the NB is that the information bits on either side of the training sequence are set to a predefined sequence in the DB. The DB is used to fill inactive time slots on the BCCH carrier, which must be transmitted continuously and at a constant power.

The RF output spectrum of the transmitted signals in a TDMA system is not only determined by the modulation process, but also by the switching transients that occur when the bursts of RF energy are transmitted. The switching transients tend to widen the spectrum of the transmitted signal, although this effect can be reduced by ramping the output power up and down when transmitting a burst, instead of just keying the transmitter on and off. The information transmitted in the burst must not be affected by the process of power ramping, which is performed at the beginning and end of the time slot. As we have already seen, the active part of an NB is 148 bit periods in duration. The useful part of a burst in all cases is one bit period shorter than the active part and it begins halfway through the first bit period. During that part of the burst when information is transmitted, the amplitude of the modulated RF signal must stay approximately constant. The power ramping mask, namely the variation in transmitted power with time, for the NB is shown in Figure 2.10(a). The amplitude of the transmitted signal must remain between the two thick lines, i.e. it must be to the non-hatched side of each thick line. This same mask also applies to the FB and SB. The AB has a similar power ramping mask; however, in this case the useful part is reduced to 87 bits. This mask is shown in Figure 2.10(b). Observe that the 70 dB power ramp-up occurs during a 28 μs period which corresponds to 7.6 bit periods, while ramp-down takes place in 18 μs, i.e. 4.9 bit periods.

2.3.5 The GSM receiver

Although the GSM specifications do not define the manner in which the transmitted information should be recovered at the BTS or MS receiver, the bursts have been specifically designed with the Viterbi equaliser in mind. In this section we will discuss the way in which the Viterbi equaliser may be used to recover the information from each GSM burst in the presence of the ISI caused by the radio channel and the GMSK modulation process. We will also examine how the performance of the GSM receiver is tested such that a minimum performance standard may be maintained across all GSM type-approved equipment.

2.3.5.1 The channel equaliser

Figure 2.11 shows the block diagram of a typical GSM baseband link. The term *baseband* is used to indicate that the RF portion has been excluded and the effects of the radio channel and the RF portions of the transmitter and receiver are modelled using an equivalent baseband channel. The figure shows that the bursts, which contain both the data and the training sequence, are passed through a baseband modulator at the transmitter and then through the baseband channel before arriving at the receiver. The received waveform will contain ISI caused by the radio transmission channel and the GMSK modulation process. At the receiver the burst is demultiplexed to give the training sequence and the data bits. The training sequence is used to estimate the impulse response of the radio channel in the channel estimator.

The entire demodulation process is accomplished using digital signal processing techniques. After the signal has passed through the RF front-end at the receiver, it is sampled to produce a complex digital representation of the baseband signal. In Figure 2.11 the flow of

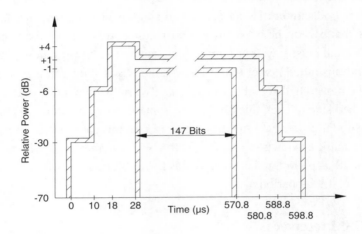

(a) The normal burst power ramping mask

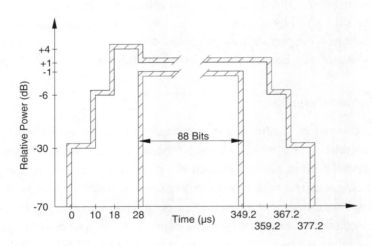

(b) The access burst power ramping mask

Figure 2.10: The power ramping masks in GSM.

complex baseband information is denoted by double arrows.

In order to explain the channel estimation process, we represent the received training sequence, $s_r(t)$, as the convolution of the transmitted training sequence, $s(t)$, and the impulse response of the baseband channel, $h_c(t)$, i.e.

$$s_r(t) = s(t) * h_c(t), \tag{2.13}$$

where $*$ denotes convolution. Passing $s_r(t)$ through a filter with an impulse response, $h_{MF}(t)$, that is matched to the training sequence yields a channel impulse response estimate, $h_e(t)$, that is given by

$$
\begin{aligned}
h_e(t) &= s_r(t) * h_{MF}(t) \\
&= s(t) * h_c(t) * h_{MF}(t) \\
&= R_s(t) * h_c(t), \tag{2.14}
\end{aligned}
$$

where $R_s(t) = s(t) * h_{MF}(t)$ is the autocorrelation function (ACF) of the training sequence. It is important to note that $R_s(t)$ is the autocorrelation of the modulated GMSK symbols that are generated by the training sequence, and not the training sequence bits themselves. The training sequences have been chosen using a computer search so that $R_s(t)$ is an approximation to a Dirac impulse-like function. The actual shape is pulse-like and its magnitude is limited to be less than the number of chips in the sounding sequence. Given that $R_s(t)$ has this property, Equation (2.14) becomes

$$
\begin{aligned}
h_e(t) &\approx \delta(t) \times h_c(t) \\
&\approx h_c(t), \tag{2.15}
\end{aligned}
$$

where $\delta(t)$ is the Dirac impulse function.

The time delay between the first and last recieved multipath components of the estimated channel may be large and there is a correspondence between the size of this delay and the complexity of the Viterbi channel equaliser. Every increase in delay corresponding to a bit duration results in a doubling of the number of states in the equaliser. There is a standardised artificial channel impulse response contrived to test the Viterbi equaliser. The channel impulse response is composed of six impulses of the same average power that are equally spaced over a time of 16 μs, i.e. the spacing between adjacent impulses is 3.2 μs. (This channel impulse response is shown later in Figure 2.13(d).) The GMSK modulator spreads each bit over a bandwidth of three bits. The delay spread of 16 μs corresponds to 16/3.69 or 4.34 bits periods. The total spreading for a 16μs excess delay spread would be 7.34 bit periods.

Recall that the training sequence consists of a 16-bit sounding sequence with five bits appended to either end. This means that channel estimates may be derived for total excess

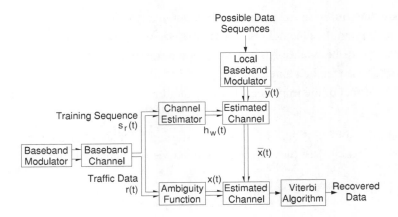

Figure 2.11: The baseband link.

delay spreads of five bit periods. If the delay spread extends beyond five bit periods, then the sounding sequence of the later paths will overlap with the information bits of the earlier paths and this will corrupt the channel estimate to a certain degree because the training sequence will be cross-correlated with information bits.

It is common practice to accommodate a channel excess delay spread of only two bits, corresponding to a duration of 7.38 μs. This is more than adequate for the specified urban and rural channels for GSM which have excess delay spreads of 5 μs and 0.5 μs, respectively. If the equaliser is designed for an excess delay spread of two bits, delays in excess of this may cause errors when it is the function of the de-interleaver and FEC decoder to remove them. Should the estimated channel impulse response, $h_e(t)$, exceed 7.38 μs, the Viterbi equaliser will only make use of 7.38 μs of it. So the question becomes, which 7.38 μs? A windowing procedure is implemented, where a two-bit duration window is moved over $h_e(t)$, and the portion of $h_e(t)$ used is where the maximum energy in $h_e(t)$ resides. At this position the windowed channel estimate, $h_w(t)$, is given by

$$h_w(t) = h_e(t) \cdot w(t), \tag{2.16}$$

where $w(t)$ is the rectangular windowing function. Observe that $h_w(t)$ will hopefully contain the most significant multipath components. However, this may not be so in hilly or mountainous terrains, where distinct reflections may result in substantial energies at large delays.

Returning to Figure 2.11, we note that all possible data sequences are generated locally within the receiver and passed through a local baseband modulator. This produces a number of GMSK symbols, $\bar{y}(t)$, which are then convolved with the estimated channel impulse response $h_w(t)$ to produce a number of *waveform templates*, $\bar{x}(t)$. If the channel estimate were perfect, then the waveform template that corresponds to the transmitted data sequence would exactly match the received waveform, $r(t)$. However, there are a number of imperfections in the channel estimation process. For example, the autocorrelation of the training sequence is not a perfect Dirac function and there are problems associated with windowing the channel estimate. The effect of the imperfect channel sounding may be mitigated to some degree by introducing the same distortions into the data path as in the channel estimation path. Accordingly we introduce an *ambiguity function*, $R_s(t)$, which is the autocorrelation function of the training sequence.

To examine the role of the ambiguity function we first consider the received data signal, $r(t)$, which is given by

$$r(t) = b(t) * h_c(t), \tag{2.17}$$

where $b(t)$ is the transmitted data sequence and $h_c(t)$ is the actual channel impulse response. The received signal is convolved by $w(t)R_s(t)$ to give

$$x(t) = b(t) * w(t)R_s(t) * h_c(t). \tag{2.18}$$

The locally generated waveform templates are produced by convolving each possible data sequence with the windowed channel estimate. This gives

$$
\begin{aligned}
\bar{x}(t) &= \bar{y}(t) * h_w(t) \\
&= \bar{y}(t) * (w(t)h_e(t)) \\
&= \bar{y}(t) * w(t)R_s(t) * w(t)h_c(t) \\
&= \bar{y}(t) * w(t)R_s(t) * h_w(t).
\end{aligned} \tag{2.19}
$$

where $\bar{y}(t)$ is the output of the local baseband modulator, as shown in Figure 2.11. Note that as $\bar{y}(t)$ is one of a number of possible $b(t)$ waveforms, Equations (2.18) and (2.19) only differ in that $x(t)$ includes $h_c(t)$, whereas $\bar{x}(t)$ is forced to have the windowed version, $h_w(t)$, and that $h_c(t)$ and $h_w(t)$ are the same when the impulse response used in forming $\bar{y}(t)$ is completely contained within the window.

Thus, by introducing the ambiguity function into the received data path, as shown in Figure 2.11, the matching between the received waveform and the correct waveform template is significantly improved.

Suppose the equaliser is to accommodate a five-bit overall spreading of the data, resulting from three-bit spreading in the GMSK modulator and two-bit spreading in the channel. So in any bit period we must consider the effect of five bits. Since five bits can yield 32 different binary patterns, we must generate 32 different waveform templates, $b\bar{x}(t)$, of one-bit duration and use them as each bit of the information $x(t)$ arrives. The same 32 values of $\bar{x}(t)$ will be used for each bit in $x(t)$. The mean square error between each waveform template, $\bar{x}(t)$, and the received waveform, $x(t)$, is computed for each bit period. These mean square error values are called incremental metrics. The waveform template that most closely matches the received waveform will produce the lowest incremental metric and this could be used to regenerate the data bit, $b(t)$. However, we refrain from regenerating a data bit. Instead this process of determining the smallest incremental metric is done for the first bit in the burst through to the last bit. Only at the end of the burst will all the bits be regenerated.

The Viterbi equaliser has 2^{v-1} states, where v in our example is five. So we have 16 states. Each state is associated with a different four-bit binary number. The states are formed into a column having 16 circles, each representing a binary state, e.g. 1101. A trellis diagram is formed of the same columns of these 16 states. Each state, say 1101, will change at the next bit interval to either 0101 or 1101 depending on whether the new bit is a logical 0 or logical 1, respectively. There are two of the above described incremental metrics for each state, so at each bit period there are 32 incremental metrics. For the first data bit we select the lower metric at each state as the favoured metric. (In GSM there are introductory logical 0 tail bits so we know how it starts and this eases the equalisation process.) The favoured metric is associated with a logical 0 or logical 1 transmitted bit. We note the logical value of the bit associated with the metric as well as the value of the metric. At the next bit interval we get another set of 32 metrics. The two incremental metrics at any state are added to the previously retained incremental metric, and the summation is referred to as a *path metric*. The smaller of the two metrics retained, its value noted, as well as the logical value of the bit with which it is associated.

As the bits are processed, one at a time, incremental and path metrics for each state are computed as well as the value of the probable transmitted bit associated with the lowest path metric. At the end of the compete burst there will be 16 metrics, one for each state. The state with the lowest path metric will be selected, and a trace will be employed to identify the bits associated with this path metric as it changed values from the beginning of the burst to when it won out as the lowest path metric. These bits are regenerated in one go as the recovered data burst, although bit de-interleaving and FEC decoding will still need to be done.

This simple description relates to digital phase modulation. When we consider GMSK, a frequency modulation method, we require four phase states 0, $\pi/2$, π and $3\pi/2$ in addition to the states previously described. Thus there are 16 states associated with 0 radians, 16 states associated with $\pi/2$ radians, and so on. There are therefore 64 instant phase states. The reader is advised to consult Steele [4], Chapter 6, for a detailed description of Viterbi equalisation of GMSK signals. We note that with GMSK either double or multiple bit errors occur, although double errors are converted to single errors if differential encoding of the data is done at the transmitter prior to GMSK modulation.

2.3.5.2 The GSM receiver performance

It is important in a system such as GSM that every piece of equipment complies with a minimum set of performance standards, regardless of its manufacturer or country of origin. The performances of the MS and BTS receivers are specified by defining a maximum allowable bit error rate (BER) for each of the different GSM logical channels for a given set of radio channel conditions. As the GSM system has been designed to operate in many different environments, from rural to dense urban, it is important that the specifications reflect this by specifying the performance of the MS and BTS receivers over a wide range of different operational environments. To this end, the GSM specifications [10] define four different *channel models* and these are used to specify the performances of the MS and BTS receivers. Each channel model consists of a number of independently fading impulses, or paths, at different time delays. In practice, the mobile radio channel cannot be separated into its different paths; however, the channel models have been defined in this manner so that they may be easily implemented in wideband channel simulators for equipment testing.

Figure 2.12 shows a block diagram of a wideband channel simulator. The input signal is passed through a delay line which has a number of taps. Each tap represents a received signal path and the signals at each tap are summed to produce the received signal. The delay blocks introduce time delays between the different paths, thereby modelling the effects of path length differences, and each tap may be weighted to model different path attenuations. The signals on the taps can also be set to fade independently with different characteristics.

The GSM channel models are defined in terms of the delays and relative amplitudes of their component paths. Each model is defined for a six-tap channel simulator; however, some channels also include a 12-tap setting for use with larger simulators. Table 2.7 gives the characteristics of the rural area (RA) model. The left-hand column is the tap number. The column entitled 'Relative time' lists the time delays for the two sets of alternative tap settings. The next two columns give the relative amplitudes of each tap for the two alternative settings and, finally, the right-hand column shows the fading characteristics for each path in terms of its Doppler spectrum. CLASS is the classical Doppler spectrum defined by

$$S(f) = A/(1 - (f/f_d)^2)^{\frac{1}{2}} \tag{2.20}$$

and RICE is the sum of a classical Doppler spectrum and one direct path defined by

$$S(f) = 0.41/2\pi f_d(1 - (f/f_d)^2)^{\frac{1}{2}}) + 0.91\delta(f - 0.7f_d), \tag{2.21}$$

where the maximum Doppler frequency is

$$f_d = v/\lambda, \tag{2.22}$$

v is the MS speed in m/s, λ is the wavelength of the carrier frequency in m, $\delta(\cdot)$ is the Dirac function, f is the frequency which is constrained between $\pm f_d$ and A is a constant.

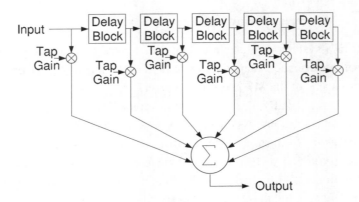

Figure 2.12: Block diagram of a wide band channel simulator.

Table 2.7: Rural area (RA) channel (six taps).

Tap number	Relative time (μs)		Average relative power (dB)		Doppler spectrum
	(1)	(2)	(1)	(2)	
1	0.0	0.0	0.0	0.0	RICE
2	0.1	0.2	−4.0	−2.0	CLASS
3	0.2	0.4	−8.0	−10.0	CLASS
4	0.3	0.6	−12.0	−20.0	CLASS
5	0.4	-	−16.0	-	CLASS
6	0.5	-	−20.0	-	CLASS

From Equation (2.22) we note that a channel's fading characteristics are governed by the speed of the MS and this information must also be included in the description of the GSM channel model. For example, an RA250 channel is an RA channel as defined in Table 2.7 with a MS speed of 250 km/h. In addition to the RA channel, the specifications also define a typical urban (TU) channel, shown in Tables 2.8 and 2.9, and a hilly terrain (HT) model, shown in Tables 2.10 and 2.11. Each of the channels mentioned so far is based on propagation measurements. The specifications also define a further channel model which is used to test the performance of the Viterbi equaliser (EQ). This model is specified in Table 2.12. The specifications state that the receiver must achieve a BER of $\leq 3\%$, without channel coding, over an EQ50 channel where the received signal power is 20 dB above the receiver's minimum sensitivity level. This means that the GSM system will operate in the presence of delay spreads up to 16 μs. Figure 2.13 shows a pictorial representation of one variation of each of the channel models.

Table 2.8: Typical urban (TU) channel (twelve taps).

Tap number	Relative time (μs)		Average relative power (dB)		Doppler spectrum
	(1)	(2)	(1)	(2)	
1	0.0	0.0	−4.0	−4.0	CLASS
2	0.1	0.2	−3.0	−3.0	CLASS
3	0.3	0.4	0.0	0.0	CLASS
4	0.5	0.6	−2.6	−2.0	CLASS
5	0.8	0.8	−3.0	−3.0	CLASS
6	1.1	1.2	−5.0	−5.0	CLASS
7	1.3	1.4	−7.0	−7.0	CLASS
8	1.7	1.8	−5.0	−5.0	CLASS
9	2.3	2.4	−6.5	−6.0	CLASS
10	3.1	3.0	−8.6	−9.0	CLASS
11	3.2	3.2	−11.0	−11.0	CLASS
12	5.0	5.0	−10.0	−10.0	CLASS

Table 2.9: Typical urban (TU) channel (six taps).

Tap number	Relative time (μs)		Average relative power (dB)		Doppler spectrum
	(1)	(2)	(1)	(2)	
1	0.0	0.0	−3.0	−3.0	CLASS
2	0.2	0.2	0.0	0.0	CLASS
3	0.5	0.6	−2.0	−2.0	CLASS
4	1.6	1.6	−6.0	−6.0	CLASS
5	2.3	2.4	−8.0	−8.0	CLASS
6	5.0	5.0	−10.0	−10.0	CLASS

Table 2.10: Hilly terrain (HT) channel (twelve taps).

Tap number	Relative time (μs)		Average relative power (dB)		Doppler spectrum
	(1)	(2)	(1)	(2)	
1	0.0	0.0	-10.0	-10.0	CLASS
2	0.1	0.2	-8.0	-8.0	CLASS
3	0.3	0.4	-6.0	-6.0	CLASS
4	0.5	0.6	-4.0	-4.0	CLASS
5	0.7	0.8	0.0	0.0	CLASS
6	1.0	2.0	0.0	0.0	CLASS
7	1.3	2.4	-4.0	-4.0	CLASS
8	15.0	15.0	-8.0	-8.0	CLASS
9	15.2	15.2	-9.0	-9.0	CLASS
10	15.7	15.8	-10.0	-10.0	CLASS
11	17.2	17.2	-12.00	-12.0	CLASS
12	20.0	20.0	-14.00	-14.0	CLASS

Table 2.11: Hilly terrain (HT) channel (six taps).

Tap number	Relative time (μs)		Average relative power (dB)		Doppler spectrum
	(1)	(2)	(1)	(2)	
1	0.0	0.0	0.0	0.0	CLASS
2	0.1	0.2	-1.5	-2.0	CLASS
3	0.3	0.4	-4.5	-4.0	CLASS
4	0.5	0.6	-7.5	-7.0	CLASS
5	15.0	15.0	-8.0	-6.0	CLASS
6	17.2	17.2	-17.7	-12.0	CLASS

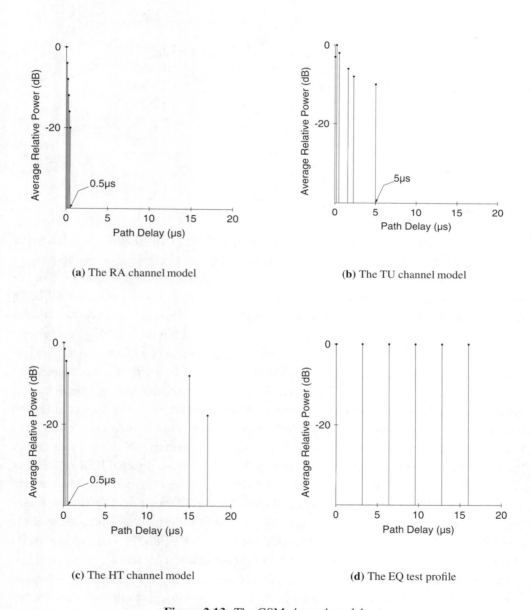

Figure 2.13: The GSM channel models.

Table 2.12: Equaliser (EQ) test profile (six taps).

Tap number	Relative time (μs)	Average relative power (dB)	Doppler spectrum
1	0.0	0.0	CLASS
2	3.2	0.0	CLASS
3	6.4	0.0	CLASS
4	9.6	0.0	CLASS
5	12.8	0.0	CLASS
6	16.0	0.0	CLASS

The specifications define a minimum performance, in terms of BER, for a given set of receiver input conditions and channel conditions. The noise performance of the receiver is defined at a reference sensitivity level that varies depending on the receiver type, as shown in Table 2.13. The receiver performance is defined as a maximum channel BER, or frame erasure rate (FER), for a given logical channel received at the reference sensitivity for various propagation conditions. We will discuss the different GSM logical channels in Section 2.3.6. Suffice to say that information of a specific type is formed into a logical channel before it is mapped onto the GSM bursts. Table 2.14 shows the reference sensitivity performance for GSM900 [10]. The DCS1800 performance specification is similar but not identical. The table shows that the performance is specified for frequency hopped (FH) and non-frequency hopped channels. Where frequency hopping is used, perfect decorrelation is assumed between bursts. The table also shows that the performance of the full-rate speech traffic channel (TCH/FS) is defined in terms of the FER and the residual BER (RBER) of the Class Ib and Class II bits. The different bit classes will be explained in a later section which deals with the speech coder. Suffice to say that the bit errors in different bit classes will have a different impact on the speech quality. The factor α has been included to allow a slight trade-off between the speech FER and the Class Ib RBER. The value of α must lie between 1 and 1.6 and it must remain constant for each channel condition.

The interference-limited performance of the receiver is defined for a number of reference interference levels which are specified relative to the wanted signal. In each case the received level of the wanted signal is set to 20 dB above the reference sensitivity level shown in Table 2.13. The reference carrier-to-interference (C/I) ratios for both co-channel and adjacent channel interference are as follows:

Table 2.13: Reference sensitivity levels.

Receiver type	Reference sensitivity (dBm)
DCS1800 MS	−100
GSM900 handhelds	−102
Other GSM900 MSs	−104
normal BTSs	−104
Micro BTSs	
GSM900 M1	−97
GSM900 M2	−92
GSM900 M3	−87
DCS1800 M1	−102
DCS1800 M2	−97
DCS1800 M3	−92

Table 2.14: Reference sensitivity performance for GSM900.

Type of channel		Propagation conditions			
		Static	TU50 (no FH)	TU50 (ideal FH)	RA250 (no FH)	HT100 (no FH)
FACCH/H	(FER)	0.1%	6.9%	6.9%	5.7%	10.0%
FACCH/F	(FER)	0.1%	8.0%	3.8%	3.4%	6.3%
SDCCH	(FER)	0.1%	13%	8%	8%	12%
RACH	(FER)	0.5%	13%	13%	12%	13%
SCH	(FER)	1%	16%	16%	15%	16%
TCH/F9.6	(BER)	10^{-5}	0.5%	0.4%	0.1%	0.7%
TCH/H4.8	(BER)	10^{-5}	0.5%	0.4%	0.1%	0.7%
TCH/F4.8	(BER)	-	10^{-4}	10^{-4}	10^{-4}	10^{-4}
TCH/F2.4	(BER)	-	2×10^{-4}	10^{-5}	10^{-5}	10^{-5}
TCH/H2.4	(BER)	-	2×10^{-4}	10^{-4}	10^{-4}	10^{-4}
TCH/FS	(FER)	0.1 α%	6α %	3 α %	2α%	7α%
Class	Ib(RBER)	0.4/α%	0.4 /α %	0.3/α %	0.2/α%	0.5/α%
Class	II(RBER)	2%	8%	8 %	7%	9%

co-channel interference	$C/I = 9$ dB
first adjacent channel interference (200 kHz)	$C/I = -9$ dB
second adjacent channel interference (400 kHz)	$C/I = -41$ dB
third adjacent channel interference (600 kHz)	$C/I = -49$ dB.

These values show that GSM is specified to operate in situations where the co-channel C/I is ≥ 9 dB and the first adjacent channel is ≤ 9 dB greater than the wanted channel. Owing to the effects of power control and shadow fading, this adjacent channel performance precludes the use of adjacent channels within the same cell. Given the input conditions shown above, the receiver performance is defined in terms of maximum FERs and BERs for each logical channel and propagation condition similar to that shown for the sensitivity performance itemised in Table 2.14. The interference performance of both GSM900 and DCS1800 is fully defined in the specifications [10] and full details have not been included here.

2.3.6 Physical and logical channels

In the previous sections we have discussed the way in which information is transmitted over the GSM radio interface in the form of bursts containing both information bits and a predefined training sequence. We have also looked at the way the information bits may be recovered in the presence of the ISI introduced by the modulation process and the radio channel. In this section we consider the manner in which the user data, e.g. speech, and the signalling data, e.g. handover commands, are mapped onto the TDMA bursts.

When an MS and a BTS communicate, they do so on a specific pair of radio frequency (RF) carriers, one for the up-link and the other for the down-link transmissions, and within a given time slot in each consecutive TDMA frame. This combination of time slot and carrier frequency forms what is termed a *physical channel*. One RF channel will support eight physical channels in time slots zero through to seven. The data, whether user traffic or signalling information, are mapped onto the physical channels by defining a number of *logical channels*. A logical channel will carry information of a specific type and a number of these channels may be combined before being mapped onto the same physical channel. Having introduced the concept of the physical and logical channels, we will now describe the function of each of the GSM logical channels.

2.3.6.1 Traffic channels

GSM defines two types of traffic channel (TCH). The *full-rate* TCH allows speech transmission at 13 kb/s and in the specifications it is termed a TCH/FS channel to show that it is a full-rate TCH carrying *s*peech information. All forms of the GSM TCH follow a similar naming convention as shown below. The full-rate TCH also allows user data transmission at the primary user rates of 9.6, 4.8 and ≤ 2.4 kb/s, referred to as TCH/F9.6, TCH/F4.8 and

TCH/F2.4, respectively. A full-rate TCH will occupy a complete physical channel, i.e. one timeslot in each TDMA frame and on each up- and down-link carrier.

The *half-rate* TCH allows speech transmission at around 7 kb/s (TCH/HS) and data at primary user rates of 4.8 and ≤ 2.4 kb/s, called TCH/H4.8 and TCH/H2.4, respectively. The half-rate channel uses one timeslot in every other TDMA frame, on average, and this means that each physical channel can support two half-rate TCHs. The half-rate channel is primarily intended to support the GSM half-rate speech coder, the design of which was finalised in January 1995.

The data rates detailed above have been chosen to coincide with those already in use in the PSTN. As the primary data rate are reduced, the power of the channel coding is increased and this results in lower decoded BERs. To demonstrate this, we present in Table 2.15 examples of BER performances taken from the specifications [10]. Table 2.15 is for a typical urban GSM900 propagation channel with ideal FH and an MS speed of 50 km/h. The TCHs always use the normal burst.

2.3.6.2 Control channels

Control channels carry signalling information between an MS and a BTS. There are several forms of control channels in GSM, and they can generally be divided into four categories according to the manner in which they are supported on the radio interface and the type of signalling information they carry.

The *broadcast channels* are used to broadcast synchronisation and general network information to all the MSs within a cell. Signalling information is carried between an MS and a BTS using *associated* control channels during a call, while *stand-alone dedicated* control channels are employed outside of a call. Finally, the *common* control channels are used by an MS during the paging and access procedures. We will now examine in detail the logical channels in each of these four categories.

Table 2.15: Typical BER performance.

Type of channel	BER
TCH/F9.6	0.3%
TCH/F4.8	0.01%
TCH/F2.4	0.001%
TCH/H4.8	0.3%
TCH/H2.4	0.01%

Broadcast channels   Broadcast channels are transmitted in the down-link direction only, i.e. they are only transmitted by the BTS.

The *frequency correction channel* (FCCH) is the simplest GSM logical channel because all its information bits are set to zero. The FCCH consists solely of frequency correction bursts which, as we have already seen, consist of an all-zero bit pattern. After GMSK modulation, these bursts produce a pure sinewave at a frequency of around 68 kHz (1625/24 kHz) above the carrier frequency. The FCCH is used by the MS in the initial stages of BTS acquisition to correct its internal frequency sources and recover the carrier phase of the BTS transmissions.

The *synchronisation channel* (SCH) contains full details of its own position within the GSM framing structure. We will examine this framing structure in the next section. Using the information supplied on the SCH, an MS can fully synchronise its frame counters with those of a BTS. The SCH information is transmitted using synchronisation bursts.

In addition to the frame synchronisation information, the SCH also contains a six-bit base station identity code (BSIC). We will see in the Section 2.4.4, which deals with handover in GSM, that an MS is required to measure the received signal strength of a special beacon carrier, termed the BCCH carrier. This carrier is transmitted at a constant power by each of the MS's neighbouring BTSs. This information is reported to the network where it is used to determine whether the MS is connected to the most appropriate BTS or whether it should be switched to a more suitable BTS. The MS will be provided with a list of the BCCH carrier frequencies used by its neighbouring BTSs and it will report a received signal level measurement for each BCCH carrier frequency. It is important to ensure that the measurement information is always associated with the correct BTS. For example, an MS may report a received signal level measurement for frequency f_1 which the network associates with a particular BS, say, BS_A. However, the MS may in fact be measuring the BCCH carrier transmissions from a different BS, say, BS_B. This confusion is resolved by requiring the MS periodically to decode the SCH of its neighbouring BTSs and extract the BSIC. When the measurement information is sent to the network it is accompanied by the BSIC for each BTS and this ensures that the network associates each measurement with the appropriate BTS. Therefore, it is important to ensure that two potential co-channel BTSs, e.g. BS_A and BS_B above, are assigned different BSICs.

The BSIC consists of a three-bit network colour code (NCC) and a three-bit BS colour code (BCC). The term 'colour' is used because the assignment of these codes may be achieved by colouring the regions on a map according to the code that is in use within a particular area. Where two networks, sharing the same frequency band, have overlapping coverage they will each be assigned a different NCC. This ensures that the two networks cannot use the same BSIC. In general, two PLMNs within the same country will use different frequency bands, and the use of the NCC only becomes important at international boundaries where the coverage of two PLMNs using the same frequency bands may over-lap. The BCC is used by the individual operators to ensure that co-channel BTSs have different BSICs. The BCC and the training sequence code (TSC) will be the same for each BTS. This means that by decoding the BSIC for a particular BTS, the MS can determine which of the eight possible training sequences to expect in the normal bursts transmitted by a particular BTS (see Table 2.6).

The *broadcast control channel* (BCCH) is used to broadcast control information to every MS within a cell. This information includes details of the control channel configuration used at the BTS (see Section 2.3.7), a list of the BCCH carrier frequencies used at the neighbouring BTSs and a number of parameters that are used by the MS when accessing the BTS.

The *cell broadcast channel* (CBCH) is used to transmit short alpha numeric text messages to all the MSs within a particular cell. These messages appear on the MS's display and a subscriber may choose to receive different messages by selecting different pages, similar to the way the teletext system works on broadcast television. The BCCH and the CBCH both use the normal burst.

Associated control channels When an MS is engaged in a call, a certain amount of sig-nalling information must flow across the radio interface in order to maintain the call. For example, an MS will continually report the received level of the BCCH carriers of its neigh-bouring BTSs. This type of signalling is supported using logical control channels which occupy the same physical channel as the traffic data. Non-urgent information, e.g. mea-surement data, is transmitted using the *slow associated control channel* (SACCH). This channel is always present when a dedicated link is active between the MS and BTS, and it occupies one timeslot in every 26. SACCH messages may be sent once every 480 ms, i.e. approximately every 2 s. More urgent information, e.g. a handover command, is sent using time slots that are 'stolen' from the traffic channel. This channel is known as the *fast asso-ciated control channel* (FACCH) because of its ability to transfer information between the BTS and MS more quickly than the SACCH. A FACCH signalling block is used to exactly replace a single (20 ms) speech block and a complete FACCH message may be sent once every 20 ms. Both the SACCH and FACCH use the normal burst and they are both up-link and down-link channels.

Stand-alone dedicated control channel In some situations, signalling information must flow between a network and an MS when a call is not in progress, e.g. during a location update. This could be accommodated by allocating either a full-rate or half-rate TCH and by using either the SACCH or FACCH to carry the information. This would, however, be a waste of the limited radio resources since the data transfer requirements of a process such as location updating are far less than those of speech transmissions. For this reason, a lower data rate channel has been defined which has around one-eighth of the capacity of a full-rate TCH. This channel is known as a *stand-alone dedicated control channel* (SDCCH). The channel is termed 'stand-alone' because it may exist independently of any TCH, and it is termed 'dedicated' because it is only used by one particular MS, i.e. it is dedicated to a particular MS.

In some ways the SDCCH is similar to a TCH since they are both used to provide a dedicated connection between a BTS and an MS. The SDCCH also has an associated SACCH. Since the SDCCH always carries signalling traffic there is no frame stealing and consequently it does not need an FACCH. Alternatively, one could argue that the SDCCH is constantly in the FACCH mode. The SDCCH operates on both the up-link and down-link and the normal burst is always used.

Common control channels   The common control channels may be used by any MS within a cell. The *paging channel* (PCH) is a down-link only channel that is used by the system to page individual MSs, e.g. in the event of an incoming call. There are two different PCHs, a full-rate PCH and a reduced rate PCH, for use in cells with a limited capacity. The normal burst is always used on the PCH.

The *access grant channel* (AGCH) shares the same physical resources as the PCH, i.e. a particular time slot may be used by either channel, though not simultaneously. As its name would suggest, the AGCH is used by the network to grant, or deny, an MS access to the network by supplying it with details of a dedicated channel, i.e. TCH or SDCCH, to be used for subsequent communications. The AGCH is a down-link only channel and it uses the normal burst.

The *random access channel* (RACH) is an up-link only channel that is used by an MS to initially access the network, e.g. at call set-up or prior to a location update. It is termed 'random' because there is no mechanism to ensure that no more than one MS transmits in each RACH time slot and there is a finite probability that two mobiles could attempt to access the same RACH at the same time. This could result in neither access attempt being successful as the two signals collide at the BTS. If an MS receives no response from the BTS, it will attempt to access the BTS again after waiting a certain period of time. If this period of time was the same for every MS, then once a collision occurs between two MSs, it will continue to occur for every subsequent access attempt. Therefore the delay between access attempts is randomised to reduce the likelihood of collisions at the BTS. The MS will always transmit access bursts on the RACH.

2.3.7 Mapping logical channels onto physical channels

The various logical channels described above may be combined in one of six different ways, before being mapped onto a single physical channel. The simplest mapping is the full-rate traffic channel (TCH/F) and its SACCH. When combined these channels fit exactly into a single physical channel. We note that the mapping between the TCH and the physical channel is the same regardless of whether the TCH is used to carry speech or user data. A single physical channel will also support two half-rate traffic channels (TCH/H) and their SACCHs or eight SDCCHs and their associated SACCHs. The remaining three logical channel combinations are a little more complicated and these are explained below.

The basic broadcast and common control channel combination consists of a single FCCH, SCH and BCCH on the down-link, along with a full-rate PCH and a full-rate AGCH. The up-link is entirely dedicated to the RACH, and for this reason we shall term this a full-rate RACH. This type of channel configuration is generally used in medium capacity or large capacity cells where the access capacity of a full-rate PCH, AGCH and RACH channel is justified. This control channel combination may only occur on time slot zero of a carrier. The carrier that supports these channels at a BTS is called the *BCCH carrier* and it will be unique within each cell, or sector, i.e. each BTS will only have one BCCH carrier.

In smaller capacity cells, i.e. cells with a smaller number of RF carriers, the capacity of the full-rate PCH, AGCH and RACH may not be justified. For this reason, a second combination of the access channels is employed. The down-link continues to support an FCCH, SCH and BCCH; however, the rate of the down-link PCH and AGCH is reduced to around one-third of their full rate. The extra slots that have been created as a result of this rate reduction on the down-link are used to support four SDCCHs and their associated SACCHs. The SDCCHs will also occupy a number of up-link slots and the number of timeslots allocated to the RACH on the up-link is reduced accordingly. This will effectively halve the number of time slots allocated to the RACH. Again, this control channel combination may only occur on time slot zero of the BCCH carrier.

The final control channel combination is defined for use in large capacity cells where the access capacity of a single PCH, AGCH and RACH is insufficient. This combination consists of a BCCH and a full-rate PCH and AGCH on the down-link and a full-rate RACH on the up-link. This channel combination may only occur on slot two, or slots two and four, or slots two, four and six of the BCCH carrier. The reasons for this restriction will be explained later when we examine the timing advance mechanism. We note that each BTS must only transmit a single FCCH and SCH and, consequently, these channels are not included in the extension channel set. Each extension set contains its own BCCH for two reasons. Firstly, the BCCH contains information that applies only to the RACH occupying the same time slot within the TDMA frame, and secondly, it is easier for the MS to monitor bursts occurring on the same physical channel.

The various channel combinations described above are summarised in Table 2.16. We note that the CBCH has been omitted. If this channel is required it may replace a single SDCCH.

2.3.8   The GSM frame structure

We have already described the basic TDMA frame structure employed in GSM, whereby each carrier supports eight timeslots, and a physical channel occupies one timeslot in each frame. The TDMA frame represents the lowest layer in a complex hierarchical frame structure. The next level in the GSM frame structure, above the TDMA frame, is the *multiframe* which consists of 26 TDMA frames in the case of the full-rate and half-rate traffic channels, or 51 frames for all of the other logical channels.

The frame structure for a full-rate traffic channel (TCH/F) occupying time slot one in each TDMA frame is shown in Figure 2.14. The first 12 time slots in each TDMA frame of the multiframe, i.e. zero to 11, are used by the TCH/F itself. The next time slot is not used for transmission, and for this reason it is termed an 'idle' time slot. The next 12 timeslots in each TDMA frame of the multiframe are used by the TCH/F, and the remaining timeslot in TDMA frame 26 is used by the SACCH. Figure 2.14 may be applied to both the up-link and down-link by noting that there is an offset of three burst periods between the frame timing on the up-link and down-link. The traffic multiframe is exactly 120 ms in duration and this defines many of the time periods that are used in GSM, e.g. 1 TDMA frame = 120 ms/26 ≈ 4.615 ms. We note that the multiframe structure shown in Figure 2.14 only applies to a TCH/F occupying the odd-numbered time slots. On even-numbered time slots, and time slot zero, the position of the idle and SACCH time slots are exchanged. To understand the reason for this we need to examine the way the information is carried on SACCH. We will see later that the SACCH messages are interleaved over four bursts, which represent a time period of four multiframes or 480 ms. This means that the BTS must receive four SACCH bursts before the information can be successfully de-interleaved and decoded, i.e. the SACCH information can only be decoded once every four frames. If the SACCH bursts occurred at the same point in the multiframe for each physical channel, then the SACCH messages from every MS would arrive within the same TDMA frame and this would put a large processing load on the BSC. In order to spread the SACCH processing load more evenly in time, the position of the SACCH bursts are changed from timeslot to timeslot and the interleaving periods are also arranged such that the BTS will have to decode a maximum of one SACCH message per TDMA frame. This also gives the BSC a period of around 12 TDMA frames to process the information in the SACCH message before another message arrives from another MS using the same carrier but in a different timeslot.

The organisation of the half-rate traffic channel (TCH/H) is a little more complicated than the TCH/F. Figure 2.15 shows an example of the organisation of two half-rate traffic

Table 2.16: Logical channel configurations that can be mapped onto a single physical channel in GSM.

Possible	Channels	
time slots	Down-link	Up-link
0–7	1 TCH/F (+SACCH)	1 TCH/F (+SACCH)
0–7	2 TCH/H (+SACCH)	2 TCH/H (+SACCH)
0–7	8 SDCCH (+SACCH)	8 SDCCH (+SACCH)
0	1 SCH + 1 FCCH + 1 BCCH	1 RACH
	+ 1 AGCH + 1 PCH	
0	1 SCH + 1 FCCH + 1 BCCH	
	+ 1 AGCH* + 1 PCH*	1 RACH*
	+ 4 SDCCH (+SACCH)	+ 4 SDCCH (+SACCH)
2,4,6	1 BCCH + 1 AGCH + 1 PCH	1 RACH

*Reduced rate channels.

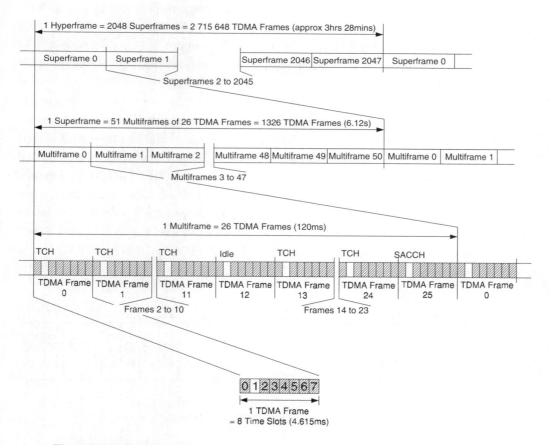

Figure 2.14: The frame structure for a full-rate traffic channel (TCH/F) on timeslot 1.

channels, TCH 0 and TCH 1. The description of a half-rate channel will include a sub-channel number of zero or one in addition to the timeslot number and the carrier frequency. The example in Figure 2.15 shows that the TCH/H (i.e. TCH 0 or TCH 1) and its associated SACCH (i.e. SACCH 0 or SACCH 1) use on average every other slot within the multiframe. We note that this diagram, in contrast to Figure 2.14, only shows the timeslots from the same physical channel; the timeslots from the remaining seven physical channels have been omitted.

Figure 2.14 shows that the next level in the frame structure above the multiframe is the *superframe*, and in the case of the TCH/F and the TCH/H consists of 51 multiframes. The duration of a superframe is 6.12 s. The final level in the frame structure is the *hyperframe* and this consists of 2048 superframes and has a duration of around 3.5 h. We shall discuss the purpose of each of these higher level frames after we have described the frame structure used on the control channels.

The frame structure for a control channel set on timeslot zero is shown in Figure 2.16. In this case the multiframe consists of 51 TDMA frames and lasts around 235 ms. All channels, except the TCH/F and TCH/H, use the 51-frame multiframe. As we have seen in Table 2.16, there are four different control channel combinations and each of these uses the 51-frame multiframe.

Figure 2.17 shows the way in which a group of eight SDCCHs are mapped onto a single physical channel. The burst mapping is based on a two-multiframe cycle, i.e. it repeats every two multiframes, and the diagram shows two multiframes with the frames numbered zero to 50 and 51 to 101. The time slots that are used by a particular SDCCH are labelled SDCCH n (e.g. SDCCH 0) and the timeslots used by its associated SACCH are labelled SACCH n (SACCH 0), where n may take any value from zero to seven. Where the control channels have been combined in this manner the specifications refer to each of them as an SDCCH/8. The figure shows that each SDCCH/8 occupies a total of eight time slots in two multiframes (102 frames). The associated SACCH, which is referred to as an SACCH/C8

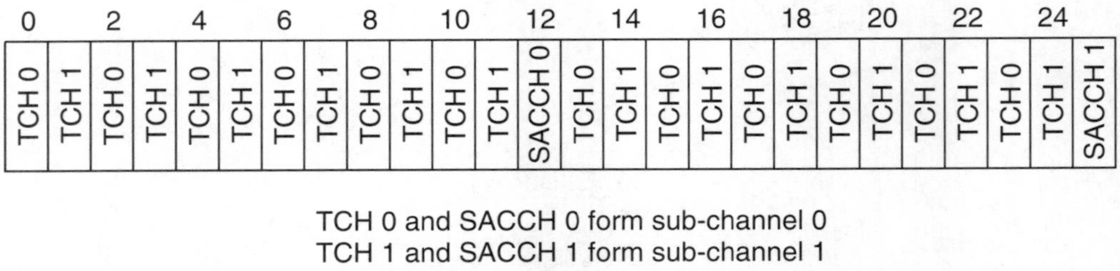

TCH 0 and SACCH 0 form sub-channel 0
TCH 1 and SACCH 1 form sub-channel 1

Figure 2.15: Half-rate channel organisation.

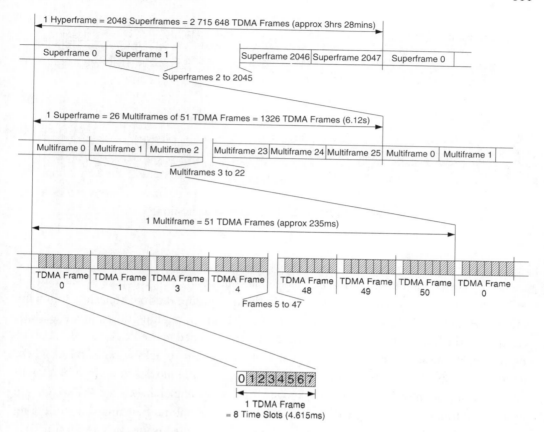

Figure 2.16: The frame structure for a control channel group on timeslot 0.

in the specifications, occupies a further four timeslots in the same period. This means that $8 \times 12 = 96$ timeslots are occupied and the remaining six are left idle. Figure 2.17 shows the situation for the down-link slots; the up-link configuration may be derived by shifting each slot 15 places to the right, i.e. SDCCH 0 will occupy slots 15 to 18 and slots 66 to 69 and its associated SACCH will occupy slots 47 to 50 on the up-link. This control channel arrangement may be used on any time slot and any carrier, except timeslot zero of the BCCH carrier.

Table 2.16 shows that there are three possible combinations of the broadcast and common channels. Figure 2.18 shows the manner in which each of these combinations is mapped onto the control channel multiframe. Figure 2.18(a) shows the basic control channel arrangement that would be present on time slot zero of the BCCH carrier. The down-link multiframe is subdivided into five groups of 10 time slots with a single idle slot at the end. The first timeslot of each of the groups is assigned to the FCCH and contains the

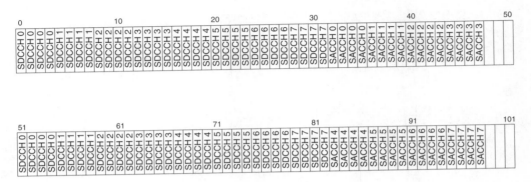

Figure 2.17: SDCCH/8 channel organisation.

frequency correction burst. Similarly, the second time slot of the group is assigned to the SCH and contains a synchronisation burst. In the first group, the four timeslots following the FCCH and SCH slots are assigned to the BCCH. The remaining down-link timeslots, except the idle burst at the end of the multiframe, are assigned to the PCH and AGCH. The PCH/AGCH timeslots may be assigned to either channel on a block-by-block basis, where a block consists of four timeslots (defined by the interleaving block size). Each block will contain sufficient information for the MS to identify the channel, i.e. PCH or AGCH. On the up-link, all timeslots are assigned to the RACH. We note that the four timeslots following the BCCH slots shown in Figure 2.18(a) may also be assigned to the BCCH. In this case these four slots will be assigned to either the BCCH, PCH or AGCH on a block-by-block basis.

Figure 2.18(b) shows the control channel arrangement for small capacity cells where the PCH and AGCH capacity is reduced to accommodate four SDCCHs. The down-link multi-frame is again subdivided into five groups of 10 slots and each group begins with an FCCH and an SCH. The first group has four slots assigned to the BCCH and the remaining four slots are used by the PCH/AGCH. The eight remaining slots in the second group are used by the PCH/AGCH. All the remaining down-link slots are used by the four SDCCHs and their SACCHs, except the idle slot at the end of the multiframe. On the up-link, 24 slots are assigned to the four SDCCHs and their SACCHs and the remaining 27 slots are used by the RACH. This control channel arrangement may only be used on slot zero of the BCCH carrier. Figure 2.19 shows the manner in which the timeslots labelled 'D' in Figure 2.18(b) are assigned to each of the four SDCCHs and their associated SACCHs. Where the SDCCH channels have been combined in this manner the specifications refer to them as SDCCH/4 and the associated control channels as SACCH/C4.

The third control channel configuration is designed for high capacity cells where a single

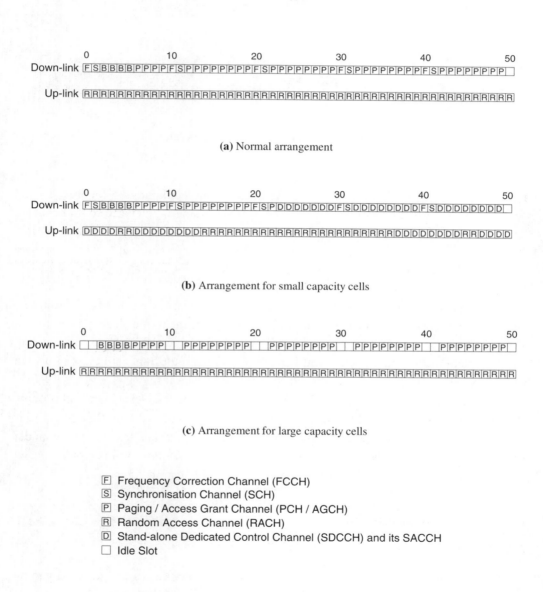

(a) Normal arrangement

(b) Arrangement for small capacity cells

(c) Arrangement for large capacity cells

F Frequency Correction Channel (FCCH)
S Synchronisation Channel (SCH)
P Paging / Access Grant Channel (PCH / AGCH)
R Random Access Channel (RACH)
D Stand-alone Dedicated Control Channel (SDCCH) and its SACCH
☐ Idle Slot

Figure 2.18: The BCCH carrier control channel arrangements.

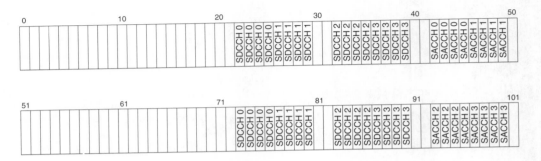

Figure 2.19: SDCCH/4 channel organisation.

PCH/AGCH and RACH are inadequate. The multiframe configuration for this channel group is shown in Figure 2.18(c). This is effectively the same as the basic control channel arrangement given in Figure 2.18(a), but with the FCCH and SCH slots replaced by idle slots. This arrangement may only be used on slot two, slots two and four, or slots two, four and six of the BCCH carrier. We also note that this extension set can only be used when there are no SDCCHs on timeslot zero of the BCCH carrier.

Returning to Figure 2.16 we see that the control channel superframe consists of 26 multiframes and is 6.12 s in duration. This is exactly the same length as the traffic channel superframe. The superframe represents the smallest time cycle for which the traffic channel and control channel relationships are repeated, and for this reason the multiframe structure in use on each physical channel may only change at superframe boundaries.

Figures 2.14 and 2.16 show that the hyperframe provides the final level in the GSM frame structure for both the traffic and control channels. It consists of 2048 superframes and lasts for around 3.5 h. Each of the 2 715 648 timeslots in the hyperframe has a unique number and this is used in the ciphering and frequency hopping algorithms. We discuss frequency hopping in Section 2.4.6 and issues relating to ciphering in Section 2.5.

2.3.9   Speech transmission

2.3.9.1   Introduction

We now examine the coding, interleaving and ciphering processes that are employed on the GSM full-rate speech channel (TCH/FS). In subsequent sections we will describe the equivalent processes for the user data channels (e.g. TCH/F9.6) and the signalling channels (e.g. SACCH). We will not consider the half-rate speech channel (TCH/HS) in this section.

Referring to Figure 2.2 we see that a speech coder is used to convert the analogue speech signal into a digital signal that is suitable for transmission over the radio interface. Forward error correction (FEC) coding is then applied to the speech data to allow all or some of the transmission errors to be corrected at the receiver. The FEC-coded data are interleaved and ciphered before being assembled into bursts ready for transmission over the radio interface. Following the equalisation and demodulation process at the receiver, the recovered data is deciphered, de-interleaved and channel decoded to remove as many transmission errors as possible. The data are then passed to the speech decoder where it is converted back into an analogue speech signal. It is important to note that the basic link shown in Figure 2.2 may be applied to both the up-link (i.e. an MS transmitter and a BTS receiver) and the down-link (i.e. a BTS transmitter and an MS receiver). We will now examine each of the coding, interleaving and ciphering processes in more detail.

2.3.9.2 Speech coding

In any wide area cellular system, such as GSM, the radio spectrum is a scarce resource and it must be used in an efficient manner. In the case of speech transmission this imposes the requirement that the speech information must be transferred over the radio interface using as low a data rate as possible whilst maintaining a satisfactory speech quality. This means that waveform encoders, such as the 32 kb/s adaptive differential pulse code modulation (AD-PCM) coder used in the digitally enhanced cordless telecommunication (DECT) system, are unsuitable for GSM because of their relatively high bit rate requirements. The transmission of toll quality speech at data rates below 16 kb/s calls for highly complex speech coders which use digital filters to model the characteristics of the vocal production apparatus. The speech information is conveyed by transmitting the filter coefficients of the vocal tract and an excitation sequence which allows the speech to be reconstructed at the receiver. A full explanation of the GSM speech coder is beyond the scope of this book and the reader is referred to Steele [4] for a detailed discussion of analysis-by-synthesis predictive coding and a comprehensive description of the full-rate GSM coder. In this section we will examine the basic principles of the GSM full-rate coder and highlight its main features.

At an MS, the user's acoustic pressure signal is converted into an analogue electrical signal using a microphone. The electrical signal is sampled at 8 kHz and each sample is converted into a 13-bit digital representation using a uniform analogue-to-digital conversion. Following this conversion process the signal is passed to the GSM speech coder. On the network side the situation is slightly more complicated because speech signals are usually coded using an eight-bit A-law pulse code modulation (PCM) format in order to be compatible with the PSTN/ISDN. This means that, before the speech signal is passed to the speech coder on the network side, it must first undergo an eight-bit A-law PCM to 13-bit uniform PCM conversion.

The GSM coder belongs to the family of regular pulse excited (RPE) linear predictive codecs (LPC). It also employs long-term prediction (LTP) in addition to the conventional short-term prediction (STP) and accordingly it is called an RPE-LTP speech coder. An RPE-LTP encoder can be divided into four parts: pre-processing, STP analysis filtering, LTP analysis filtering and RPE computation. We will now briefly examine each of these processes.

Pre-processing   The sampled speech signal is initially passed through a notch filter to remove any dc offsets that may be present and passed through a first-order finite impulse response (FIR) pre-emphasis filter. The pre-emphasis filter is used to emphasise the low power, high frequency part of the speech spectrum and this provides better numerical precision in the subsequent computations. The speech signal is then divided into non-overlapping frames consisting of 160 samples, each having a duration of 20 ms.

STP analysis filtering   Speech data inherently contain a high level of redundancy and this means that it is possible to predict a future speech sample from previous speech samples. In mathematical terms we can say that a speech sample may be approximated as the linear combination of a number of past speech samples (e.g. 8–16 samples) such that the predicted speech sample at an instant n is given by

$$\tilde{s}(n) = \sum_{k=1}^{p} a_k s(n-k), \tag{2.23}$$

where a_k are the predictor coefficients, $s(n)$ is the speech sample at sampling instant n and p is the number of predictor coefficients or the *predictor order*. The prediction error, $e(n)$, is defined as

$$
\begin{aligned}
e(n) &= s(n) - \tilde{s}(n) \\
&= s(n) - \sum_{k=1}^{p} a_k s(n-k).
\end{aligned}
\tag{2.24}
$$

Taking the z-transform of Equation (2.24) gives

$$E(z) = S(z)A(z), \tag{2.25}$$

where

$$A(z) = 1 - \sum_{k=1}^{p} a_k z^{-k}. \tag{2.26}$$

The coefficients a_k are computed by minimising the mean-squared error over a 20 ms segment of the speech waveform. The inverse of $A(z)$, i.e. $H(z) = 1/A(z)$, is an all-pole digital filter which models the spectral envelope of the speech waveform.

In practice, the predictor coefficients, a_k, are not calculated directly, but instead, a number of reflection coefficients are derived from the autocorrelation coefficients of the speech block. Uniform quantisation of the reflection coefficients is inefficient and, for this reason, they are transformed into another set of coefficients known as log area ratios (LARs). There are eight LARs per 20 ms speech block and they are quantised using six bits for LAR(1) and LAR(2), five bits for LAR(3) and LAR(4), four bits for LAR(5) and LAR(6) and three bits for LAR(7) and LAR(8), which results in 36 bits per speech block. These bit allocations are determined according to the dynamic range and the distribution of the LAR values.

LTP analysis filtering Filtering the speech signal using the inverse filter, $A(z)$, tends to remove much of the redundancy by subtracting from each speech sample its predicted value using the past p samples. The resulting signal is known as the short-term prediction residual and it will generally exhibit a certain amount of periodicity related to the pitch period of the original speech when it is voiced. This periodicity represents a further level of redundancy which can be removed using a pitch predictor or a long-term predictor. The general form of the long-term predictor filter is given by

$$\frac{1}{P(z)} = \frac{1}{1 - P_l(z)},$$

(2.27)

where

$$P_l(z) = \sum_{k=-m_1}^{m_2} G_k z^{-(\alpha+k)}$$

(2.28)

is the long-term predictor, m_1 and m_2 determine the number of predictor taps, α is the LTP delay and G_k is the LTP gain. In the case of the GSM full-rate speech coder, $m_1 = m_2 = 0$, resulting in a single-tap predictor. The parameters α and G_0 are determined by minimising the mean-squared residual error after both short-term and long-term prediction over a period of 40 samples, i.e. 5 ms. For the single-tap predictor, this residual error, $e(n)$, at an instant n, is given by

$$e(n) = r(n) - Gr(n - \alpha),$$

(2.29)

where $r(n)$ is the residual that is produced following the short-term prediction. The mean-squared residual, E, is given by

$$E = \sum_{n=0}^{39} e^2(n) = \sum_{n=0}^{39} [r(n) - Gr(n - \alpha)]^2.$$

(2.30)

The parameters G and α are quantised and coded using two and seven bits, respectively, resulting in nine bits per 5 ms sub-block and 36 bits per 20 ms block.

RPE computation Removing the redundancy from the speech signal produces a residual signal. In the speech decoder this residual is used to excite the reconstructed LTP and STP filters. The GSM system uses the regular pulse excited (RPE) approach to encode this residual efficiently. For each 5 ms subsegment, the excitation signal is assumed to consist of 13 pulses spaced apart by three samples. The magnitude and initial starting position of the first pulse are computed to minimise the error between the speech and its locally reconstructed version. Given the pulse spacing of three samples, there are three possible grid positions for the first excitation pulse and this information can be encoded using just two bits. The pulse magnitudes are normalised to the highest magnitude for the block and quantised using three bits. Finally, the block maxima is quantised using six bits. This results in 47 bits per 5 ms sub-block.

The full bit allocation for the GSM voice coder is given in Table 2.17. Each 20 ms block of speech is encoded using 260 bits and this produces a bit rate of 260/20 ms = 13 kb/s.

2.3.9.3 Speech decoding

Decoding the speech data is a much less complex task than the encoding process. The decoder performs the opposite operations, namely RPE decoding, LTP synthesis filtering, STP synthesis filtering and post-processing.

RPE decoding   In the decoder, the grid position, the subsegment excitation maxima and the excitation pulse amplitudes are recovered from the received data and the actual pulse amplitudes are computed by multiplying the decoded amplitudes by their corresponding block maxima. The LTP residual model is recovered by properly positioning the pulse amplitudes according to the initial offset grid position.

LTP synthesis filtering   The LTP filter parameters, G and α, are recovered from the received data and they are used to derive the LTP synthesis filter. Then the recovered LTP excitation model is used to excite this LTP synthesis filter to recover a new subsegment of the estimated STP residual.

Table 2.17: Summary of the RPE-LTP bit-allocation scheme.

Parameter to be encoded	No. of bits
8 STP coefficients	36
4 LTP gains G	$4 \times 2 = 8$
4 LTP delays D	$4 \times 7 = 28$
4 RPE grid positions	$4 \times 2 = 8$
4 RPE block maxima	$4 \times 6 = 24$
$4 \times 13 = 52$ pulse amplitudes	$52 \times 3 = 156$
Total number of bits per 20 ms	260
Transmission bit rate	13 kb/s

STP synthesis filtering   The STP filter is reconstructed and excited by the reconstructed STP residual signal to regenerate the speech.

Post-processing   The speech signal is de-emphasised using the inverse of the pre-emphasis filter employed in the encoding.

The data bits produced by the speech coder will each have a different impact on the speech signal if they are received in error. For this reason, each bit is ranked in order of importance and forward error correction is applied accordingly. In GSM the most important bits are protected by a parity check and a powerful half-rate convolutional code, whereas the least important bits are left unprotected. The specifications define three classes of bits, based on subjective testing, and these are termed Class Ia, Class Ib and Class II. The Class Ia bits are the most important and receive the most protection. In fact, if an error is detected in the Class Ia bits, then the entire speech frame is discarded and interpolation techniques are used to reconstruct the speech from the frames on either side of the discarded block. The Class Ib bits are slightly less important and receive correspondingly less protection. The Class II bits are the least important and they are transmitted unprotected.

2.3.9.4   Voice activity detection

During a typical telephone conversation, a person will generally speak for around 40% of the time and remain silent for the other 60%. The interference levels may be reduced and the MS battery recharging cycle increased by turning the transmitter off during periods of silence. This technique is known as *discontinuous transmission* (DTX) and relies on the accurate detection of periods of silence in the user's speech. This is achieved using voice activity detection (VAD) where the energy in the speech signal is computed for each speech block and a decision is made using an adaptive threshold as to whether the block contains speech or background noise. It is important that the system quickly recognises the beginning of a talk spurt to prevent initial talk spurt clipping. In addition, a hangover delay is incorporated at the end of a talk spurt to prevent clipping occurring. The hangover delay of four speech frames (80 ms) also prevents the VAD system from reacting to very short periods of silence, e.g. between syllables or within a word.

Subjectively, the periods of complete silence introduced by the DTX process at the receiver are annoying since they give the impression that the link has been lost. It was found that this situation could be improved by inserting 'comfort noise' to fill the periods of silence. The comfort noise is derived from the spectral envelope of the background noise using the LARs and it is conveyed in the form of a silence descriptor frame (SID). The SID is the same size as a speech frame and is updated once every four multiframes, i.e. every 480 ms.

2.3.9.5 Channel coding

Channel coding is employed in an effort to reduce the system BER to acceptable levels and hence improve the overall performance. However, FEC coding introduces a level of redundancy into the transmitted data, thereby increasing the transmitted data rate and channel bandwidth. For this reason, channel coding is applied in a selective manner in GSM. We have seen in Table 2.17 that the speech coder delivers 260 bits every 20 ms, which is equivalent to a data rate of 13 kb/s, and that these bits are divided into three classes depending on the impact on the received speech quality if they are received in error. Figure 2.20(a) shows the format of the 260-bit speech frame at the output of the speech coder. The Class Ia bits are labelled $\{d_0...d_{49}\}$, the Class Ib bits are labelled $\{d_{50}...d_{181}\}$ and the Class II bits are labelled $\{d_{182}...d_{259}\}$.

As we have already explained, the Class Ia bits are so important that the speech frame must be discarded if any of these bits are received in error. Therefore it is important for the receiver to be able to detect when errors in the Class Ia bits have leaked through the FEC process. This is achieved using a weak error detecting block code. The code used is a shortened cyclic code (53,50,2) with a generator polynomial given by [12]

$$g(D) = D^3 + D + 1. \tag{2.31}$$

This coding process generates three parity bits, $\{p_0, p_1, p_2\}$, which are appended to the end of the Class Ia bits, as shown in Figure 2.20(b). The Class Ib bits and four 'all zero' tail bits are then added to the Class Ia bits and the parity bits, and all the bits are reordered according to the following relationship:

$$
\begin{aligned}
u_k &= d_{2k}, & \text{for } k = 0, 1, ..., 90, \\
u_{184-k} &= d_{2k+1}, & \text{for } k = 0, 1, ..., 90, \\
u_k &= 0(\text{tail bits}), & \text{for } k = 185, 186, 187, 188,
\end{aligned}
$$

where the reordered bits are given by $\{u_0...u_{188}\}$. The reordering process groups the even-numbered data bits, $\{d_0...d_{180}\}$, at the beginning of the frame, the odd-numbered bits, $\{d_{181}, d_{179}, ..., d_1\}$, at the end of the frame followed by the four tail bits $\{d_{185}...d_{188}\}$. The parity bits, $\{p_0, p_1, p_2\}$, are inserted in the centre of the reordered frame as shown in Figure 2.20(c). The resulting 189-bit block, $\{u_0...u_{188}\}$, is then convolutionally encoded using a half-rate code with the following generator polynomials:

$$
\begin{aligned}
G0 &= 1 + D^3 + D^4, \\
G1 &= 1 + D + D^3 + D^4.
\end{aligned}
$$

This produces a coded block of $189 \times 2 = 378$ bits to which the uncoded 78 Class II bits are added to produce a block of 456 bits labelled $\{c_0, c_1, c_2, ..., c_{455}\}$ as shown in Figure 2.20(d).

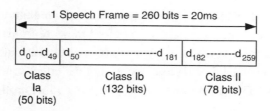

(a) The GSM speech frame

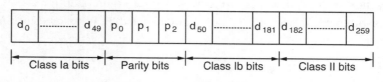

(b) The addition of the parity bits

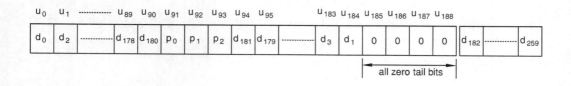

(c) The bit reordering process

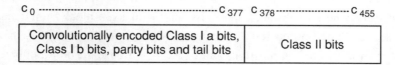

(d) The FEC coded speech frame

Figure 2.20: The FEC coding process for the full-rate GSM speech frame.

2.3.9.6   Interleaving

In the mobile radio environment, the errors in the transmitted bits tend to occur in bursts as the MS moves into and out of deep fades. The convolutional error correcting code described above is most effective when the errors are randomly distributed throughout the bit stream. For this reason the coded data are interleaved before they are transmitted over the radio interface. At the receiver, the de-interleaving process tends to distribute the error bursts randomly throughout the received data thereby increasing the effectiveness of the subsequent convolutional decoding. The GSM system employs two distinct levels of interleaving and these are described below.

Block diagonal interleaving   For a full-rate traffic channel (TCH/FS) carrying speech information, the 456-bit coded speech block, shown in Figure 2.20(d), is partitioned into eight 57-bit sub-blocks, B_0, B_1, ..., B_7 by assigning coded bit c_k to sub-block B_i, based on the following relationship:

$$i = k \bmod 8, \tag{2.32}$$

i.e. every eighth bit is assigned to the same sub-block. This process is shown in Figure 2.21(a). Each sub-block then forms one half of eight consecutive transmission bursts on the radio interface. The remaining half of each burst is occupied by sub-blocks from either the previous speech frame or the next frame, as shown in Figure 2.21(b), where B_i^n refers to sub-block i of speech frame n. The burst also contains two stealing flags, h_l and h_u, which are used to indicate whether either half-burst has been stolen by the fast associated control channel (FACCH).

Inter-burst interleaving   In addition to the block diagonal interleaving described above, the data bits are interleaved within the burst. One sub-block will occupy either the odd or even bit positions within the burst. Where a sub-block from a speech frame shares its burst with a sub-block from the previous speech frame, it will use the even-numbered bit positions. Conversely, where a sub-block shares its burst with a sub-block from the next speech frame, it will use the odd-numbered bit positions. Alternatively, we may say that B_0, ..., B_3 will use the even-numbered bit positions and B_4, ..., B_7 will use the odd-numbered bit positions. The bits within the sub-blocks are also reordered to increase the maximum distance between consecutive bits. The interleaving process is fully defined in the specifications [12] and there is also a comprehensive explanation in Reference [4].

2.3.9.7   Ciphering

The GSM system has the ability to encrypt the information on the radio path to reduce the security threat posed by eavesdroppers. The encryption process involves performing

456 bits, $\{c(0), ..., c(455)\}$. Puncturing is used to precisely tailor the rate of a convolutional code to the requirements of a transmission link. At the receiver, the convolutional decoder will effectively treat the deleted bits as errors and they will be corrected in the conventional way within the Viterbi decoder.

TCH/F4.8   The data are delivered to the coding unit at an intermediate bit rate of 6 kb/s, or to be more precise, one 60-bit block every 10 ms. Each block is extended to 76 bits by the addition of 16 all zero bits, which are inserted in blocks of four, once every 15 bits. Two of these blocks are then concatenated to form a single 152-bit block which is one-third rate convolutionally encoded using the following generator polynomials:

$$
\begin{aligned}
G1 &= 1+D+D^3+D^4, \\
G2 &= 1+D^2+D^4, \\
G3 &= 1+D+D^2+D^3+D^4.
\end{aligned}
$$

This results in a coded block of 456 bits.

TCH/H4.8   The data are delivered to the coding unit in blocks of 60 information bits every 10 ms, i.e. 6 kb/s. Four of these blocks are grouped together to produce a block of 240 bits. The subsequent coding processes are identical to those described for the TCH/F9.6 channel above.

TCH/F2.4   The data are delivered to the coding unit at an intermediate bit rate of 3.6 kb/s in the form of 36-bit blocks every 10 ms. The coding unit operates on blocks of 72 bits formed by the concatenation of two 36-bit blocks. Initially four all zero tail bits are added at the end of the block to produce a block of 76 bits. This block is then one-sixth rate convolutionally encoded using the following generator polynomials:

$$
\begin{aligned}
G1 &= 1+D+D^3+D^4, \\
G2 &= 1+D^2+D^4, \\
G3 &= 1+D+D^2+D^3+D^4, \\
G1 &= 1+D+D^3+D^4, \\
G2 &= 1+D^2+D^4, \\
G3 &= 1+D+D^2+D^3+D^4.
\end{aligned}
$$

This results in a coded data block of 456 bits.

TCH/H2.4   The information data are delivered to the coding unit at a rate of 3.6 kb/s in the form of one 36-bit block every 10 ms. Two blocks are concatenated to form a single 72-bit block and each block is expanded by the addition of four all zero tail bits at the end of the block. Two of these 76-bit blocks are then concatenated to form a single 152-bit block for the purposes of channel coding. This block is one-third rate convolutionally encoded to produce a coded data block of 456 bits using the convolutional code specified for the TCH/F4.8 channel.

2.3.10.2   Interleaving

The interleaving process may be partitioned into two levels: block diagonal and inter-burst interleaving. The TCH/H2.4 channel uses the same interleaving scheme as the TCH/FS described above. The remaining channels use a complex interleaving scheme which is described below for the TCH/F9.6 channel.

We recall that the convolutional encoding process results in a 456-bit block after puncturing. Since a data channel can, in general, accept a greater overall end-to-end delay than a speech channel, the block diagonal interleaving depth is increased to 22 bursts, thereby increasing the system's ability to randomise burst errors. Unfortunately, 456 bits cannot be evenly divided between 22 bursts so, consequently, the interleaving scheme becomes complex. The specifications present two different ways of viewing the block interleaving process [12].

1. The 456-bit block is sub-divided into four 114-bit sub-blocks, each of which is evenly distributed over 19 bursts with six bits in each. The sub-blocks are block diagonally interleaved with a shift of one burst between each sub-block.

2. The 456-bit block is sub-divided into two blocks of six bits, which are placed in bursts 0 and 21; two blocks of 12 bits, which are placed in bursts 1 and 20; two blocks of 18 bits, which are placed in bursts 2 and 19; and 16 blocks of 24 bits, which are placed in bursts 3 to 18. Therefore, each 456-bit block is interleaved over 22 bursts. The data blocks are also diagonally interleaved with a new data block beginning every fourth burst.

We also note that the data bits are reordered within each burst [12].

2.3.10.3   Ciphering

The bursts are encrypted in the same manner as that described in the previous section for the full-rate speech traffic channel.

2.3.11   Control data transmission

In this section we examine the channel coding and interleaving processes that are performed on the control data transferred between the MS and the network. Each logical control channel will be dealt with separately. At the end of this section we will also discuss the type of ciphering that is applied to the control channels.

2.3.11.1 Slow associated control channel

The SACCH data are delivered to the coding unit in fixed blocks of 184 information bits. The initial encoding is performed using a shortened binary cyclic code defined by the following generator polynomial:

$$g(D) = (D^{23} + 1)(D^{17} + D^3 + 1). \tag{2.33}$$

This type of code is commonly referred to as a fire code and is used to detect 'bursty' residual errors that are not corrected by the convolutional decoder. The result of this coding process is the generation of 40 parity bits that are appended to the end of the data block to form a 224-bit block. This block is extended to 228 bits with the addition of four all zero tail bits at the end of the block. This data block is then convolutionally encoded using a one-half rate convolutional code with the following generator polynomials:

$$\begin{aligned} G0 &= 1 + D^3 + D^4, \\ G1 &= 1 + D + D^3 + D^4. \end{aligned}$$

The result is a block of 456 coded bits. The bits are reordered and divided into eight, 57-bit sub-blocks in the same way as the 456-bit speech block on the TCH/FS, i.e. some blocks occupy the even-numbered bits and some blocks occupy the odd-numbered bits. However, in the case of the SACCH block, the block interleaving occurs over four full bursts with each burst containing bits from the same block in both the odd and even bit positions. This technique is termed *block rectangular interleaving* and a new data block begins once every fourth burst and is interleaved over four bursts. We note that the SACCH bursts occur once every 26 bursts or 120 ms, in the case of the full-rate traffic channel, which means that the delay caused by the interleaving and channel coding processes will be $4 \times 120\,\text{ms} = 480\,\text{ms}$. In the case of a SACCH burst the stealing flags are always set to one.

2.3.11.2 Fast associated control channel

The FACCH information is delivered to the channel coder in blocks of 184 bits which are block and convolutionally encoded using the same codes as employed for the SACCH. The interleaving scheme is different for FACCHs on full-rate and half-rate channels. For the full-rate channel, the interleaving scheme is identical to that used for the full-rate 456-bit coded speech frames, i.e. the block is divided into eight 57-bit sub-blocks and block diagonally interleaved over eight consecutive bursts with the first four sub-blocks occupying only the even-numbered bit positions and the last four sub-blocks occupying the odd-numbered bit positions. Where the even-numbered bits of a burst are stolen by the FACCH, the h_u flag is set to a logical 1, and the h_l flag is set to a 1 when the odd-numbered bits have been stolen. The similarity between the FACCH and full-rate speech interleaving means that a 456-bit

FACCH block will completely replace a 456-bit coded speech block on a one-for-one basis. Consequently, the insertion of an FACCH block will result in the loss of a 20 ms speech block.

Since the TCH/F2.4 also uses the same interleaving scheme, the insertion of an FACCH block results in the loss of a single 456-bit coded information block, or a 72-bit information block. In the case of the TCH/F9.6, the difference between the two interleaving schemes means that the insertion of the FACCH results in the loss of a maximum of 24 coded bits in each 114-bit block, and in the case of the TCH/F4.8, a maximum of 48 coded bits will be lost in every 228 bits.

When an FACCH is inserted on the half-rate channel, the 184-bit FACCH block is block and convolutionally encoded in the same manner as the SACCH block to produce a 456-bit coded data block. The block is interleaved over six bursts in the following manner. The 456-bit block is divided into eight sub-blocks with the first four sub-blocks, $\{B(0), ..., B(3)\}$, occupying the even-numbered bit positions and the last four blocks, $\{B(4), ..., B(7)\}$, occupying the odd-numbered bit positions. Sub-blocks $B(2)$ and $B(3)$ are combined with sub-blocks $B(4)$ and $B(5)$ to fill two complete bursts and the remaining sub-blocks fill half-bursts. The blocks are therefore effectively block diagonally interleaved over six bursts with a new data block beginning once every fourth burst. Accordingly an FACCH block steals the even-numbered bits of the first two bursts of the TCH/H (h_u set to 1), all of the bits of the next two bursts (both h_u and h_l set to 1), and the odd-numbered bits of the next two bursts (h_l set to 1).

The effects of FACCH bit stealing on a half-rate speech channel (TCH/HS), will be the loss of two consecutive speech frames. In the case of the TCH/H4.8 and TCH/H2.4 channels, a maximum of 24 in every 114 coded bits will be lost.

2.3.11.3 Random access channel (RACH)

The short RACH bursts transmitted on the up-link contain only eight information bits. Six parity bits are generated using a simple systematic cyclic code with the following feedback polynomial:

$$G(D) = D^6 + D^5 + D^3 + D^2 + D + 1. \tag{2.34}$$

The six parity bits are then added, bit-wise modulo-2, to the six-bit base station identity code (BSIC) of the BTS for which the RACH message is intended. This process is included to ensure that two BTSs with the same BCCH carrier frequency do not both decode and respond to a RACH frame from a single MS. Only the BTS with the same BSIC as that used in the RACH burst generation will be able successfully to decode the information. This process results in a 14-bit block to which four all zero tail bits are added to form an 18-bit block. This block is then one-half rate convolutionally encoded using the same generator polynomials as those used in the TCH/FS producing a 36-bit coded block. Referring Figure 2.9, we note that this block fits exactly into a single RACH burst. We emphasise that there is no interleaving on the RACH channel.

2.3.11.4   Synchronisation channel (SCH)

The synchronisation bursts, each containing 25 information bits, are transmitted on time slot zero on the down-link BCCH carrier. The data transmitted include the BSIC and the frame number of the current frame within the hyperframe. Ten parity bits are generated using the following polynomial:

$$G(D) = D^{10} + D^8 + D^6 + D^5 + D^4 + D^2 + 1, \tag{2.35}$$

and four all zero tail bits are added to produce a 39-bit data block. The block is then one-half rate convolutionally encoded using the same code as the TCH/FS to produce a coded data block of 78 bits. Referring to Figure 2.9, this information block fits exactly into a single burst and we note that interleaving is not used on the SCH.

2.3.11.5   Other control channels

The broadcast control channel (BCCH), the paging channel (PCH), the access grant channel (AGCH), the cell broadcast channel (CBCH) and the stand-alone dedicated control channel (SDCCH) all use the same coding and interleaving scheme as the SACCH, described above.

2.3.12   Ciphering of control data

The SCH, BCCH, PCH, AGCH and CBCH are never encrypted because they must be available to every MS within a cell, and the GSM security mechanisms are designed such that the encrypted information may only be decoded by a single known user. Also, the RACH is never encrypted because the BTS must be able to decode this burst with no prior knowledge of the identity of the MS. The only control channels that may carry encryption are the SACCH, FACCH and SDCCH. The data on these channels are encrypted by EXOR-ing the 114 bits in the burst with a 114-bit encryption word, thereby scrambling the data in a known way. We will examine the manner in which the encryption word is generated in Section 2.5, which deals with the security aspects of GSM.

2.4   Control of the radio resource

In this section we consider the manner in which the radio resource is allocated and used in a GSM system. We will discuss many of the operational aspects associated with a cellular

tialisation and cell selection, paging, access and handover. We
of the other features of GSM including power control and timing

2.4.1   Cell selection

When an MS is switched on, its first task is to locate a suitable BTS through which it
can gain access to the network, if required. This is achieved by searching the relevant
frequency band for BCCH carriers and then decoding the information they carry to select an
appropriate BTS. The MS may implement one of two different search algorithms depending
on its knowledge of the BCCH carriers in use.

The first algorithm is applied when the MS has no knowledge of the BCCH carriers deployed in a particular PLMN. Initially the MS searches the entire down-link frequency band
(i.e. 124 carriers for primary GSM900, 174 carriers for E-GSM900 and 374 carriers for
DCS1800) and measures the received signal strength of each carrier. The received signal
level for each carrier is determined from the average of at least five measurements spread
evenly over a time period of 3 to 5 s [13]. The MS then retunes to the strongest carrier and
waits for an FCCH burst, i.e. a burst of pure sine wave. If an FCCH burst, which occurs
every 10 or 11 time frames on time slot zero of a BCCH carrier, is not detected, then the MS
retunes to the next strongest carrier and repeats the process. Once the MS identifies a BCCH
carrier by means of the FCCH burst, it synchronises to the BTS and attempts to demodulate
the synchronisation information. The FCCH burst is used by the MS to correct its internal
time base to ensure that its carrier frequency is accurate to within 0.1 part per million (ppm)
compared with the signal received from the BTS. The MS employs its internal time base to
generate both the local versions of the RF carriers for demodulation, and the clock signals
for its internal counters.

Having applied the relevant frequency correction, the mobile attempts to decode the synchronisation burst contained in the SCH time slot. This slot is easily located because it
always follows immediately after the FCCH time slot on the same physical channel, i.e.
eight time slots later. The synchronisation burst contains sufficient information for the MS
to identify its position within the complete GSM frame structure. The burst contains 25 bits
of information prior to channel coding and six of these bits are used to transmit the base
station identity code (BSIC). The remaining 19 bits are used to transmit the reduced TDMA
frame number (RFN) of the time slot containing the synchronisation burst. The RFN consists of three parameters, $T1$ (11 bits), $T2$ (5 bits) and $T3'$ (3 bits), which are determined
using the full frame number (FN) unique to each TDMA frame within the hyperframe. The

FN ranges from 0 to 2 715 647 and the RFN parameters are defined as follows:

$$T1 = \text{FN div } (26 \times 51), \quad \text{(11 bits),} \quad \text{range 0 to 2047,}$$
$$T2 = \text{FN mod } 26, \quad \text{(5 bits),} \quad \text{range 0 to 25,}$$
$$T3' = (T3 - 1) \text{ div } 10, \quad \text{(3 bits),} \quad \text{range 0 to 4,}$$

where $T3$ is a number in the range 0 to 50 and is given by

$$T3 = \text{FN mod } 51. \tag{2.36}$$

The 'mod' and 'div' operators return the integer result and the remainder of an integer division, respectively. This shows that $T1$ provides the position of the superframe containing the synchronisation burst within the hyperframe, while $T2$ provides the position of the multiframe within the superframe. The control channel multiframe contains 51 TDMA frames and the position of the frame containing the synchronisation burst within the multiframe is given by $T3$, which requires six bits. However, the synchronisation burst can only occupy one of five different positions within the multiframe and consequently this information is transmitted using three bits as $T3'$. We note that the synchronisation burst always occupies time slot zero of the TDMA frame and there is no requirement to transmit this information.

In addition to the FN, the mobile must also maintain the time slot number (TN) and quarter-bit number (QN) counters. The QN counter is set using the extended training sequence located in the middle of the synchronisation burst and is incremented every $12/13 \, \mu s$. The QN counts the quarter-bit periods and its value ranges from 0 to 624, i.e. 0 to 156-bit periods. The TN counter is set to zero when the synchronisation burst is received and it is incremented each time the QN count changes from 624 to 0. The TN counter is used to hold the position of the time slot within the TDMA frame and its value ranges from 0 to 7. As the value of TN changes from 7 to 0 the FN is incremented.

Having successfully synchronised to the BS, the mobile may proceed to decode the system information contained on the BCCH. The BCCH is easily located since it always occupies the same position within the control channel multiframe. This channel contains a number of parameters that influence the cell selection, including the maximum power that the MS may transmit while accessing the BTS (parameter $MS\_TXPWR\_MAX\_CCH$) and the minimum received power at the MS for access (parameter $RXLEV\_ACCESS\_MIN$). These parameters are combined with the received power of the base station, R, and the maximum output power of the MS, P, to produce a radio parameter known as $C1$ given by [13]

$$C1 = A - B, \quad \text{for } B > 0,$$
$$C1 = A, \quad \text{for } B \leq 0,$$

where

$$A = R - RXLEV\_ACCESS\_MIN,$$
$$B = MS\_TXPWR\_MAX\_CCCH - P,$$

and all values are expressed in dBm. If $C1$ for a given BTS is greater than zero, then the MS is considered to have the ability to access that BTS, if required. Also, the BTS with the highest $C1$ is considered to be the most suitable BTS as far as the radio interface is concerned.

Another parameter that influences the cell selection process is the *CELL_BAR_ACCESS* flag. As its name suggests this flag indicates whether the cell has been barred for access. It is used to prevent normal subscribers from accessing an active cell, e.g. for the purposes of testing. Special engineering MSs will ignore this flag.

Using the cell selection algorithm, the MS will choose the unbarred cell with the highest positive $C1$ value in its current location area. For cells outside the current location area, a handicap factor is introduced into the selection algorithm to prevent an MS from switching back and forth between two cells in different location areas, as this results in an unacceptable location updating load. The maximum time allowed for synchronisation to a BCCH carrier is 0.5 s, and the maximum time allowed to read the BCCH data is 1.9 s, once the MS is synchronised. When the MS has selected the most appropriate cell, it will 'camp on' that cell and enter the idle mode.

A second cell selection algorithm exists for situations where the MS has a knowledge of the BCCH carriers used within the network. In this case the MS will initially search only the carriers stored in its BCCH carrier list and apply the cell selection algorithm. In the event that this process does not produce a valid BCCH carrier, the MS will revert to the full cell selection algorithm described above. The storage of BCCH carrier information in the MS is an optional feature and the manner in which the list is derived may vary. For example, the MS may store the list of BCCH carriers in use when it was last switched on. However, this may prove inappropriate if the MS is switched off and moved to an entirely different location. It is encumbent upon an MS to select the fourth strongest cell (where the three strongest cells are inappropriate) and be able to respond to paging messages within 30 s, assuming the BCCH carriers are all received at an adquate C/I.

2.4.2   Idle mode

Having selected an appropriate cell, the MS enters the 'idle mode' where it must monitor the BTS paging channel and continue to ensure that it is camped on the most appropriate cell. Each of these processes will now be described in detail.

2.4.2.1 Paging

A paging call, transmitted on the paging channel (PCH), is used to alert an MS to an incoming call. Consequently, an MS must continually monitor the relevant PCH for paging calls containing its unique address. The MS may determine the full control channel organisation on the BCCH carrier using the data contained within the BCCH. In cells where the paging channels are located on more than one timeslot, the MS will determine the relevant PCH using the last three digits of its unique international mobile subscriber identity (IMSI). In this way the MSs are 'randomly' distributed between the radio resources that have been allocated to the PCH.

The GSM system also supports a slotted paging mode, whereby the PCH is divided into a number of paging sub-blocks and the MS is only required to listen to the channel during its assigned paging sub-block. The MS may be 'powered-down' during the periods it is not monitoring the PCH and this leads to an increase in the MS battery recharging cycle, a technique known as *discontinuous reception* (DRX).

The messages sent on the PCH will include the IMSI of the MS to be paged. Once a MS recognises its own IMSI, it will answer the paging call by entering the access mode. Unanswered paging messages will be repeated to allow for variations in the radio channel; however, the exact repetition policy is a matter for the network operator.

2.4.2.2 Cell reselection

Whilst in idle mode, the MS must continue to monitor the down-link signal strength of surrounding BTSs to ensure that it remains camped on the most appropriate BTS. The MS will continue to measure the received signal strength from the serving BTS and also the received signal strength of the six strongest BCCH carriers contained in the neighbour list message carried on the BCCH of the serving BTS. At least every 5 s, the MS is required to apply the cell selection algorithm to identify whether an idle mode handover is required to a more appropriate cell. Provided the new cell is contained within the current location area, the MS may simply switch to the new cell and begin to decode the PCH data. Where the new cell belongs to a different location area, the MS must perform a location update before it is in a position to receive an incoming call.

2.4.3 Access mode

The access mode is used to describe the process by which an MS in idle mode gains access to the network and receives a dedicated channel. An MS will enter the access mode to perform a location update, or to answer a paging call, or as a result of a subscriber-originated call. An MS initially accesses a BTS using the RACH on the up-link of the BCCH carrier. Although an MS may only transmit an access request in the up-link slots assigned to the RACH, there is no restriction on which slots it may use. Consequently, a collision of two or more MSs' RACH signals may occur at the BTS. In this situation, either one or none of the attempted accesses may be decoded and answered. This means that all unanswered access requests must be repeated to ensure a reasonable chance of success. The repetition strategy must be such that it minimises the load on the RACH whilst maintaining a satisfactory level of success for access requests. We also note that the time between retransmissions cannot be a fixed parameter, otherwise two colliding MSs will continue to collide during successive access attempts. For this reason, the repetition period is effectively randomised, although the average time between retransmissions and the maximum number of retransmissions is controlled through parameters that are regularly broadcast on the BCCH.

The RACH load may also be controlled by forbidding an MS from attempting to access the system for a given period of time after an access attempt has failed. Alternatively the RACH load may be reduced by prohibiting fractions of the entire MS population from attempting to access the network. Specifically, the entire subscriber population is divided into 10 equal subgroups using information stored in the SIM card. A subgroup may then be prohibited from accessing the network, except in specific cases (e.g. emergency calls). In this way the RACH load may be reduced in steps of 10% in times of congestion.

The RACH burst is very short compared with the other bursts, and it contains only eight bits of information prior to channel coding; this is insufficient to transmit the MS's unique address, i.e. its IMSI. The main requirement of the RACH burst is that it contains sufficient information for the MS to identify a response to its RACH burst from the network. Accordingly, an MS transmits a five-bit random number chosen by itself, along with a three-bit number that informs the network of the reason for the access attempt, e.g. an answer to a paging call, emergency call, etc. The MS transmit power used for the access burst is determined using a system parameter transmitted on the BCCH. Some of the bits in the access burst are EXORed with the BSIC of the BTS for which the message is intended. This ensures that only the intended BTS will decode the message correctly and attempt to respond.

On receiving an access burst on its RACH channel, the BTS will relay the message to the BSC along with an estimate of the transmission delay between the arrival of the access burst and the BTS timing schedule. This parameter is used to initialise the timing advance at the MS and to avoid collisions with data from other MSs at the BTS once the MS enters the dedicated mode. At this stage the BSC will allocate a free channel on the BTS, and once the BTS has acknowledged that the channel has been successfully activated, an initial assignment message is constructed. This message contains details of the allocated channel, the initial timing advance to be applied by the MS, the initial maximum transmission power, and an address allowing MSs to identify the destination of the message.

The timing advance parameter is transmitted as a six-bit number providing 64 timing advance steps each corresponding to the period of one bit, namely 3.69 μs. This means that the system may apply a maximum timing advance of $63 \times 3.69\,\mu s \approx 232\,\mu s$, which corresponds to a round trip propagation distance of approximately 70 km and a maximum distance from a BS to the cell boundary of some 35 km. Large cells may be accommodated by enlarging the time guard bands surrounding each time slot by only using every other time slot (i.e. time slots 0, 2, 4 and 6). In this way the up-link transmission can spill into the neighbouring time slot and the cell radius may be increased to more than 120 km. This approach will result in a decrease in the cell capacity and is deployed in special circumstances, e.g. with coastal BTSs conducting maritime communications. This technique also explains the use of time slots 0, 2, 4 and 6 for common control channels, as these are the only time slots that are provided in these large cells.

The initial assignment message from the BTS is addressed to the appropriate MS by including the exact contents of the access burst transmitted by the MS to the BTS along with a time reference describing the time at which the access burst was received. The initial assignment message is sent to the MS using the access grant channel (AGCH) which shares the same physical resources as the PCH. The MS continually monitors the AGCH until it detects a message containing both the contents of its initial access burst and a time reference corresponding to the time at which the message was sent. We note that at this point the MS has not been identified unambiguously, since there is a finite possibility that two MSs could have transmitted identical access bursts in the same RACH time slot.

Once the MS receives an appropriate initial assignment message, it retunes to the new dedicated channel and sends an initial message using the timing advance contained in the initial assignment message. The format of this message will depend on the reason for the access, e.g. a response to a paging call.

The initial message contains details of the MS including its classmark and its unique IMSI. Once this message has been received by the BTS its contents are sent back to the MS without modification [6]. If an MS receives a message that is different from the one it transmitted it will leave the channel immediately and restart the access process. This ensures that any remaining ambiguity is removed and only the correct MS accesses the dedicated channel. Having established a link on a dedicated channel the MS will enter the dedicated mode.

2.4.4 Handover

A handover is the process by which an MS relinquishes its connection with one BTS while establishing a new connection with another BTS while ensuring that an existing call is maintained. Handovers are also performed to decrease interference, or to relieve traffic congestion. Although the GSM specifications provide an example handover decision algorithm based on radio criteria, these solutions are not mandatory and the equipment manufacturer is free to implement its own algorithms. The complete handover process may be sub-divided into two distinct phases, preparation and execution [6].

2.4.4.1   Handover preparation

The decision to perform a handover and the identification of the most suitable new BTS is based on several different measurements, performed both at the MS and at the BTS. It is also based on a number of static parameters and the distribution of the traffic load within the network. The measurements utilised in the handover process are as follows.

1. Measurements performed at the BTS.

   - The level of the up-link signal received from the MS (parameter $RXLEV\_UL$).

   - The quality (or BER) of the up-link signal received from the MS (parameter $RXQUAL\_UL$).

   - The distance between the MS and the BTS based on the adaptive timing advance parameter.

   - The interference level in unallocated time slots.

2. Measurements performed at the MS.

   - The level of the down-link signal received from the serving cell (parameter $RXLEV\_DL$).

   - The quality (or BER) of the down-link signal received from the serving cell (parameter $RXQUAL\_DL$).

   - The level of the down-link signal received from the nth neighbour cell (parameter $RXLEV\_NCELL(n)$).

The $RXQUAL$ and $RXLEV$ parameters are coded using the schemes shown in Tables 2.18 and 2.19, respectively.

The handover process also takes into account the maximum transmission powers of the MS, the serving BTS and the neighbouring BTSs. A handover may also be used to distribute the traffic more evenly throughout the network by handing MSs from congested BTSs onto less congested neighbouring BTSs, provided the neighbours are able to support the MS. Consequently, the handover process will also take into account the traffic load within the network.

Table 2.18: The RXQUAL parameter and the corresponding channel BERs (for both up-link and down-link).

RXQUAL 0	BER < 0.2%
RXQUAL 1	BER = 0.2% to 0.4%
RXQUAL 2	BER = 0.4% to 0.8%
RXQUAL 3	BER = 0.8% to 1.6%
RXQUAL 4	BER = 1.6% to 3.2%
RXQUAL 5	BER = 3.2% to 6.4%
RXQUAL 6	BER = 6.4% to 12.8%
RXQUAL 7	BER >12.8 %

Table 2.19: The RXLEV parameter and the corresponding received signal strength values (for up-link, down-link and neighbouring cells).

	Received signal level (dBm)
RXLEV 0	< −110
RXLEV 1	−110 to −109
RXLEV 2	−109 to −108
RXLEV 3	−108 to −107
.	.
.	.
.	.
RXLEV 62	−49 to −48
RXLEV 63	> −48

MS measurement schedule   The complicated frame structure employed in GSM is, in many ways, dictated by the requirement that the MS must not only measure the down-link signal from its serving BTS but also the down-link signals from surrounding BTSs. Each BTS transmits a neighbour list which contains the BCCH carrier frequencies of its neighbouring BTSs. The MS will measure the received signal strength for each BTS in the neighbour list. Examining the transmission and reception schedule at the MS reveals three 'windows' during which these measurements might take place. There is a gap of two time slots (minus the timing advance) between the reception of the down-link burst and the transmission of the up-link burst. With the timing advance at its maximum, this window is 920 μs in duration and is too short for measurement purposes. The second window occurs between the transmission of the up-link burst and the reception of the down-link burst, and its minimum duration is four time slots or 2.3 ms (with no timing advance) as shown in

Figure 2.22. This window is used by the MS to measure the down-link signal strength of the BCCH carriers in neighbouring cells. During this time, the MS must retune to the BCCH carrier to be measured, take the measurement, and then retune to the current down-link frequency in time to receive the next burst. This tight schedule does not allow the MS the time to wait for an active burst, and consequently every slot on the BCCH carrier must remain active. This is achieved by using dummy bursts to fill slots that would normally be inactive. There is also a requirement that the BCCH carrier is transmitted at full power and therefore down-link power control must *not* be applied. We also note that DTX may *not* be used on any slots on the BCCH carrier.

The third measurement window is produced by the idle frame included in each traffic multiframe (i.e. 26 TDMA frames). This window is a minimum of 12 timeslots, or around 7 ms, as shown in Figure 2.23. This measurement window is used to ensure that the BCCH carrier measurements described above are always associated with the correct BTS. As the BCCH carrier frequencies are reused by different BTSs throughout the network, we cannot guarantee that an MS is measuring a particular BTS simply by knowing the carrier frequency on which the measurements were taken. For this reason, the MS is required periodically to check the identity of the BTS by decoding the synchronisation burst transmitted on the BCCH carrier and extracting the base station identity code (BSIC). This is a six-bit colour code and will be assigned to the BTSs such that two BTSs using the same BCCH carrier frequency will have different BSICs if there is a possibility of co-channel interference. The MS reports the BSIC to the network along with the BCCH carrier measurements and this allows the network to ensure that the measurements are associated with the correct BTS.

During the long measurement window, the MS is required to retune to the BCCH carrier, identify and decode a valid synchronisation burst, and then retune to its current down-link frequency to receive the next burst from the serving BTS. This obviously assumes that the synchronisation burst on the neighbouring BCCH carrier falls within the 12-slot measurement window. The synchronisation bursts occur at intervals of 10 or 11 TDMA frames within the 51-frame control multiframe. Consequently, there is no guarantee that the mea-

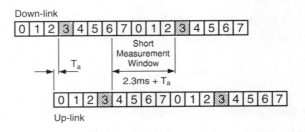

Figure 2.22: The short measurement window for communications on time slot 3.

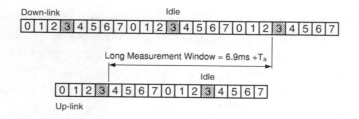

Figure 2.23: The long measurement window for communications on timeslot 3.

surement window will align with a synchronisation burst every time. However, the time frame structure means that the idle frame window in the 26-frame traffic multiframe gradually slips past the 51-frame control multiframe ensuring that the idle frame window will coincide with a synchronisation burst.

The specifications require that the BSIC of a BCCH carrier is decoded at least once every 10 s. The MS will also decode and store the synchronisation information carried in the synchronisation burst, i.e. the multiframe, superframe and hyperframe parameters, $T1$, $T2$ and $T3'$, and it may use this to schedule the decoding of the BSIC. This synchronisation information may also be employed to reduce the switching times at handover.

MS measurement averaging The MS measurement results are reported to the network via the slow associated control channel (SACCH) in the form of a measurement report message. The MS may report on up to six neighbouring cells in the same measurement report, in addition to the serving cell. The information carried on the SACCH is interleaved over four bursts and this represents a delay of 4×120 ms $= 480$ ms. This time period is referred to as a *reporting period* and the measurements taken at the MS are averaged over this period before they are reported to the network. Further averaging will take place once the measurements arrive at the base station subsystem (BSS). The BSS must be able to store at least 32 measurement samples, where a sample is defined as the value evaluated by the MS during the measurement reporting period of 480 ms. The BSS may perform an unweighted average of the 32 samples, or it may weight the samples to attach more importance to the more recent measurements [13].

2.4.4.2   Handover execution

Once the decision has been taken to initiate a handover and the most suitable new cell has been identified, the MS and network enter the handover execution phase whereby the connection with the old BTS is relinquished and the connection with the new BTS is established. We will now provide a general overview of the handover process from the network and the MS point of view. For a more detailed description of the GSM handover process, the reader is referred to Reference [6]. The handover execution process is heavily influenced by the position of the switching point in the infrastructure, e.g. whether it is an intra-BSC, intra-MSC or inter-MSC handover, and on whether the old and new BTSs are synchronised. At this stage we will simplify the handover description by adopting the notation used in Reference [6], where the terms BTS, BSC and MSC are suffixed with the word 'old' if they form part of the communication path prior to the handover, e.g. BTS-old, and the word 'new' if they from part of the communication path after the handover, e.g. BTS-new.

The position of the switching point for different scenarios is shown in Figure 2.24. Where BTS-old and BTS-new are controlled by the same BSC, the switching point will be the BSC, see Figure 2.24(a). If BTS-old and BTS-new are controlled by different BSCs (i.e. BSC-old $\neq$ BSC-new) but they are both connected to the same MSC, then the MSC will be the switching point, as shown in Figure 2.24(b). Figures 2.24(c), (d) and (e) apply if the BTS-old and BTS-new are connected to different MSCs and in these cases the *anchor-MSC* will provide the switching point.

The anchor-MSC is the MSC to which the MS was connected at call initiation and it remains in control of the communication for the duration of the call. Any handover between BTSs connected to different MSCs results in MSC-new being added to the communication chain along with the anchor-MSC, see Figure 2.24(c). In this case MSC-new is sometimes referred to as the relay-MSC [6]. Where the connection already includes a relay-MSC, and MSC-new is not the anchor-MSC, a different relay-MSC, i.e. MSC-new, is included in the communication path, as shown in Figure 2.24(d). Where MSC-new is not MSC-old and MSC-new is the anchor-MSC, the relay-MSC is removed from the communication path, see Figure 2.24(e). The position of the switching point determines the type of messages that are transmitted between the various machines on the infrastructure side.

The first step of the handover execution phase is for BSC-new to be informed by BSC-old that a handover is required. Except where BSC-new and BSC-old are the same, this message is transmitted via the switching point. At this point the new communication path is established between the switching point and BSC-new. Having been informed of the requirement for a handover, BSC-new attempts to allocate an appropriate channel on BTS-new. If this process is successful, then BSC-new sends details of the new channel back to BSC-old via the switching point (in the case where BSC-new $\neq$ BSC-old). The switching point now generates a handover command message and this is sent via BSC-old and BTS-old to the MS. The handover command message contains details of the new channel on BTS-new and some essential parameters relating to the new cell, e.g. BCCH carrier frequency.

An MS is completely unaware of the impending handover until it receives the handover command message. At this stage the MS retunes to the new channel. From this point onwards the action of the MS is governed by the timing relationship between BTS-old and

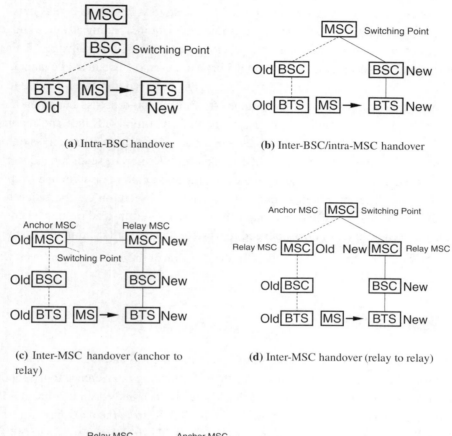

(a) Intra-BSC handover

(b) Inter-BSC/intra-MSC handover

(c) Inter-MSC handover (anchor to relay)

(d) Inter-MSC handover (relay to relay)

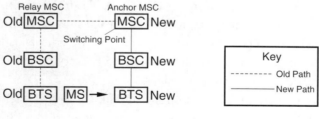

(e) Inter-MSC handover (relay to anchor)

Figure 2.24: The different types of handover in GSM.

BTS-new. The handover command message contains an indication as to whether or not BTS-old and BTS-new are synchronised. If they are synchronised, then the MS will have been able to determine the timing advance to be used at BTS-new beforehand. In this situation, once tuned to the new channel, the MS transmits a few short access bursts with no timing advance to enable BTS-new to determine the required timing advance and it then commences normal transmission using the previously computed timing advance. This is the only time where bursts other than the normal bursts are used on a dedicated channel.

In the case where BTS-old and BTS-new are unsynchronised, the MS is unable to compute the new timing advance and it continues to transmit short access bursts on the new channel. The BTS uses these bursts to determine the new timing advance and reports this to the MS. When the MS receives this message it commences normal transmission on the new channel. Once the MS has established a link with BTS-new, it sends a handover completion message, and this is the cue for the communication path to be switched from BTS-old to BTS-new. The interruption time during handover execution can be reduced from roughly 200 ms to 100 ms if a synchronous handover is used instead of an asynchronous handover [6] because the MS does not need to wait for the network to compute the required timing advance.

2.4.5   Power control

The GSM system employs power control to ensure that the MS and BTS only transmit sufficient power to maintain an acceptable link, thereby reducing interference to neighbouring cells and improving spectral efficiency. We have already seen in Section 2.2 that an MS has the ability to decrease its transmitted power in steps of 2 dB from the maximum for its class down to 5 dBm for GSM900 and 0 dBm for DCS1800. The transmission power of the MS is controlled by the network conveying messages over the slow associated control channel (SACCH). After receiving a power control command, an MS adjusts its transmitted power to the requested power level at a maximum rate of 2 dB every 60 ms. Thus a transmitter power change of 30 dB will take around 900 ms.

The power control algorithm is based on the up-link signal measurements taken at the BTS, and although the specifications include an example algorithm, the implementation will be manufacturer/operator specific. The BTS must be able to dynamically adjust its power in at least 15 steps of 2 dB. The operator has the option of whether or not to use both up-link and down-link power control, and it may be applied independently on either link. However, down-link power control may not be applied to any slots on the BCCH carrier as it must be transmitted at a constant power because it is measured by the MSs in surrounding cells for the purposes of handover preparation.

2.4.6 Frequency hopping

GSM employs slow frequency hopping (SFH) to mitigate the effects of multipath fading and interference. Each burst belonging to a particular physical channel will be transmitted on a different carrier frequency in each TDAM frame. Thus the hopping rate is equal to the frame rate (i.e. $\simeq 217$ frames/s). The only physical channels that are not allowed to hop are the broadcast and common control channels (i.e. the FCH, SCH, BCCH, PCH and AGCH). This is because an MS must be able easily to locate these channels on power-up and this process would become more complex if frequency hopping were allowed on these channels. The radio carrier on which these broadcast and common control channels are transmitted is called the BCCH carrier. The BCCH carrier is measured by the MSs in neighbouring cells to determine whether the BTS is a suitable handover candidate. As a result of this, the BCCH carrier must be transmitted continuously and at a constant power. This means that down-link power control and discontinous transmission (DTX) may not be used on the BCCH carrier. A traffic channel or dedicated control channel may use the BCCH carrier frequency as part of its hopping sequence, but it must obey the above rules while it is using that carrier, i.e. power control and DTX are not permitted. Let us now examine the two significant gains that acrue from deploying SFH in GSM.

The effect of frequency hopping on interference In a non-frequency hopping GSM system, an MS will tend to experience interference from the same set of MSs in neighbouring co-channel cells for the duration of a call, provided the MS is not handed over. The level of this interference will vary depending on the relative positions of each of the MSs. For example, the interference levels will be high if all the MSs are situated at the edges of their respective cells, whereas it will be low if the MSs are situated close to their serving BTS. This means that, in situations where an MS experiences a high level of interference, it is likely that it will continue to suffer for the duration of a call. In a frequency hopped system, the hopping patterns (i.e. the sequence of transmission frequencies) are different in co-channel cells and the MS will experience interference from a different set of MSs on each burst. This effectively randomises the interference and each MS will experience an average level of interference. This situation is preferable to the non-frequency hopped case where it is possible for some MSs to experience very high levels of interference while others experience very low levels of interference.

The effect of frequency hopping on multipath fading The mobile radio propagation channel imposes large amplitude variations onto the received signal as a number of different propagation paths are either constructively or destructively summed at the receiving antenna. The position of the signal fades is heavily dependent on the environment, e.g. positions of buildings and vehicles, and the operating frequency. At high mobile speeds the rate of signal fading is high and the length of the fades will be short, and the resulting error bursts will be combated by the symbol interleaving and channel coding. At slower speeds the MS will spend longer periods of time in signal fades and the effectiveness of the interleaving and channel coding schemes is less. Frequency hopping may be used to ensure that an MS does not spend large periods of time in signal fades. By ensuring that the frequency change between two successive frequency hops is beyond the coherence bandwidth of the channel an MS is unlikely to hop from one fade directly into a second one. Consequently, frequency hopping may be used to mitigate the effects of fast fading at slow MS speeds.

2.4.6.1 Frequency hopping sequences

The hopping sequence defines the order in which the different carrier frequencies are used on the up-link and the down-link. Since the up-link and down-link frequencies always remain separated by the duplex channel spacing (i.e. 45 MHz for GSM900 and 95 MHz for DCS1800) only a single hopping sequence is required to describe the complete duplex link.

The *mobile allocation* parameter prescribes the carrier frequencies that may be used by each MS in its hopping sequence. For 124 possible TDMA carriers the mobile allocation parameter requires a minimum of 124 bits to uniquely describe every possible carrier combination. We recall that initial assignment messages are sent on the common access grant channel (AGCH) where the message size should be kept short to preserve the access capacity of the system. To avoid transmitting the full mobile allocation parameter at initial assignment a two-step approach is used. Each BTS transmits details of all the carriers it is using in the form of a cell channel description message, carried on the BCCH. This message takes the form of a 124-bit map where each bit represents a carrier and a '1' or '0' is inserted to indicate whether each particular carrier is in use at a BTS. The MS decodes and stores this information whilst it is in its idle mode. On initial assignment, the mobile allocation is described as a subset of the cell allocation, thus reducing the signalling overhead on the AGCH.

Having established the list of carrier frequencies assigned to the frequency hopping channel, the MS must also determine the sequence in which each frequency is to be used. The hopping sequence is described by two parameters: the hopping sequence number (HSN) and the mobile allocation index offset (MAIO). The HSN selects one of 64 predefined 'random' hopping sequences, while the MAIO selects the start point within the sequence. The MAIO may take as many values as there are frequencies in the mobile allocation. The value HSN=0 chooses a cyclic sequence where the frequencies in the mobile allocation are used one after the other.

Frequency hopping channels with the same HSN, but having different MAIOs, will never use the same frequency simultaneously because they are orthogonal. Consequently, all frequency hopping channels within a cell employ the same HSN but have different MAIOs.

Where two frequency hopping channels use different HSNs, they will interfere for $1/n$ of the bursts, and consequently frequency hopping channels in co-channel cells will use different HSNs [6]. For a complete description of the frequency hopping algorithm the reader is referred to the specifications [11].

2.5 Security Issues

2.5.1 Introduction

The first generation cellular systems, e.g. NMT, TACS and AMPS, had few security features, and this resulted in significant levels of fraudulent activity, which harms both subscribers and network operators. A number of high profile incidents highlighted the susceptibility of analogue phones to eavesdropping on the radio paths. The GSM system has a number of security features which are designed to afford both the subscriber and the network operator a greater level of protection against fraudulent activity. Authentication mechanisms ensure that only bona fide subscribers possessing bona fide equipment, i.e. not stolen or non-standard, are granted access to the network. Once a connection has been established, the information on the radio path is encrypted to deter eavesdropping. Each subscriber's privacy is protected by ensuring that details of both her identity and location are protected. This is achieved by assigning each user with a temporary mobile subscriber identity (TMSI) number which changes from call to call. In this way there is no requirement to transmit the user's IMSI over the radio interface making it very difficult for an eavesdropper to identify and locate a user. We will now address each of these security techniques in more detail.

2.5.2 PIN code protection

The first and most basic level of protection against fraudulent use of the mobile phone is the personal identification number (PIN) code, designed to protect against the fraudulent use of stolen SIM cards [8]. In the SIM, the PIN takes the form of a four to eight decimal digit code. The user may have the option of disabling this level of PIN protection. The SIM also has the facility to store a second four to eight digit PIN known as PIN2 to protect certain features that are accessible to a subscriber. Once the PIN, and where appropriate PIN2, are entered correctly, the ME will have access to the data held in the SIM.

The specifications also define the procedures that should be followed when a PIN is entered incorrectly. After three consecutive incorrect attempts to enter a PIN, the SIM becomes blocked and any further attempts to enter the PIN are ignored, even if the SIM is removed from the ME. The SIM may be unblocked by entering a further eight-digit code known as the PIN unblocking key (PUK) which is also stored in the SIM. After 10 incorrect

attempts to enter the PUK, the unblocking key itself becomes blocked and there is no means
to unblock the SIM.

2.5.3   Authentication

The authentication procedure is triggered when the MS attempts one of the following actions:

- a change of subscriber-related information stored in either the HLR or VLR; this will
  include a location update which involves the subscriber's details being stored in a new
  VLR and network registration;

- accessing the network for the purposes of making or receiving a call;

- on the first access following the restart of an MSC/VLR.

Authentication is initiated by the network in the form of an *authentication request* message sent to the MS. This message contains a 128-bit random number, called *RAND*. At the
MS this number is used as one input to a secret algorithm known as A3. The other input to
A3 is the subscriber's secret key, K_i. Both A3 and K_i are stored in the SIM. The security
of GSM hinges around the secrecy of K_i and for this reason the key is stored in the SIM
under heavy protection. The secret key cannot be read from the SIM itself and it is only
accessed when the SIM is personalised under the control of the network operator. During
the security procedures that occur within GSM, K_i is only used internally within the SIM.
K_i may be of any format and any length. We also note that the A3 algorithm is kept secret
to provide additional security. The result of applying the A3 algorithm to K_i and *RAND*
is another number, *SRES* (Signed RESult), which must be 32 bits in length. Once it has
been computed by the MS, *SRES* is returned to the network in the form of an *authentication
response* message. On the network side, the AuC also stores the user's secret key, K_i, and
the A3 algorithm, and it generates a version of *SRES* in an identical manner to that at the
MS. The HLR then sends *SRES* to the visited MSC/VLR where the two versions are compared. If they match, then the MS is deemed to be authentic. The secret A3 algorithm has
the property that it is a relatively simple task to generate *SRES* from *RAND* and K_i, but it is
very difficult to determine K_i from *SRES* and *RAND* or pairs of *SRES* and *RAND*.

The A3 algorithm is not unique and each operator has the freedom to choose its own
version, although an example algorithm is available on a restricted basis. This obviously
has a major implication when inter-network roaming is considered between operators using
different A3 algorithms. The problem is overcome by requiring that the computation of
SRES is performed in the HLR of the subscriber's home network and sent to the visited
MSC/VLR where the authentication check is made. The MSC/VLR stores a number of

RAND and *SRES* pairs for each MS to reduce the number of information transfers between the HLR and MSC/VLR. The HLR and VLR may be in different countries.

2.5.4 Encryption

Once a subscriber has been authenticated, thereby protecting both the subscriber and the network operator from the effects of fraudulent access, the user must be protected from eavesdroppers. This is achieved by encrypting the data on the radio interface using a second key, K_c, and the secret algorithm, A5. K_c is generated during the authentication phase using K_i, *RAND* and the secret algorithm A8, which is also stored in the SIM. In common with the A3 algorithm, A8 is not unique and it may also be chosen by the individual operators. The K_c key for each user is computed in the home network's AuC to overcome the problems of inter-network roaming. In some implementations the A3 and A8 algorithms are combined into a single algorithm A38 which uses *RAND* and K_i to generate K_c and *SRES*.

In contrast to A3 and A8, which may not be the same for each individual operator, A5 will be chosen from a list of different candidates, which will not exceed seven. Prior to encryption being enabled a negotiation phase will occur whereby the MS and the network decide which version of A5 to use. Where the network and the MS do not have any versions of A5 in common, the connection must continue in unciphered mode, or it must be discontinued. The A5 algorithm takes the 64-bit long K_c key and a 22-bit long representation of the TDMA frame number and produces two 114-bit long encryption words, *BLOCK*1 and *BLOCK*2, for use on the up-link and the down-link, respectively. The encryption words are EXORed with the 114 data bits in each burst. Because the encrypted data are computed using the TDMA frame number, the words change from burst to burst and are not repeated over the hyperframe cycle (around 3.5 h). The authentication and encryption process is summarised in Figure 2.25, where $\oplus$ represents the EXOR function.

2.5.5 The temporary mobile subscriber identity (TMSI)

Some transmissions on the radio path cannot be protected by encryption. For example, following an initial assignment, the MS must transmit its identity to the network before encryption can be activated. This would obviously allow an eavesdropper to determine the location of a subscriber by intercepting this message. This problem is reduced in GSM by the introduction of the temporary mobile subscriber identity (TMSI) which is an 'alias' assigned to each MS by the VLR. The TMSI is transmitted to an MS during a previous encrypted connection and it is used by the MS and the network in any future paging and access procedures. The TMSI will only be valid within the location area served by the particular VLR.

◇   ◇

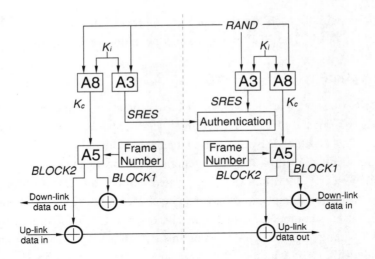

Figure 2.25: The GSM authentication and encryption process.

Having described the basic philosophy, organisation and functionality of GSM, we now move to Chapter 3 where we will quantify the performance of GSM for specific network scenarios.

Bibliography

[1] Steele R., Digital cellular mobile radio scene in Europe, *ISELDECS-87*, Dec. 16–18, 1987, India.

[2] Langewellpott U., M. Aldinger and H.P. Kuchenbecker, Perfect analysis of radio transmission in the fully digital cellular radio system CD900, *Proc. Nordic. Semin. Digital Land Mobile Radio Commun.*, 1985, Espoo, Finland.

[3] Lam W.H. and R. Steele, The error performance of CD900-like cellular mobile radio system, *IEEE Trans. Veh. Technol.* **40**(4), Nov. 1991, 671–685.

[4] Steele R. (Ed.), *Mobile Radio Communications*, Pentech Press, London 1992.

[5] Chia S., The universal mobile telecommunication system, *IEEE Commun. Mag.* Dec. 1992, 54–62.

[6] Mouly M. and M-B. Pautet, *The GSM System for Mobile Communications*. Published by the authors, 1992.

[7] Redl S.M., M.K. Weber and M.W. Oliphant, *An Introduction to GSM*, Artech House, 1995.

[8] GSM 02.17 (ETS 300 509), *European Digital Cellular Telecommunications System (Phase 2); Subscriber Identity Modules (SIM) Functional Characteristics.*

[9] GSM 05.04 (ETS 300 576), *European Digital Cellular Telecommunications System (Phase 2); Modulation.*

[10] GSM 05.05 (ETS 300 577), *European Digital Cellular Telecommunications System (Phase 2); Radio Transmission and Reception.*

[11] GSM 05.02 (ETS 300 574), *European Digital Cellular Telecommunications System (Phase 2); Multiplexing and Multiple access on the Radio Path.*

[12] GSM 05.03 (ETS 300 575) *European Digital Cellular Telecommunications System (Phase 2); Channel Coding.*

[13] GSM 05.08 (ETS 300 578) *European Digital Cellular Telecommunications System (Phase 2); Radio Subsystem Link Control.*

Chapter 3

Capacity of GSM Systems

3.1 List of Mathematical Symbols

a_i represents path loss and shadow fading effects for the ith MS

B TDMA carrier frequency

dA area occupied by an MS

da area occupied by an MS, or an area of ring centred around a BS at a distance r_j and having thickness dr_j

d_j distance between an MS in jth cell and the zeroth BS

$d\phi$ infinitesimal change in ϕ

D distance between a cell site and the nearest co-channel cell site

$E[\Upsilon_j]$ average interference experienced at the zeroth BS due to all MSs in the jth cell in the presence of frequency hopping

$E[(\cdot)]$ expectation of $(\cdot)$

$f(a)$ probability density function (PDF) of an MS's location area within a cell

f_i ith carrier frequency

I_j interference power from an MS in the jth cell at the zeroth BS

I'_j I_j in the presence of power control errors

I_T total interference from J co-channel cells

J number of co-channel cells

k slot number in the TDMA frame

M	cluster size
N_f	number of carriers allocated to each BS
N_{ts}	time slots per carrier
$P_j(r_j)$	transmitted power from the jth BS
P_T	MS transmit power
$Q(k)$	Q-function .
r	distance between an MS and a BS in the same cell
R	cell radius
S	power received from an MS at a BS that is just sufficient to maintain good communications
SIR	signal-to-interference ratio
SIR_{min}	minimum SIR
S'	received power at the BS in the presence of power control errors
v_j	voice activity variables, 1 or 0
W	width of the streets in street microcells
x	normalised distance $(= r/R)$
X	distance from a microcell BS to the street microcell boundary
X_b	path loss break-distance in street microcells
α	path loss exponent (dB)
χ	fraction of channels for signalling
δ	random variable of the error in S' (dB)
δ_0	random variable of the power control error in the received interference power
ε	random variable with standard deviation $\sqrt{2}\,\sigma_e$
λ	shadow fading random variable for path r
λ_0	shadow fading random variable for path d_j
μ	voice activity factor (VAF) $(= \mathrm{E}[v_j])$
ϕ	angle between a line from an MS to its BS in the jth cell, and a line between the zeroth and jth BSs (see Figure 3.3)
σ	standard deviation of λ and λ_0
σ_ε	standard deviation of δ
Υ_j	average interference power at the zeroth BS during slot-k, frequency f_i, from MSs over the jth cell in the presence of frequency hopping
Υ'_j	Υ_j in the presence of power control errors
ζ	rv $(\lambda_0 - \lambda)$ with standard deviation $\sqrt{2}\sigma$

$\overline{(\cdot)}$ average of $(\cdot)$

3.2 Introduction

A precise analysis of the capacity of GSM is exceedingly complex, but by making reason-able assumptions we can provide good estimates of the capacity for a variety of conditions. We will confine ourselves to voice traffic channels since this represents the bulk of the traf-fic in GSM networks [1–8]. While cognisant that radio cells are fundamentally irregular in shape, depending on terrain, buildings, road topologies, and so forth, we will estimate the capacity for hexagonal macrocells and for microcells in a rectilinear grid pattern of roads. These types of cells have the virtue of simplicity and they enable the performance of other systems to be compared on the same cellular basis.

The methodology used in the analysis of macrocellular GSM networks is as follows. We will calculate the signal-to-interference ratios (*SIR*s) for different cluster sizes and identify the cluster size (*M*) that will support the minimum acceptable *SIR*, namely SIR_{min}. Knowing *M* we will determine the number of channels per MHz per cell site, allowing for signalling channels. The traffic carried by the network can then be computed for a given blocking probability.

The *SIR* needs to be calculated for both up-link (reverse) and down-link (forward), and for omnicells and sectorised cells. In computing the *SIR* we will make the following as-sumptions: that the power control is applied; that frequency hopping (FH) is used where the carrier in each frame hops beyond the coherence bandwidth; and that discontinuous transmission (DTX) enables transmissions to be suspended on a link when the user is not speaking [9–11]. All the traffic channels are considered to be occupied, and initially we will ignore the signalling channels. The GSM radio link is assumed to be able to combat the ef-fect of fast fading by means of its channel coding, bit interleaving, channel equalisation and signal processing sub-systems. The radio channel is subjected to log-normal shadow fading, and path losses that increase with distance raised to the power α.

Section 3.3 provides *SIR* calculations for transmissions in macrocellular GSM networks, dealing with omnidirectional and sectorised cells, and examining the effect of power control errors. The analysis is repeated in Section 3.4 for down-link transmissions. Armed with the SIR calculations and the knowledge of cluster size *M* required to ensure SIR_{min}, the capacity of the hexagonal macrocellular network is determined in Section 3.5, along with the effect of sectorisation on the teletraffic performance.

Section 3.6 is concerned with a street microcellular GSM network. The models used are cross-shaped microcells formed by placing the base stations (BSs) at street intersections, and rectangular-shaped microcells where the BSs are mid-way along the sides of the city blocks. The cluster sizes for these two models are two and four, respectively. The other parameters, e.g. DTX and FH, used in the macrocellular analysis apply. After calculating the *SIR* the capacity of the microcellular GSM network is computed. Section 3.6 concludes with a discussion of irregularly-shaped microcells.

3.3   Macrocellular GSM Network: Up-link Transmissions

3.3.1   The *SIR* for omnidirectional macrocells

We commence by determining the *SIR* for the up-link using omnidirectional antennas at the cell sites [9]. Figure 3.1 shows the pattern of co-channel hexagonal cells in a macrocellular network. The other cells are not shown, but we assume that there is a continuum of tessellated hexagonal cells. Observe that around any cell there is an inner ring of co-channel cells followed by outer rings of co-channel cells. Only the inner rings of co-channel cells contribute significantly to the co-channel interference. Each ring has six co-channel cells, except for a cluster size of two. This is because a hexagon has six sides. The distance between a cell site and the nearest cochannel cell site is D, and the distance between a cell site and the apex of a hexagonal cell is R. A hexagonal cell may be approximated by a circle of radius R as shown in Figure 3.2. Using this approximation we show in Figure 3.3 two cells, the zeroth cell, and one of its co-channel cells, the jth cell.

We consider a mobile station (MS) in Figure 3.3 occupying an area da, at a distance r from the jth BS. The MS transmits a power P_T such that the received power at the jth BS is S, a power just sufficient to maintain good communications. The received signal decreases as the MS moves away from the cell site because of an increase in the path loss. The received power at a distance r from the BS will be different at different angles because of the variations in the terrain and the distribution of buildings and streets. We allow for these variations by introducing a random variable λ having zero mean and standard deviation σ. Often λ is referred to as a *shadowing random variable* because it is associated with the electromagnetic shadows cast by buildings and terrain variations. The received power in dBs is

$$10\log_{10} S \;=\; 10\log_{10} P_T \;-\; \alpha 10\log_{10} r \;+\; \lambda, \tag{3.1}$$

where the path loss is $\alpha\, 10\log_{10} r$, and α is called the path loss exponent. Note that λ is in dBs. Measurements of λ show that it is normally distributed between -4σ and 2σ [2]. Since λ is in dBs, it is said to be log-normally distributed between -4α and 2σ. In order to make S a constant, P_T is varied using a closed-loop power control system. From Equation (3.1) the power transmitted by the MS is

$$P_T \;=\; S\, r^{\alpha} 10^{-\lambda/10} \tag{3.2}$$

and this power causes interference at the zeroth BS. The actual interference depends on the

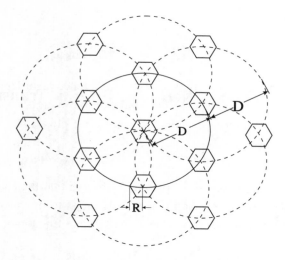

Figure 3.1: Co-channel cells in a mosaic of tessellated hexagonal cells. Non-co-channel cells are not displayed.

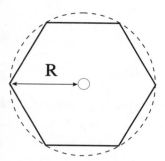

Figure 3.2: An hexagonal cell and its circular representation.

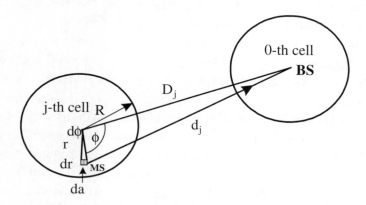

Figure 3.3: Up-link: an MS in the *j*th cell interfering with the zeroth BS.

distance d_j between the MS and the zeroth BS, namely

$$d_j = \sqrt{D_j^2 + r^2 - 2D_j r \cos\phi}, \qquad (3.3)$$

where D_j and ϕ are shown in Figure 3.3. The interference also depends on the shadow fading affecting the MS's transmissions over the path between it and the zeroth BS. The interference due to the MS in the jth cell at the zeroth cell site is

$$I_j = P_T \, d_j^{-\alpha} \, 10^{\lambda_0/10}, \qquad (3.4)$$

where λ_0 is the shadowing variable for path d_j. Observe that the shadowing variables λ and λ_0 have the same variance σ^2, and that α is assumed to be universal over the entire geographical area. Substituting P_T from Equation (3.2) into Equation (3.4) yields

$$I_j = S \left(\frac{r}{d_j} \right)^{\alpha} 10^{(\lambda_0 - \lambda)/10} \qquad (3.5)$$

and

$$\zeta = \lambda_0 - \lambda \qquad (3.6)$$

is a random variable having a normal distribution with standard deviation $\sqrt{2}\sigma$, with

$$-4\sqrt{2}\sigma < \zeta < 2\sqrt{2}\sigma. \qquad (3.7)$$

Unlike the digital enhanced cordless telecommunication (DECT) system [12], GSM cannot hop onto a different radio carrier at each time slot. The radio carrier only hops to a new frequency every GSM frame. In first generation GSM equipment the carriers are assigned to a cell site (or sector) and hopping only occurs between these carriers. This is termed *baseband frequency hopping*. Second generation equipment facilitates hopping over all GSM frequencies and this is termed *RF frequency hopping*. We will consider the special case of the beacon carrier at a later stage.

Let each BS be allocated N_f carriers, say $f_1, f_2, f_3, \ldots, f_{Nf}$, and each carrier supports N_{ts} time slots, resulting in $N_{ts}N_f$ traffic channels per cell. The carriers are hopped on a frame-by-frame basis. Each user is assigned a specific slot in a time frame and it stays in this slot as frequency hopping occurs and the call progresses. Consider the kth timeslot supported by carrier frequency, f_i. In the jth interfering cell, although all the users change their carrier frequency during each TDMA frame, there is always a user who occupies the kth timeslot of the carrier f_i. Owing to frequency hopping at each TDMA frame the interference associated with the channel specified by the kth timeslot and f_i carrier can come from users in different locations within the jth cell, although the interference is only from one user during any frame. Eventually frequency f_i will be used by a subset of N_f users who occupied the kth timeslot in the jth cell. If there is a sufficiently large number of carriers, and the MSs are uniformly distributed over the jth cell, then the probability that an MS will be using the kth slot on carrier f_i is $f(a)da$, where $f(a)$ is the probability density function (PDF). This probability is also $da/(\pi R^2)$, giving

$$f(a) = \frac{1}{\pi R^2}. \tag{3.8}$$

The average interference power at the zeroth BS during the kth slot on frequency f_i is

$$\Upsilon_j = \int\int_{\text{cell area}} I_j f(a)da, \tag{3.9}$$

where I_j is the interference from location da. Substituting I_j from Equation (3.5) into Equation (3.9), and using Equation (3.6) yields

$$\Upsilon_j = \frac{S}{\pi R^2} \int_0^R \int_0^{2\pi} \left(\frac{r}{d_j}\right)^\alpha 10^{\zeta/10} r \, dr \, d\phi \tag{3.10}$$

where da has been replaced by $r \, dr \, d\phi$, and dr and $d\phi$ are defined in Figure 3.3. The variable ζ depends on the paths between the MS and the jth and zeroth BSs while communicating using the kth slot on carrier f_i. Owing to hopping, the MS using the kth slot on carrier f_i is located in a different part of the jth cell during each TDMA frame and thereby has a different shadow variable ζ. Consequently we must take the expectation or average of Υ_j to get the average of $10^{\zeta/10}$ and hence obtain the total interference experienced at the zeroth cell site due to all the mobiles in the jth cell, namely

$$E[\Upsilon_j] = \frac{S}{\pi R^2} \int_0^R \int_0^{2\pi} \left(\frac{r}{d_j}\right)^\alpha E[10^{\zeta/10}] r \, dr \, d\phi, \tag{3.11}$$

where $E[(\cdot)]$ means the expectation of $(\cdot)$. Normalising the distance as

$$x = r/R, \ 0 \leq x \leq 1,$$

$$\frac{E[\Upsilon_j]}{S} = \frac{1}{\pi} \int_0^1 \int_0^{2\pi} \left(\frac{x}{d_j/R}\right)^\alpha E\left[10^{\zeta/10}\right] x \, dx \, d\phi. \tag{3.12}$$

We now need to determine $E\left[10^{\zeta/10}\right]$.

3.3.1.1 Expectation of $E\left[10^{\zeta/10}\right]$

We have stated that the shadow random variable λ is a normal random variable, but bounded from -4σ to $+2\sigma$. These limits have been experimentally observed [2]. Since ζ is the difference between two independent random variables λ, ζ is bounded as given by Inequality

(3.7). Let us commence to find $E[10^{\zeta/10}]$ on the basis that ζ is from $-\infty$ to $+\infty$, and then examine the case when $-4\sqrt{2}\sigma < \zeta < 2\sqrt{2}\sigma$.

Because ζ is a normal random variable,

$$E[10^{\zeta/10}] = \int_{-\infty}^{\infty} 10^{\zeta/10} \cdot \frac{\exp\left(-\frac{\zeta^2}{4\sigma^2}\right)}{(4\pi\sigma^2)^{1/2}} \cdot d\zeta. \tag{3.13}$$

Setting

$$10^{\zeta/10} = \exp(z)$$

or

$$z = \frac{\zeta}{10}\ln(10)$$

then

$$
\begin{aligned}
E\left[10^{\zeta/10}\right] &= \int_{-\infty}^{\infty} \frac{\exp\left(\frac{\zeta}{10}\ln(10) - \frac{1}{2}\frac{\zeta^2}{2\sigma^2}\right)}{(4\pi\sigma^2)^{1/2}} \cdot d\zeta \\
&= \int_{-\infty}^{\infty} \frac{\exp\left\{\left(\frac{\sigma\ln(10)}{10}\right)^2 - \frac{1}{2}\left(\frac{\zeta}{\sqrt{2}\sigma} - \frac{\sqrt{2}\sigma\ln(10)}{10}\right)^2\right\}}{(4\pi\sigma^2)^{1/2}} \cdot d\zeta \\
&= \exp\left(\frac{\sigma\ln(10)}{10}\right)^2 \int_{-\infty}^{\infty} \frac{\exp\left\{-\frac{1}{2}\left(\frac{\zeta}{\sqrt{2}\sigma} - \frac{\sqrt{2}\sigma\ln(10)}{10}\right)^2\right\}}{(4\pi\sigma^2)^{1/2}} \cdot d\zeta
\end{aligned}
$$

and the integral is unity because it represents a normal distribution of mean $\sqrt{2}\sigma ln(10)/10$. The expectation is, therefore,

$$E[10^{\zeta/10}] = \exp\left(\frac{\sigma\ln(10)}{10}\right)^2. \tag{3.14}$$

If we truncate the normal distribution of ζ at $-4\sqrt{2}\sigma$ and $2\sqrt{2}\sigma$, then we rewrite Equation (3.13) as

$$E[10^{\zeta/10}] = \frac{1}{Q(-4) - Q(2)} \int_{-4\sqrt{2}\sigma}^{2\sqrt{2}\sigma} 10^{\zeta/10} \cdot \frac{\exp\left(-\frac{\zeta^2}{4\sigma^2}\right)}{(4\pi\sigma^2)^{1/2}} \cdot d\zeta, \tag{3.15}$$

where

$$Q(k) = \frac{1}{\sqrt{2\pi}} \int_{k}^{\infty} \exp(-\lambda^2/2) \cdot d\lambda,$$

$$E[10^{\zeta/10}] = 1.023\exp\left(\frac{\sigma\ln(10)}{10}\right)^2$$

$$\times \int_{-4\sqrt{2}\sigma}^{2\sqrt{2}\sigma} \frac{\exp\left[-\frac{1}{2}\left\{\frac{\zeta}{\sqrt{2}\sigma} - \frac{\sqrt{2}\sigma\ln(10)}{10}\right\}^2\right]}{(4\pi\sigma^2)^{1/2}} \cdot d\zeta. \tag{3.16}$$

Let

$$x = \frac{\zeta}{\sqrt{2}\sigma} - \frac{\sqrt{2}\sigma \ln(10)}{10}, \tag{3.17}$$

then

$$dx = \frac{d\zeta}{\sqrt{2}\sigma}; \tag{3.18}$$

$$
\begin{aligned}
E[10^{\zeta/10}] &= 1.023 \exp\left(\frac{\sigma \ln(10)}{10}\right)^2 \\
&\quad \times \frac{\sqrt{2}\sigma}{(4\pi\sigma^2)^{1/2}} \int_{-4-\frac{\sqrt{2}\sigma \ln(10)}{10}}^{2-\frac{\sqrt{2}\sigma \ln(10)}{10}} \exp\left(-\frac{x^2}{2}\right) \cdot dx \\
&= 1.023 \exp\left(\frac{\sigma \ln(10)}{10}\right)^2 \left[Q\left\{-4 - \frac{\sqrt{2}\sigma \ln(10)}{10}\right\}\right. \\
&\quad \left. -Q\left\{2 - \frac{\sqrt{2}\sigma \ln(10)}{10}\right\}\right]. \tag{3.19}
\end{aligned}
$$

3.3.1.2   Discontinuous transmission (DTX)

A voice activity detection (VAD) circuit is used to detect when a user is speaking. When a user is silent, the transmissions to and from this user are essentially stopped, although background noise at a low bit rate is sent, as described in Section 2.8.1. The VAD therefore supports discontinuous transmissions (DTX), i.e. transmissions that only occur when a user is speaking. To allow for DTX we introduce a voice activity variable

$$v_j = \begin{cases} 1 \,, \text{ with probability } \mu, \\ 0 \,, \text{ with probability } 1 - \mu, \end{cases} \tag{3.20}$$

and μ is called the voice activity factor (VAF). The mean of v_j is

$$E[v_j] = \mu. \tag{3.21}$$

When we include DTX, the total interference from the MSs in the jth cell is decreased to $E[v_j]\,E[\Upsilon_j]$. Extending the situation to include J co-channel cells, we have a total interference of

$$I_T = \sum_{j=1}^{J} E[v_j]\,E[\Upsilon_j], \tag{3.22}$$

and with the aid of Equation (3.21) we have

$$SIR = \frac{S}{\mu \sum_{j=1}^{J} E[\Upsilon_j]} \tag{3.23}$$

and normally J is set to six.

3.3.1.3   Computing the *SIR*

We need to compute the graph of *SIR* as a function of cluster size M (or reuse factor), and then note the minimum value of M that provides an *SIR* above SIR_{min}. By using this value of M the radio links will operate with an acceptable bit error rate (BER). Figures 3.4 and 3.5 show examples of tessellated cells in three and seven cell clusters. The first step is to note that the distance between co-channel cell sites, D, is given by

$$D = R\sqrt{3M}, \tag{3.24}$$

where R is the cell radius and M is the cluster size. The path loss exponent α is set to 4 (although others might prefer 3.5). The standard deviation σ of the shadow fading is 8 dB, and $E[10^{\zeta/10}]$ is calculated using Equation (3.19). Three values of the VAF, i.e. μ, are used: 1.0, when all users are speaking, which is a worse case scenario, and 1/2 and 3/8. In normal conversation a speaker on average may speak for only 40% or so of the time. We observe that in any one slot there are only six significant interferers in a fully loaded system, i.e. the received packet in the kth slot on carrier f_i has interference from six other MSs. We will assume that the BSs have sufficient carriers to ensure that DTX has statistical meaning.   Using the above parameters and on computing by numerical methods, the *SIR* of Equation (3.23) is calculated for different values of cluster size M. The variation in *SIR* as a function of M is displayed in Figure 3.6. Note the low gradient of the curves. This is unfortunate as we would prefer the *SIR* to increase rapidly with cluster size, producing a substantial increase in *SIR* for a small increase in M. Instead, considerable increases in M yield a relatively small increase in *SIR*. This feature is common to both frequency division multiple access (FDMA) and time division multiple access (TDMA) systems using fixed channel allocation (FCA) techniques. The SIR_{min} for GSM is said to be 9 dB [2], although operators employ a higher figure, sometimes 12 dB or more. Using the best figure of 9 dB, Figure 3.6 shows that for a VAF of 3/8 a three-cell cluster could be used, but if a conservative design is done and VAF is ignored, i.e. VAF = 1, then the minimum cluster size is five for up-link transmissions.

3.3.2   The *SIR* for sectorised macrocells

Sectorisation of cells is generally employed in current macrocellular systems. This is because macrocellular cell sites are often on the tops of tall buildings and as the terrain may vary in different directions, and because of the high rents for these sites, it is common to employ directional antennas with each antenna covering a particular sector. Although six and four sectors are used, the most common is the three-sector cell.

We will now consider sectorising our cells. Ideal sectorisation will be assumed. This means that the radiation pattern will be precisely the area of a sector with no backlobes.

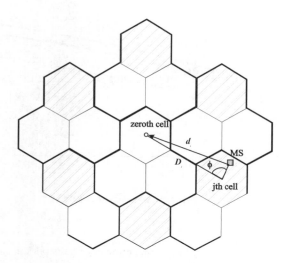

Figure 3.4: Up-link: three cells per cluster, omnidirectional sites. Co-channel cells are shown shaded.

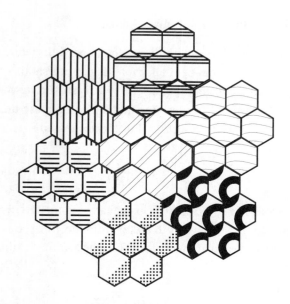

Figure 3.5: Tessellated clusters with seven cells per cluster, omnidirectional sites.

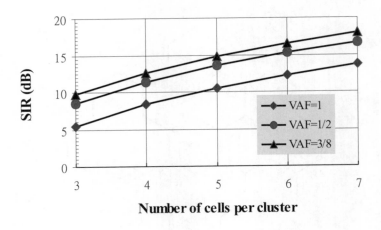

Figure 3.6: Up-link *SIR* versus number of cells per cluster.

The *SIR* will be calculated for different cluster sizes where each cell has three sectors. This will eventually lead to us understand how sectorisation affects the traffic carried by the network.

3.3.2.1 Sectorisation for three- and four-cell clusters

Figure 3.7 shows a three-cell cluster with three sectors per cell, while Figure 3.8 displays a four-cell cluster with three sectors per cell. Note that sectorisation produces sectors that are essentially smaller cells, and that the spectrum at each site has been partitioned into three parts. The most apparent effect of sectorisation is to decrease the number of first tier co-channel cells from six to just two sectors. For the sectorised situation, Equation (3.12) becomes

$$\frac{E[\Upsilon_j]}{S} = \frac{3}{\pi} \int_0^1 \int_0^{2\pi/3} \left(\frac{x}{d_j/R}\right)^\alpha E[10^{\zeta/10}]\, x\, \mathrm{d}x\, \mathrm{d}\phi \qquad (3.25)$$

as ϕ is from 0 to $2\pi/3$ and $f(a) = 3/\pi R^2$. This equation is the interference-to-signal ratio (ISR) for the total interference in the zeroth sector from the jth sector. Since there are two significant interfering sectors, and allowing for DTX, the *SIR* for the three cell clusters with three sectors per cell is

$$SIR = \frac{S}{2\mu E[\Upsilon_j]}. \qquad (3.26)$$

Employing Equation (3.19), the *SIR* for the four cell clusters is 17.8, 20.8 and 22 dB for $\mu = 1$, 1/2 and 3/8, respectively. The corresponding *SIRs* for the three-cell clusters are

14.9, 17.9 and 19.1 dB, respectively. The improvement in *SIR* due to sectorisation is 9.5 dB as a result of the lower $E[\Upsilon_j]$.

3.3.2.2 Sectorisation for the two-cell cluster

Figure 3.9 shows tessellated two-cell clusters where each cell has three sectors. This cellular topology yields asymmetrical interference conditions. In sectors 1 and 2 there are four co-channel interfering sectors on the central cell site, where two of the co-channel cells are at a distance D_j of $2\sqrt{3}R$, one is $3R$ and the closest is at $\sqrt{3}R$. Sector 3 has five interfering sectors on the central cell site. The closest is at $3R$, while two are at $2\sqrt{3}R$ and the other two at $4.72R$. The values of D_j for a particular sector are inserted into Equation (3.3) and d_j computed as a function of $x = r/R$ and ϕ. Next $E[\Upsilon_j]$ is found for this interfering sector using Equation (3.25). The interference is computed for mobiles (which use omnidirectional antennas) from each sector, and the *SIR* is computed using Equation (3.23), where J is either 4 or 5.

The computed *SIR* values for sectors 1 and 2 are 7.7, 10.7 and 11.9 dB for VAFs, i.e. values of 1, 1/2, and 3/8, respectively. The *SIR* of sector 3 is higher because although there are five co-channel sectors, they are all spaced by at least two cells. The corresponding *SIR* values are 13.9, 16.9 and 17.9 dB. However, note that the interfering sectors 3 are in a contiguous line, and mobiles near the corner of one sector are likely to interfere with mobiles in an adjacent sector. Consequently, the *SIR* values for sectors 3 are somewhat optimistic. Indeed, this contiguous group of five sectors 3 may invalidate the two-cell cluster option.

3.3.2.3 Sectorisation for the single-cell cluster

Since the spectral efficiency in fixed channel allocation networks is inversely proportional to the cluster size, it follows that it would be highly desirable if the entire bandwidth allocation could be deployed at every cell site. Let us determine if sectorisation will facilitate single-cell frequency reuse by calculating the *SIR*. Figure 3.10 shows the cellular arrangement. The reuse distances D_j for sectors E and F are $\sqrt{3}R$. The next closest sectors are D, B and G at $D_j = 3R$, while the remaining sectors A and C are at $2\sqrt{3}R$. The procedure outlined in Section 3.3.2.2 is employed again, and we display our *SIR* values, along with the others evaluated in Section 3.3.2, in Table 3.1. We also include the *SIR* values for a three-cell cluster without sectorisation as a bench marker. Observe that as with the two-cell cluster, the single-cell cluster has contiguous co-channel sectors. The interference between adjacent sectors is likely to exceed the interference to the zeroth cell sector.

If the *SIR* for GSM is taken as 9 dB, then the three cells per cluster only works with omnicells when VAF = 3/8. It is therefore better to use sectorisation when the *SIR* is 14.9 dB even when VAF = 1. Single-cell clusters will not work, but a cluster size of two with

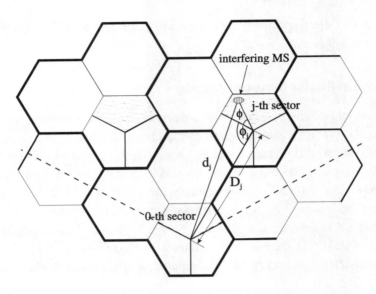

Figure 3.7: Up-link: three-cell cluster with three sectors per cell. The two most significant interfering sectors are shown shaded.

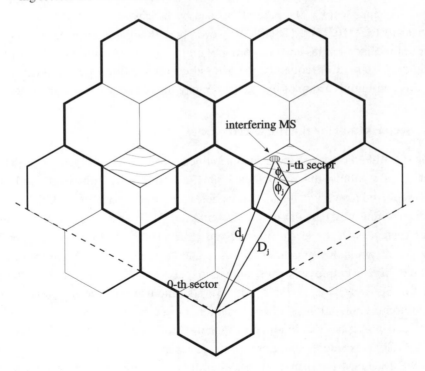

Figure 3.8: Up-link: four-cell clusters with three sectors per cell. The two most significant interfering sectors are shown shaded.

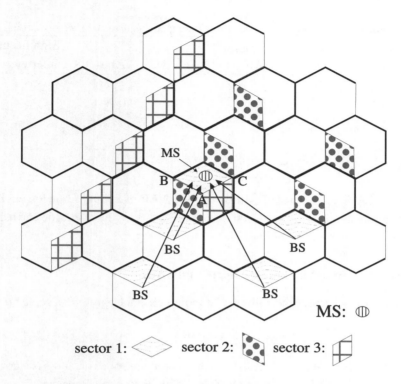

Figure 3.9: Up-link: two-cell clusters with three sectors per cell.

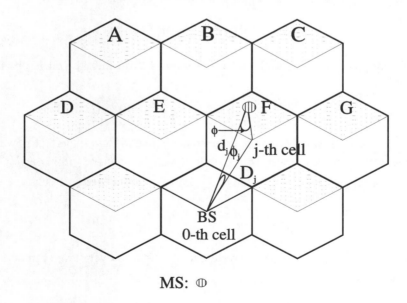

Figure 3.10: Up-link: single-cell cluster with three sectors per cell.

Table 3.1: Up-link: *SIR* values for different cluster sizes and sectorisation.

Cluster size	Number of sectors	VAF = 1	VAF = 1/2	VAF = 3/8	Significant interferers
3	0	5.4	8.4	9.6	6
3	3	14.9	17.9	19.1	2
2	3	7.7	10.7	11.9	4
1	3	4.6	7.6	8.9	7

three ideal shaped sectors per cell will work for a VAF of about 1/2 or lower. Before we discuss the ramifications of these cluster sizes and sectors on the teletraffic carried, we must examine the *SIR* values on the down-link.

3.3.2.4 Effect of imperfect power control in the *SIR*

Power control errors occur, especially in fast fading channels, and the received power is

$$S' = S 10^{\frac{\delta}{10}} , \tag{3.27}$$

rather than the target power, S, where $10^{\frac{\delta}{10}}$ is the power control error, and δ is a normally distributed random variable having a standard deviation of σ_e [13]. Because of power control errors the mobile transmitted power in Equation (3.2) becomes

$$P_T = S' r^\alpha 10^{\frac{-\lambda}{10}} = S r^\alpha 10^{\frac{-\lambda}{10}} 10^{\frac{\delta}{10}} . \tag{3.28}$$

The interfering power at the zeroth cell BS from an MS in the jth cell is obtained from Equation (3.5) as

$$I'_j = S' \left(\frac{r}{d_j}\right)^\alpha 10^{\frac{\zeta}{10}} = S \left(\frac{r}{d_j}\right)^\alpha 10^{\frac{\zeta}{10}} 10^{\frac{\delta_0}{10}}, \tag{3.29}$$

where δ_0 is the random variable of the interference power due to imperfect power control. The average of the received power from mobiles located anywhere within the jth cell using the kth timeslot can be found by inserting $10^{\frac{\delta_0}{10}}$ into Equation (3.10), namely

$$\Upsilon'_j = \frac{S}{\pi R^2} \int_0^R \int_0^{2\pi} \left(\frac{r}{d_j}\right)^\alpha 10^{\frac{\zeta}{10}} 10^{\frac{\delta_0}{10}} r dr d\phi . \tag{3.30}$$

From Equations (3.27) and (3.30), the interference-to-signal power ratio for imperfect power controlled up-link is

$$\frac{\Upsilon'_j}{S} = \frac{1}{\pi} \int_0^1 \int_0^{2\pi} \left(\frac{x}{d_j/R}\right)^\alpha 10^{\frac{\zeta}{10}} 10^{\frac{\varepsilon}{10}} x dx d\phi , \tag{3.31}$$

where ε is a random variable

$$\varepsilon = \delta_0 - \delta \qquad (3.32)$$

having a normal distribution with a standard deviation of $\sqrt{2}\sigma_e$. Since ζ and ε are two mutually independent random variables, the mean value of Equation (3.31) is

$$\mathrm{E}\left[\frac{\Upsilon'_j}{S'}\right] = \mathrm{E}\left[10^{\frac{\zeta}{10}}\right]\mathrm{E}\left[10^{\frac{\varepsilon}{10}}\right]\frac{1}{\pi}\int_0^1\int_0^{2\pi}\left(\frac{x}{d_j/R}\right)^\alpha xdxd\phi . \qquad (3.33)$$

Compared with the interference-to-signal ratio for perfect power control given by Equation (3.12), there is an increase of $\mathrm{E}\left[10^{\frac{\varepsilon}{10}}\right]$ in the interference-to-signal ratio due to imperfect power control. The corresponding decrease in the *SIR* is 0.2 dB, 0.9 dB, and 2.1 dB for δ having a standard deviation of 1 dB, 2 dB and 3 dB, respectively, irrespective of the voice activity factor and cluster size.

3.4 Macrocellular GSM Network: Down-link Transmissions

3.4.1 The *SIR* for omnidirectional macrocells

The BSs in the tessellated hexagonal structures adjust the transmitted power to their mobiles such that each mobile receives a power S. This statement implies perfect power control. Again frequency hopping beyond the coherence bandwidth is assumed, and DTX is applied. Mobiles receive interference from nearby BSs, and not from other mobiles. Thus, as the MSs move around the cell the interference they experience varies, but the sources of interference are static. The *SIR* on the down-link is dependent upon the location of an MS, and this leads us to consider the average *SIR* for the mobiles in a cell, or the *SIR*s for particular mobiles.

Figure 3.11 shows an MS at a distance r from its BS in the zeroth cell. The MS occupies an area dA. In the figure the source of interference is from the jth BS in the jth cell. The distance between the two base sites is D_j. Suppose that the MS receives information from its zeroth BS in the k-slot on carrier f_i. The interference is caused by the jth BS when it transmits to an MS in its cell on the k-slot on carrier f_i, and the magnitude of this interference depends on where this MS is located. In Figure 3.11 we display a ring distance r_j from the jth cell site. The area of this ring is

$$da = \pi(r_j + dr_j)^2 - \pi r_j^2 \approx 2\pi r_j dr_j. \qquad (3.34)$$

All MSs in this ring receive a power S, and this means that the jth BS transmitted power is

$$P_j(r_j) = Sr_j^\alpha 10^{-\lambda_j/10}, \qquad (3.35)$$

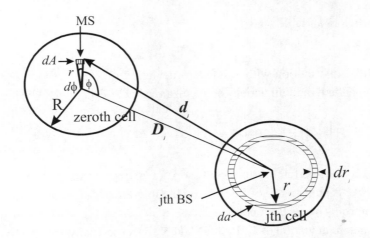

Figure 3.11: Down-link: two-cell cluster with three sectors per cell.

where α is the path loss exponent and λ_j is the shadow fading random variable. For MSs uniformly distributed over the jth cell, the average transmitted power per channel from the jth BS is

$$\overline{P}_j(r_j) = \int\limits_{\text{cell}} \int\limits_{\text{area}} P_j(r_j) f(a) \mathrm{d}a, \tag{3.36}$$

where $f(a)$ is given by Equation (3.8). Consequently,

$$\overline{P}_j(r_j) = \int_0^R S r_j^\alpha \, 10^{-\lambda_j/10} \cdot \frac{1}{\pi R^2} \cdot 2\pi r_j \mathrm{d}r_j$$

$$= \frac{2}{\alpha + 2} S R^\alpha \, 10^{-\lambda_j/10}, \tag{3.37}$$

where R is the cell radius. This average transmitted power from the jth BS is the source of interference to MSs in the zeroth cell. The particular MS at a distance d_j from the jth BS receives an interference power of

$$I_j = \overline{P}_j(r_j) \, 10^{\lambda_0/10} \, d_j^{-\alpha}, \tag{3.38}$$

where λ_0 is the shadowing random variable for path d_j. Substituting $\overline{P}_j(r_j)$ from Equation (3.37) into Equation (3.38) yields

$$I_j = \frac{2}{\alpha + 2} S \, 10^{\zeta/10} \left(\frac{R}{d_j} \right)^\alpha, \tag{3.39}$$

where

$$\zeta = \lambda_0 - \lambda_j \tag{3.40}$$

has a standard deviation of $\sqrt{2}\sigma$, and where σ is the standard deviation of λ_0 and λ_j. The distance d_j is given by Equation (3.3).

3.4.1.1 Average *SIR*

The N_f carriers in the zeroth cell, associated with N_f users assigned to the kth slot, are uniformly distributed over the area of the cell. It is important to realise that for a uniform distribution to have validity, N_f must be sufficiently large. This in turn means that the operator has a wide bandwidth at his disposal. We will now consider the average interference experienced by mobiles in the zeroth cell from the jth BS. The area dA shown in Figure 3.11 is rdϕdr, and the PDF $f(A)$ is again $1/\pi R^2$, and hence the interference average over the zeroth cell is, with the aid of Equation (3.39),

$$\bar{\Upsilon}_j = \int_0^{2\pi} \int_0^R \left[\frac{2S}{\alpha + 2} \cdot 10^{\zeta/10} \cdot \left(\frac{R}{d_j} \right)^\alpha \right] \frac{1}{\pi R^2} \cdot r\,d\phi\,dr \tag{3.41}$$

and on normalising r by introducing $x = r/R$,

$$\bar{\Upsilon}_j = \frac{2S}{(\alpha + 2)\pi} \int_0^{2\pi} \int_0^1 10^{\zeta/10} \left(\frac{R}{d_j} \right)^\alpha x\,d\phi\,dx . \tag{3.42}$$

Note that

$$\left(\frac{d_j}{R} \right)^2 = \left(\frac{D_j}{R} \right)^2 + x^2 - 2\left(\frac{D_j}{R} \right) x\cos\phi$$

and that the number of cells per cluster from Equation (3.23) is

$$M = \frac{1}{3}\left(\frac{D_j}{R} \right)^2 , \tag{3.43}$$

giving

$$\left(\frac{d_j}{R} \right)^2 = 3M + x^2 - 2\sqrt{3M}\, x\cos\phi. \tag{3.44}$$

Owing to frequency hopping, the expectation of $\bar{I}_j$ is made with respect to the random variable ζ, because ζ is different for each hop. This expectation is

$$\mathrm{E}\left[\bar{\Upsilon}_j \right] = \frac{2S}{(\alpha+2)\pi} \int_0^{2\pi} \int_0^1 \mathrm{E}\left[10^{\frac{\zeta}{10}} \right] \left(\frac{R}{d_j} \right)^\alpha x\,d\phi\,dx , \tag{3.45}$$

where $\mathrm{E}\left[10^{\frac{\zeta}{10}} \right]$ is given by Equation (3.19). The *SIR* can be found from Equations (3.23) and (3.45), namely

$$\frac{S}{I} = \frac{S}{\sum_{j=1}^J \mathrm{E}[v_j]\mathrm{E}\left[\bar{\Upsilon}_j \right]}$$

$$= \frac{1}{\frac{2\mu}{(\alpha+2)\pi} \sum_{j=1}^J \int_0^{2\pi} \int_0^1 \mathrm{E}\left[10^{\frac{\zeta}{10}} \right] \left(\frac{R}{d_j} \right)^\alpha x\,d\phi\,dx} . \tag{3.46}$$

For a hexagonal pattern of cells there are six close interferers, as shown in Figure 3.12. We compute $\mathrm{E}\left[10^{\zeta/10}\right]$ in Equation (3.19) for a path loss exponent of $\alpha = 4$, and for shadow fading with a standard deviation of $\sigma = 8$. This expectation value is inserted into Equation (3.46) and the *SIR* found. When the number of cells per cluster M is changed, d_j changes (as evident in Equation (3.44)), which in turn changes the *SIR*. Graphs of the *SIR* as a function of the number of cells per cluster are displayed in Figure 3.13 for three different voice activity factors (VAFs) of 1, 1/2, and 3/8. For three cells per cluster, the average *SIRs* for VAFs of 1, 1/2, 3/8 are 5.9 dB, 8.9 dB, and 10.1 dB, respectively. Compared with the *SIR* on the up-link system, the average *SIR* on the down-link system is 0.5 dB larger.

3.4.1.2　The *SIR* of mobiles in different locations

We now examine the *SIR* of the three-cell cluster arrangement for MSs at different locations within its cell, from the immediate vicinity of the BS ($r = 0$) to the cell boundary, $r = R$. As a consequence, those MSs near their BS will receive less interference power than MSs at locations away from their BS. From Equations (3.23) and (3.39), the *SIR* is

$$\frac{S}{I} = \frac{1}{\sum_{j=1}^{J} \mathrm{E}\left[v_j\right] \mathrm{E}\left[\frac{\Upsilon_j}{S}\right]} = \frac{1}{\frac{2\mu}{\alpha+2} \sum_{j=1}^{J} \mathrm{E}\left[10^{\frac{\zeta}{10}}\right] \left(\frac{R}{d_j}\right)^{\alpha}} , \tag{3.47}$$

where $\mathrm{E}\left[10^{\frac{\zeta}{10}}\right]$ is given by Equation (3.19). For three cells per cluster, the graphs of the *SIR* for MSs along a line from the zeroth BS site to a corner of the zeroth cell, is shown in Figure 3.14. The worst *SIR* occurs when mobiles are located at the corner ($x = 1$) where the *SIRs* are 4.8 dB, 7.8 dB, and 9.2 dB for VAFs of 1, 1/2, and 3/8, respectively. The worst case *SIR* is about 1 dB less than the average *SIR*, while the *SIR* difference between the best and worst case is about 2 dB.

3.4.2　The SIR for sectorised macrocells

By dividing each cell into three sectors, as we did for the up-link in Section 3.3, we have the arrangement shown in Figure 3.15. By averaging over the sector area, the average SIR can be found from Equation (3.46), namely

$$\frac{S}{I} = \frac{S}{\sum_{j=1}^{J} \mathrm{E}\left[v_j\right] \mathrm{E}\left[\overline{\Upsilon}_j\right]} = \frac{1}{\frac{6\mu}{(\alpha+2)\pi} \sum_{j=1}^{J} \int_0^{\frac{2\pi}{3}} \int_0^1 \mathrm{E}\left[10^{\frac{\zeta}{10}}\right] \left(\frac{R}{d_j}\right)^{\alpha} x\,d\phi\,dx} . \tag{3.48}$$

From Equation (3.47), the *SIR* for mobiles at different locations is

$$\frac{S}{I} = \frac{S}{\sum_{j=1}^{J} \mathrm{E}\left[v_j\right] \mathrm{E}\left[\overline{\Upsilon}_j\right]} = \frac{1}{\frac{2\mu}{\alpha+2} \sum_{j=1}^{J} \mathrm{E}\left[10^{\frac{\zeta}{10}}\right] \left(\frac{R}{d_j}\right)^{\alpha}} . \tag{3.49}$$

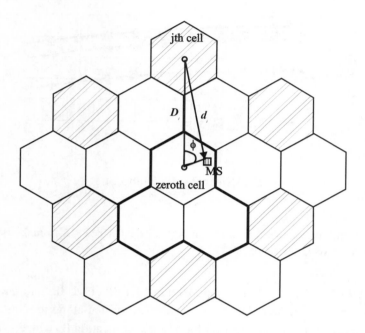

Figure 3.12: Down-link: three-cell cluster, omnidirectional sites.

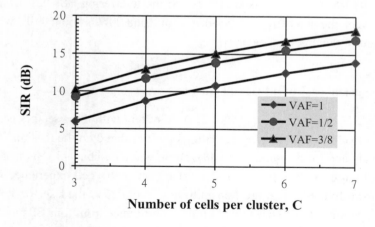

Figure 3.13: Down-link *SIR* as a function of cluster size for different VAFs. No sectorisation.

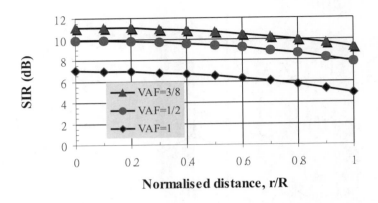

Figure 3.14: Down-link: the *SIR* for mobiles at normalised distance r/R from the BS, for different VAFs. No sectorisation.

The average *SIR*s for VAFs of 1, 1/2, and 3/8 are 14.1 dB, 17.1 dB, and 18.4 dB, respectively, for the three-cell cluster. The *SIR*s for mobiles located along the line AB in Figure 3.15 are shown in Figure 3.16. The *SIR*s for mobiles at location B (where $r/R = 0.87$) are 15.7 dB, 18.7 dB, and 19.9 dB for VAF values of 1, 1/2, and 3/8, respectively. The worst *SIR* conditions occur when the mobiles are located near the BS in Figure 3.15, where the *SIR*s are 11.7 dB, 14.7 dB, and 15.9 dB for VAF of 1, 1/2, and 3/8, respectively. This is because the power control means that the mobile still receives the same power *S*, in spite of being near its BS, while the distance between the interfering BSs has decreased. The difference between the two extreme cases of r/R of 0 and 0.87 is about 4 dB, which is 2 dB larger than that for the unsectorised system.

3.4.2.1   The special case of a two-cell cluster

The down-link *SIR* of the two-cell cluster TDMA system with three sectors per cell is calculated for the two different interference situations considered on the up-link. The average *SIR* can be evaluated from Equation (3.48), but now the co-channel separation D_j is different. From Figure 3.17, mobiles in sector 1 of the centre cell experience intercellular interference from four BSs that are transmitting to mobiles in their sector 1. Mobiles in sector 2 of the centre cell also have intercellular interference from four BSs, but mobiles in sector 3 of the central cell are subjected to interference from BSs in five sectors positioned along a line as shown in Figure 3.17. Following a similar procedure as used above, we find that the average *SIR* for mobiles in sectors 1 and 2 are 7.7 dB, 10.7 dB, and 11.9 dB for a VAF of 1, 1/2, and 3/8, respectively, while for mobiles in sector 3 the average *SIR* values

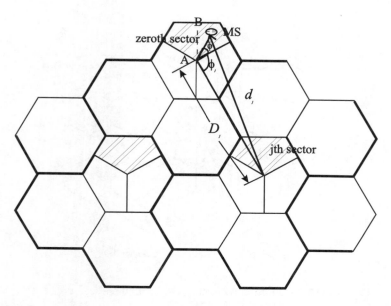

Figure 3.15: Down-link: three cells per cluster with three sectors per cell. Primary interfering sectors are shown shaded.

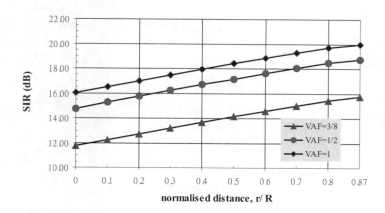

Figure 3.16: Down-link *SIR* for mobiles as a function of normalised distance r/R from the BS to the boundary along the line AB in Figure 3.15 for different VAFs; three sectors per cell and three cells per cluster.

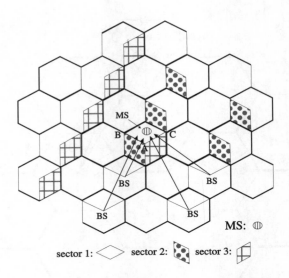

Figure 3.17: Down-link: two-cell cluster with three sectors per cell.

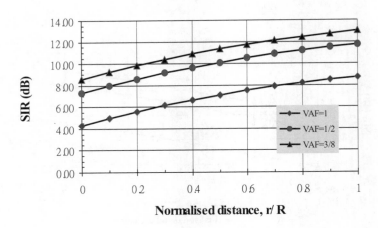

Figure 3.18: Down-link: the *SIR* for mobiles as a function of normalised distance *r/R* along the boundaries AB and AC for different VAFs; three sectors per cell and two cells per cluster.

are 13.9 dB, 16.9 dB, and 17.9 dB for a VAF of 1, 1/2, and 3/8, respectively.

The *SIR* for MSs at different locations in a sector is given by Equation (3.49). Mobiles located along the cell boundaries AB and AC in Figure 3.17 receive a greater interference power than MSs located elsewhere in a sector. The *SIR*s for MSs located along these boundaries are shown in Figure 3.18. The worst *SIR* is when mobiles are near the BS where the interference is the strongest. The *SIR*s for VAFs of 1, 1/2, and 3/8 are 4.2 dB, 7.2 dB, and 8.4 dB, respectively. From Figure 3.18, the difference between the two extreme values of the *SIR* due to different mobile locations is about 3.5 dB.

3.4.2.2   The special case of a one-cell cluster

The down-link *SIR* of the one-cell cluster TDMA system with three sectors per cell is calculated using the same assumptions as for the up-link case. The average *SIR* is calculated from Equation (3.48), but now with different co-channel separation D_j. From Figure 3.19, there are seven significant co-channel interfering BSs. Sectors E and F have a reuse distance of $\sqrt{3}R$, sectors B, D and G have a reuse distance of $3R$, and sectors A and C have a reuse distance of $2\sqrt{3}R$. We find that the average *SIR* is 4.4 dB, 7.4 dB, and 8.6 dB for a VAF of 1, 1/2, and 3/8, respectively.

The *SIR* for MSs at different locations in its sector is given by the Equation (3.49). Mobiles located along the sector boundaries extending from the zeroth cell site in Figure 3.19 receive more interference power than MSs elsewhere in the sector. The *SIR*s for MSs located along these boundaries are shown in Figure 3.20. The lowest *SIR* is when mobiles are near the BS where the interference is the strongest. The *SIR*s for VAFs of 1, 1/2, and 3/8 are 1.2 dB, 4.2 dB, and 5.5 dB, respectively.

3.4.2.3   Imperfect power control on the down-link

Similar to the derivation of the effect of imperfect power control on the up-link *SIR*, as addressed in Section 3.3.4, the effect of imperfect power control on the down-link *SIR* can be calculated by inserting $E\left[10^{\frac{\varepsilon}{10}}\right]$ into the denominator of Equations (3.46) and (3.47) for the average *SIR* and the *SIR* for mobiles at different locations in a cell, while for sectorised cells it can be calculated by inserting it into the denominator of Equations (3.48) and (3.49). The decrease in *SIR* is 0.2 dB, 0.9 dB, and 2.1 dB for a standard deviation of power error of 1 dB, 2 dB, and 3 dB, respectively, irrespective of VAF, cluster size and sectorisation.

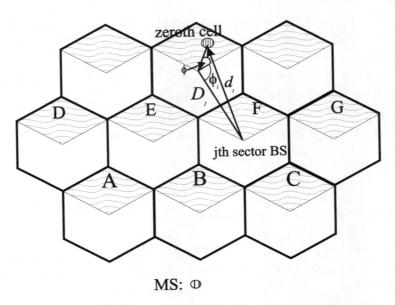

Figure 3.19: Down-link: single-cell cluster with three sectors per cell.

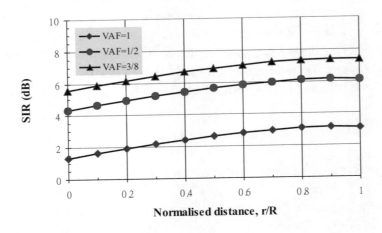

Figure 3.20: Down-link: the *SIR* for mobiles as a function of normalised distance r/R along the sector boundaries extending from the zeroth cell site for different VAFs; three sectors per cell and one cell per cluster.

3.5 Macrocellular GSM Network: Capacity

The capacity of a TDMA system in terms of channels per cell per MHz can be defined as

$$\kappa \triangleq \frac{N_{ts}(1-\chi)}{MB},\tag{3.50}$$

where N_{ts} is the number of time slots per carrier, χ represents the fraction of channels allocated for signalling, M is the cluster size, and B is the TDMA carrier spacing in MHz on each duplex link. From Equation (3.50), the capacity of a TDMA system is inversely proportional to the cluster size, and cluster size depends on SIR_{min}. Consequently, to operate with an acceptable bit error rate (BER) the SIR must exceed a threshold that is system dependent.

Without considering the shadowing effect, the SIR for a conventional TDMA system without power control, frequency hopping (FH), DTX, and sectorisation is found by noting that the received power for an MS at the boundary is proportional to $R^{-\alpha}$ while the interference for six co-channel cells is proportional to $6D^{-\alpha}$, giving [14]

$$\frac{S}{I} = \frac{1}{6}\left(\frac{D}{R}\right)^{\alpha}.\tag{3.51}$$

For tessellated omnidirectional hexagonal cells there are always six close co-channel interferers for a cluster size other than two.

We can still employ the simple S/I Equation (3.51) to allow for shadow fading, the absence of power control, frequency hopping, DTX and sectorisation in the following way. Because the shadow fading is different for different paths associated with the co-channel cells, we modify S/I by a factor which is the average of the random shadow fading variable, ζ. Since ζ is in dBs, we use $E\left[10^{\frac{\zeta}{10}}\right]$ given by Equation (3.19). Hence, an approximate expression for the SIR in the presence of shadow fading is

$$\frac{S}{I} \simeq \frac{1}{6}\left(\frac{D}{R}\right)^{\alpha}\frac{1}{E\left[10^{\frac{\zeta}{10}}\right]}.\tag{3.52}$$

This approximate representation applies for clusters other than two. For a cluster size of two, there are not just six near co-channel interferers; the situation is more complex as shown in Figure 3.21, where there are 10 significant co-channel cells. For this situation three different values of D must be used, namely $D_j = \sqrt{3}R$, $3R$ and $2\sqrt{3}R$, and the equation for S/I becomes

$$\frac{S}{I} = \frac{1}{E\left[10^{\frac{\zeta}{10}}\right]}\sum_{j=1}^{10}\left(\frac{D_j}{R}\right)^{\alpha}.\tag{3.53}$$

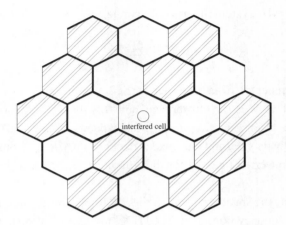

Figure 3.21: Down-link: two-cell cluster without sectorisation. Co-channel cells are shown shaded.

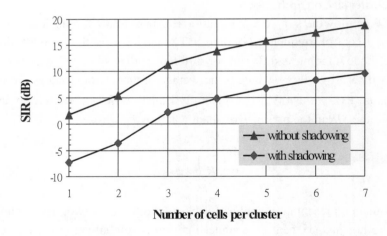

Figure 3.22: *SIR* versus number of cells per cluster with and without shadow fading. Neither power control, nor frequency hopping, nor DTX nor sectorisation are used.

For an α of 4, the *SIRs* without and with shadowing are computed for the basic TDMA system that does not have DTX, power control or FH, and the results are displayed in Figure 3.22.

Suppose the minimal required *SIR* is 9 dB, then if there is no power control, no frequency hopping, no DTX, and no sectorisation, then the cluster size must be at least seven. By dividing a cell into three sectors, a gain of 4.8 dB can be attained, and the cluster size reduces to four. With perfect power control, frequency hopping, and DTX with a VAF of $1/2$, the system can operate in a four-cell cluster using omnicells. By dividing each cell into three sectors, a three-cell cluster is theoretically feasible, even without DTX. Although the average *SIR* of the two-cell cluster with sectorisation is above 9 dB for a VAF of $1/2$, as shown in Figure 3.18, the worst *SIR* on the down-link is below 9 dB for mobiles located near the BS ($r < 0.3R$). The minimal cluster size for an *SIR* of 9 dB is four cells per cluster without sectorisation or three cells per cluster with sectorisation. The minimum *SIR* value of 9 dB is often quoted for GSM, although operators tend to design for a higher figure, such as 12 dB or even as high as 16 dB.

For a carrier spacing of $B = 200$ kHz, $N_{ts}=8$ channels per carrier, and $\chi = 0.1$, the system capacity is 12, 9 and 5 channels per MHz per cell for omnicell sites and 3, 4, and 7 cells per cluster, respectively. For a three-cell cluster with three sectors per cell, the capacity is 15 channels per MHz per cell or five channels per sector per MHz. Without sectorisation, the minimal cluster size is three and four, yielding a capacity of 12 and 9 channels per MHz per cell for a VAF of $3/8$ and $1/2$, respectively.

3.5.1   Effect of sectorisation on teletraffic

For ideal directional antennae with an infinite front-to-back ratio, the introduction of sectorisation in a TDMA system provides a gain of 4.8 dB and 9.5 dB in the *SIR* for systems without and with both power control and frequency hopping, respectively. For a TDMA system without frequency hopping and power control, the 4.8 dB gain in the *SIR* due to three sectors per cell can reduce the cluster size from seven to four. Hence the capacity increases by a factor of up to $7/4$. By using frequency hopping, power control, and DTX, the system is able to function with four cells per cluster without sectorisation. Even when we apply frequency hopping, power control, sectorisation, and with or without DTX, the optimal cluster size is three. However, the number of channels per sector is decreased by three times compared with an unsectorised cell. With three sectors per cell, there are $N_f N_{ts}/3$ channels per sector, where N_f is the number of carriers per cell site. The offered traffic per sector is Π_s, and the total offered traffic per cell site is $3\Pi_s$. Without sectorisation there are $N_{ts}N_f$ channels per cell site, and the offered traffic is Π. For a given blocking probability of 2% and $N_{ts} = 8$ time slots per carrier, the offered traffic versus number of carriers per cell site for systems with and without sectorisation is shown in Figure 3.23. There is a loss of offered traffic due to sectorisation.

Thus for any number of cells per cluster, introducing sectorisation decreases the teletraffic in the network, although there is an improvement in the *SIR*. Yet sectorisation is widely deployed. Sectorisation does increase the teletraffic if it enables fewer cells per cluster to be used. For example, if we consider a seven-cell cluster using omnidirectional antennae, and by using three sectors per cell we can operate in the four-cell cluster, then there is a gain in carried traffic. Representing this gain as the ratio of the carried traffic in the four-cell cluster with three sectors per cell, i.e. four-cell/three-sector, to the carried traffic in the seven-omnicell, we obtain the curve shown in Figure 3.24. Note that if we have a very large number of carriers to deploy per sector then the loss in teletraffic due to sectorisation becomes negligible, and in the limit of an infinite number of carriers the gain in teletraffic approaches 7/4. Also shown in Figure 3.24 are the gains when going from the seven-omnicell to three-cell/three-sector (limits at 7/3 with infinite number of carriers) and the three-omnicell to three-cell/three-sector. The latter is of particular interest because if the three cells per cluster omnidirectional cell arrangement is optimal, then there is no point in sectorisation being used because the teletraffic gain will always be smaller than unity. Of course there may be other more significant factors, such as sectorising to avoid renting new cell sites and accommodating special terrain features.

3.5.2   Summary of the performance of the macrocellular GSM network

The average *SIR* values and the worst *SIR* values for down-link transmissions are summarised in Tables 3.2 and 3.3, respectively, for smaller cluster sizes of 1, 2, and 3. The difference between the worst *SIR* for the TDMA system without sectorisation and the average *SIR* is about 1 dB, while the differences between the average *SIR* and the worst case *SIR* with sectorisation are 2.4 dB, 3.5 dB, and 3.2 dB for three cells per cluster, two cells per cluster, and a single cell per cluster, respectively. For good system design, the worst conditions should be adopted for a sectorised TDMA system.

Our interest in the three-cell cluster is because it is quoted to be the minimum size for GSM [2]. We also note that SIR_{min} is said to be 9 dB. Our analysis of both the up-link and down-link is based on perfect power control, frequency hopping and DTX. GSM, with its beacon carriers, will not attain the *SIR* values we have calculated. If we design for the worst case *SIR* values for the down-link we use the values in Table 3.3, since we note that they are lower than the ones in Table 3.1 for the up-link. Consequently, for all the arrangements listed in Table 3.3, only the three-cell cluster with three sectors per cell is appropriate for the GSM scenario posed here. However, a different TDMA system that could operate satisfactorily with lower *SIR* values could use some of the other options.

The capacity of macrocellular TDMA systems having frequency hopping, perfect power control, DTX, tessellated three-cell hexagonal clusters with three sectors, an inverse fourth

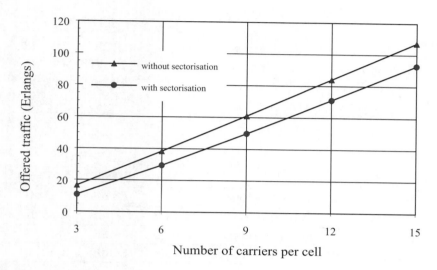

Figure 3.23: Offered traffic per cell as a function of the number of carriers per cell site, N_f, for systems with and without sectorisation.

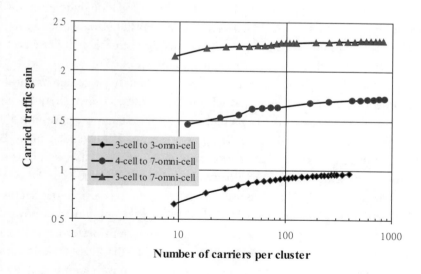

Figure 3.24: The carried traffic gain of the four-cell/three-sector to the seven-omnicell, three-cell/three-sector to the seven-omnicell, and three-cell/three-sector to 3-omnicell.

Table 3.2: Average *SIR* values for the down-link.

	VAF=1	VAF=1/2	VAF=3/8
Three-cell cluster without sectors	5.9 dB	8.9 dB	10.1 dB
Three-cell cluster with three sectors	14.1 dB	17.1 dB	18.4 dB
Two-cell cluster with three sectors	7.7 dB	10.7 dB	11.9 dB
Single-cell cluster with three sectors	4.4 dB	7.4 dB	8.6 dB

Table 3.3: Worst *SIR* values for the down-link.

	VAF=1	VAF=1/2	VAF=3/8
Three-cell cluster without sectors	4.8 dB	7.8 dB	9.2 dB
Three-cell cluster with three sectors	11.7 dB	14.7 dB	15.9 dB
Two-cell cluster with three sectors	4.2 dB	7.2 dB	8.4 dB
Single-cell cluster with three sectors	1.2 dB	4.2 dB	7 5.5 dB

path loss law, and log-normal fading has an SIR greater than 9 dB, which corresponds a system capacity of 12 channels per cell per MHz. Without sectorisation, the minimal cluster size is three and four, yielding a capacity of 12 and 9 channels per cell per MHz for a VAF of 3/8 and 1/2, respectively. The effect of sectorisation on teletraffic will yield gains in capacity, provided it results in a decrease in the cluster size, i.e. the reuse factor. A three-cell cluster with omnidirectional sites gives the highest capacity, while sectorisation decreases the carried teletraffic but improves link quality.

In GSM there may be only one TDMA carrier at a BS, and because one of the slots must contain the broadcast control channel (BCCH) neither power control, frequency hopping nor DTX can be used on any of the remaining seven traffic channels. We note that while the single TDMA carrier BS will be deployed as a form of BS extender, many BSs may only have two or three TDMA carriers, particularly in the network start-up phase. Systems having the luxury of wide bandwidth allocation, such as GSM1800 (i.e. DCS1800) and some GSM1900 (i.e. PCS1900) systems, may have up to 10 carriers per sector if there is sufficient user demand. In our analysis all TDMA carriers enjoy perfect power control, frequency hopping or DTX giving an upper bound on system performance. The lower bound is when there are many carriers in each sector, but the TDMA network does not employ power control, frequency hopping, nor DTX. We have examined these extreme conditions. Of course other viable scenarios could be considered, such as dynamic channel allocation (DCA) [15] and packet reservation multiple access (PRMA) [16]. However, no cellular TDMA network currently employs these techniques, and so they have not been considered here. What we have attempted to show is the importance of power control, frequency hopping and DTX on system performance, and the role of sectorisation.

3.6   Microcellular GSM Network

High densities of teletraffic in city centres can be accommodated by a microcellular personal communications network (PCN) [17–23]. Small base stations (BSs) located in the streets with their antennae below the urban sky-line provide radio coverage that is largely dependent on the road topology, position and cross-sectional area of the buildings and the local terrain and vegetation. For cities with irregular street patterns, as exemplified by the city of London, the microcells also are irregularly shaped. Cities, such as Manhattan, that have grid patterns of streets and block shape buildings yield regularly shaped microcells. In this section we analyse the latter microcellular patterns because of their relative simplicity. However, our theory can be modified to accommodate less regular microcellular structures.

We consider two microcellular arrangements. Figure 3.25 shows tessellated cross-shaped city microcells formed by positioning the BSs at street intersections. Rectangular-shaped microcells that occupy a side of a city block are shown in Figure 3.26. When time division multiple access (TDMA) systems are deployed, each cluster reuses the allotted spectrum, with the spectrum partitioned between the microcells within the cluster. We assume that this partitioning results in the equal bandwidth assignment for each microcell. Since microcells in different clusters use the same bandwidth we must separate them by sufficient distances to yield a level of co-channel interference that can be tolerated by the TDMA radio link. In Figures 3.25 and 3.26 the cluster sizes are two and four, respectively. The consequences of the differences in these cluster sizes on capacity will be evaluated.

The city street microcells shown in Figures 3.25 and 3.26 are formed by controlling the radiated power levels to provide acceptable communications at the microcell boundaries. Although there will be penetration of electromagnetic energy into buildings this does not affect the shape of the street microcell, assuming the buildings are of a similar type. While the microcell boundary is considered to be half a city block from its BS, the radiation from both BS and mobile stations (MSs) will extend well beyond this microcellular boundary and thereby produce cochannel interference.

The capacity of cellular radio networks is co-channel interference limited. Co-channel interference will decrease if the transmitted power to and from all mobiles is decreased while ensuring that the received power level is sufficiently high to ensure acceptable link quality. Similarly, discontinuous transmission (DTX), where users do not transmit when they are not speaking, causes co-channel interference to decrease. As users are assigned a particular slot in a TDMA frame, the interference comes from other users communicating in the same slot and on the same frequency in cells in the neighbouring clusters. During any particular time slot the interference is decreased if some of the co-channel users are not

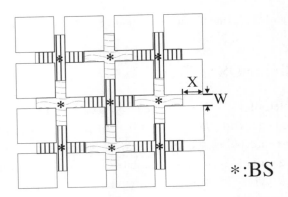

Figure 3.25: Cross-shaped city street microcells in a two-cell per cluster arrangement.

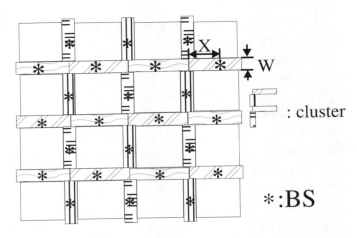

Figure 3.26: Rectangular-shaped city street microcells in a four-cell per cluster arrangement.

speaking. To fully derive the benefit of DTX we may employ frequency hopping (FH) in conjunction with DTX, whereby the user's carrier changes every TDMA frame to provide an average co-channel interference that is similar for each user.

Since the *SIR*s of the up-link and the down-link are similar for TDMA systems, in the analysis to follow we will calculate the *SIR* for a TDMA up-link in our city street micro-cells in the presence of perfect power control, FH and DTX. We will compare these *SIR*s with those of a conventional TDMA system that does not employ power control, FH, and DTX. By knowing the *SIR* and the minimum (SIR_{min}) required to ensure acceptable link performance, we can determine the number of microcells per cluster and then the capacity. To simplify our analysis, we assume that all the traffic channels in a microcell are occupied by users.

3.6.1   Path loss and shadow fading in city street microcells

Radio propagation in microcells is often characterised by line-of-sight (LOS) paths [19, 24]. For bit rates where flat fading is appropriate, the LOS paths cause the fading to be Rician [25, 26]. The Rician parameter K may vary significantly with distance, sometimes being close to Gaussian, and at the other extreme near to Rayleigh fading when no dominant path is present. We consider here path loss and shadow fading. Fast fading can be combated by the use of channel coding, symbol interleaving and signal processing [27].

Figure 3.27 shows both LOS and out-of-sight (OOS) paths, where the latter is caused by diffraction around the corners of the buildings [28]. The LOS path loss in dBs as a function of distance r can be characterised by a curve having two straight segments of slope given by the path loss exponent

$$\alpha = \begin{cases} 2 \text{ , if } r \leq X_b, \\ 4 \text{ , if } r > X_b, \end{cases} \tag{3.54}$$

where X_b is the path loss break-distance, which is the distance from the BS antenna to where the curve changes its gradient [25, 26]. For OOS path loss, there is a significant loss in signal level (15 dB to 25 dB in some 10 m) when an MS turns a corner. The slope of path loss with distance is even steeper farther away from the turning point, therefore the path loss exponent α has a value larger than 4. In our cellular arrangements of Figures 3.25 and 3.26 the interference from MSs and BSs on OOS paths is negligible compared with that for MSs and BSs on the LOS paths.

The path loss profile experiences variations about its mean value due to local features, such as irregular building-lines, parked vehicles, etc. We account for this by modelling these variations using a log-normal distribution having a standard deviation of 4 dB and 3.5 dB for LOS and OOS paths, respectively [26]. The effect of path loss and shadow fading on the received signal power for the ith mobile located at a distance, r_i, from its BS is

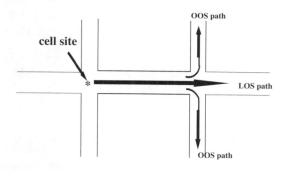

Figure 3.27: The propagation paths in street microcells.

represented by

$$a_i^2 = \begin{cases} (1/r^2)\,10^{\lambda_i/10}, & \text{if } r \leq X_b, \\ (X_b^2/r^4)\,10^{\lambda_i/10}, & \text{if } r > X_b, \end{cases} \qquad (3.55)$$

where λ_i is a normally distributed random variable with a standard deviation of σ.

3.6.2   Up-link SIR values for a cross-shaped microcellular network

Consider the zeroth BS receiver experiencing interference from mobiles in the jth co-channel cell. For perfect power control on the up-link, the received signal power at all BSs from mobiles in their cells is the constant target power, S. Suppose the power control compensates for the path loss and shadowing that affects each mobile. For a particular mobile at a distance r from the jth BS operating over a radio channel having path loss and shadowing as given in Equation (3.55), the transmitted power from this mobile is

$$P_T = \begin{cases} Sr^2 10^{\frac{-\lambda}{10}}, & \text{for } r \leq X_b, \\ SX_b^{-2} r^4 10^{\frac{-\lambda}{10}}, & \text{for } r > X_b. \end{cases} \qquad (3.56)$$

From Reference [2], the detectable range of the shadowing variable, λ, is between -4σ and 2σ. Even for the normal distribution, this range (-4σ to 2σ) occurs 98% of the time.

Figure 3.28 shows tessellated clusters of microcells with two cross-shaped microcells per cluster. For the cross-shaped microcellular pattern there are $J = 8$ close interfering microcells for the zeroth microcell. The interference from mobiles in the four interfering microcells marked A can be neglected as their interference must pass through an OOS path to reach the zeroth BS and thereby experience a power loss of about 20 dB at street corners. The significant interfering microcells are the four microcells marked B in Figure 3.28, where the interference can reach the zeroth BS via LOS paths. The microcells marked B are split into two smaller areas, B1 and B2, where B1 areas are in the LOS streets, and B2 areas are in the OOS streets. Consider the jth interfering microcell. On the up-link, the significant interference at the zeroth BS does not come from MSs in the B2 areas because of the OOS paths. The majority of the interference is generated by mobiles located in the B1 areas where the LOS conditions prevail. Consequently, the interfering power at the zeroth microcell BS from a MS in the jth microcell is

$$I_j = P_T X_b^2 r_j^{-4} 10^{\frac{\lambda_0}{10}}, \qquad (3.57)$$

where

$$r_j = D_j - r, \qquad (3.58)$$

D_j is the distance between the zeroth and jth base sites, r is the distance between the mobile in the jth cell and the jth BS, and λ_0 is the shadowing random variable over the path

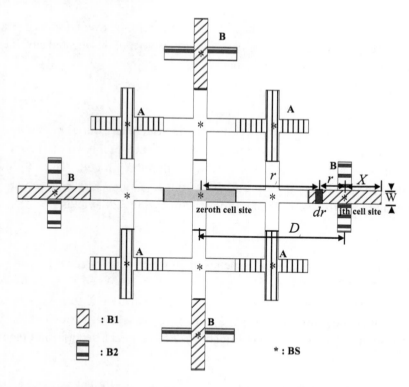

Figure 3.28: Up-link interfering microcells, shown shaded, for the centre zeroth cross-shaped microcell. There are two microcells per cluster.

between the zeroth BS and the interfering MS in the jth microcell. Substituting P_T from Equation (3.56) into Equation (3.57), yields

$$I_j = \begin{cases} \frac{SX_b^2 r^2}{r_j^4} 10^{\frac{\zeta}{10}}, & \text{for } r \le X_b, \\ \frac{Sr^4}{r_j^4} 10^{\frac{\zeta}{10}}, & \text{for } r > X_b, \end{cases} \tag{3.59}$$

where

$$\zeta = \lambda_0 - \lambda \tag{3.60}$$

is a random variable having a normal distribution with a standard deviation of $\sqrt{2}\sigma$. Note that ζ can vary from $-4\sqrt{2}\sigma$ to $2\sqrt{2}\sigma$.

Carrier f_i will be used by a subset of N_f users who occupied the kth timeslot in the jth cell. Assuming the N_f users of the kth timeslot are uniformly distributed over the entire cell, and that N_f is large, then the PDF of the kth timeslot user in the area da is

$$f(a) = \frac{1}{4XW - W^2} \cong \frac{1}{4XW} \tag{3.61}$$

for $W \ll X$, where X and W are defined in Figure 3.28. The average interference power received at the zeroth BS is the average of the powers from mobiles located anywhere within the jth cell using the kth timeslot, namely

$$\Upsilon_j = \int\int_{\text{cell area}} I_j f(a) da. \tag{3.62}$$

Note that whereas I_j in Equation (3.59) is the interference from a mobile at a distance r_j to the zeroth BS, Υ_j is the interference in the presence of FH where at each hop the interference may arise from a different location within the jth cell. Thus Υ_j involves an averaging over the cell area. Substituting $f(a)$ from Equation (3.61) into Equation (3.62), and noting that $da = W dr$, with W and dr defined in Figure 3.28, yields

$$\Upsilon_j = \frac{1}{4X}\int_{-X}^{X} I_j dr. \tag{3.63}$$

From Equation (3.59), the value of I_j depends upon whether $r \le X_b$ or $r > X_b$. The integration in Equation (3.63) is done for $X_b \le X$ and $2X > X_b > X$. When $X_b \le X$, the integration is divided into three sections, as shown in Figure 3.29, namely $-X$ to $-X_b$, $-X_b$ to X_b, and X_b to X. In the first and third sections, $r > X_b$, while in the second section $r \le X_b$. Substituting the appropriate value of I_j given in Equation (3.59) for each section, and then applying these values of I_j to Equation (3.63) to yields

$$\Upsilon_j = \frac{S}{4X}\left\{\int_{-X_b}^{X_b}\left(\frac{r}{r_j}\right)^4 10^{\frac{\zeta}{10}}dr + \int_{-X}^{-X_b}\frac{X_b^2 r^2}{r_j^4}10^{\frac{\zeta}{10}}dr + \int_{X_b}^{X}\left(\frac{r}{r_j}\right)^4 10^{\frac{\zeta}{10}}dr\right\}. \tag{3.64}$$

When $2X > X_b > X$, as shown in Figure 3.30, $|r| \le X_b$, because r is integrated from $-X$ to X. Consequently, we have

$$\Upsilon_j = \frac{S}{4X}\int_{-X}^{X}\frac{X_b^2 r^2}{r_j^4}10^{\frac{\zeta}{10}}dr. \tag{3.65}$$

Because of FH, as the interferer changes in each hop, the shadowing variable ζ associated with the interferer changes. The average of Υ_j for the jth cell is

$$E[\Upsilon_j] = \frac{S}{4X}E\left[10^{\frac{\zeta}{10}}\right]\left\{\int_{-X}^{-X_b}\left(\frac{r}{r_j}\right)^4 dr + \int_{-X_b}^{X_b}\frac{X_b^2 r^2 dr}{r_j^4} + \int_{X_b}^{X}\left(\frac{r}{r_j}\right)^4 dr\right\} \tag{3.66}$$

and

$$E[\Upsilon_j] = \frac{S}{4X}E\left[10^{\frac{\zeta}{10}}\right]\left\{\int_{-X}^{X}\frac{X_b^2 r^2}{r_j^4}dr\right\} \tag{3.67}$$

for $X_b \le X$ and $2X > X_b > X$, respectively, where $E\left[10^{\frac{\zeta}{10}}\right]$ is given in Equation (3.19). By rearranging Equations (3.66) and (3.67) and on letting $x = r/X$, where $-1 \le x \le 1$, we have

$$\frac{E[\Upsilon_j]}{S} = \frac{1}{4}E\left[10^{\frac{\zeta}{10}}\right]\left\{\int_{-1}^{-\frac{X_b}{X}}\left(\frac{x}{\frac{r_j}{X}}\right)^4 dx + \int_{-\frac{X_b}{X}}^{\frac{X_b}{X}}\frac{(\frac{X_b}{X})^2 x^2 dx}{(\frac{r_j}{X})^4} + \int_{\frac{X_b}{X}}^{1}\left(\frac{x}{\frac{r_j}{X}}\right)^4 dx\right\} \tag{3.68}$$

and

$$\frac{E[\Upsilon_j]}{S} = \frac{1}{4}E\left[10^{\frac{\zeta}{10}}\right]\int_{-1}^{1}\frac{\left(\frac{X_b}{X}\right)^2 x^2 dx}{\left(\frac{r_j}{X}\right)^4} \tag{3.69}$$

for $X_b \leq X$ and $X_b > X$, respectively.

By applying voice activity detection and thereby DTX, the mobiles transmit only when speech is present. The voice activity variable v_j is given by Equation (3.20), and its mean by Equation (3.21). The average interfering power from mobiles in the jth co-channel cell is given by $E[v_j]E[\Upsilon_j]$. So far we have considered only one interfering cell, namely the jth cell. Extending the argument to J interfering cells, yields an *SIR* of

$$\frac{S}{I} = \frac{1}{\frac{\mu}{5}\sum_{j=1}^{J}E[\Upsilon_j]}. \tag{3.70}$$

Since the length of the microcells is a few hundred metres, we arbitrarily assign the distance, X, to be 200 m. For the path loss exponent α given by Equation (3.54), a standard deviation of the shadowing fading of 4 dB, we compute $E\left[10^{\frac{\zeta}{10}}\right]$ in Equation (3.19). This expectation value is inserted into Equations (3.68) and (3.69) and the *SIR* is found from Equation (3.70). The *SIR* versus the normalised path loss break-distance, X_b/X, for the two-cell cluster cross-shape microcell system for different values of VAFs is shown in Figure 3.31. The *SIR* is seen to be highest when the microcellular dimension X is smaller than the path loss break-distance, i.e. $X_b/X \leq 1$, where the *SIR* is over 21 dB, 24 dB, 26 dB for a VAF of 1, 1/2, 3/8, respectively. When X_b/X is equal to 1.25 (a maximal X_b of 250 m), the *SIR* is 20 dB, 23 dB, and 24.2 dB for VAFs of 1, 1/2, and 3/8, respectively. As the break-distance X_b increases, the path loss exponent approaches a value of 2. These *SIR*s are far in excess of those required by a modern GSM system, even for the worst condition ($X_b = 250$ m). Note that with hexagonal cellular structures the whole area may be occupied by mobiles, and there are six dominant interfering cells, whereas in cross-shaped microcells the mobiles are confined to the streets and there are only four significant interfering microcells, and in each microcell only half the users offer significant interference.

3.6.3 Up-link *SIR* values for rectangular-shaped microcells

The up-link *SIR*s for the GSM system operating with rectangular-shaped city street microcells can be calculated by considering the position of the interfering cells shown in Figure 3.32. There are six co-channel microcells in Figure 3.32. Co-channel microcells B are in the LOS street, while the other four co-channel microcells A are in the OOS parallel streets. Since the interference created from the mobiles in the four OOS co-channel microcells must be diffracted and reflected around two corners before they reach the zeroth BS, experiencing an attenuation at each corner of about 20 dB, then only the two co-channel microcells B in the LOS street need to be considered.

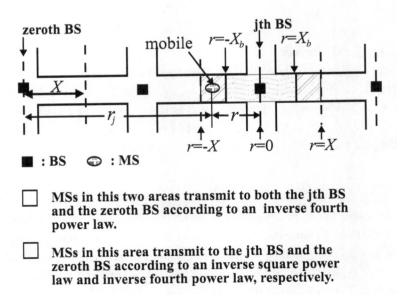

Figure 3.29: Path loss exponent for the interfering MSs in co-channel microcells when $X_b \leq X$.

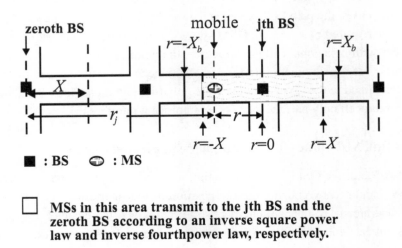

Figure 3.30: Path loss exponent for the interfering MSs in co-channel microcells when $X_b > X$.

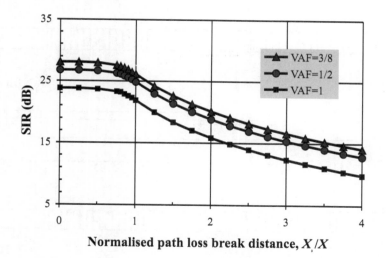

Figure 3.31: *SIRs* versus the normalised path loss break-distance, X_b/X, for the two-cell cluster cross-shaped microcellular system and for different values of VAFs.

The PDF of the kth timeslot user in an area da for a rectangular-shaped microcell is

$$f(a) = \frac{1}{2XW} \tag{3.71}$$

rather than $1/(4XW)$ for the cross-shaped microcells. By replacing $f(a)$ in Equation (3.61) with Equation (3.71), we have

$$\Upsilon_j = \frac{1}{2X} \int_{-X}^{X} I_j \mathrm{d}r. \tag{3.72}$$

From Equations (3.63)–(3.69), the interference-to-signal power ratio for the jth cell now becomes

$$\frac{\mathrm{E}[\Upsilon_j]}{S} = \frac{1}{2}\mathrm{E}\left[10^{\frac{\zeta}{10}}\right]\left\{\int_{-1}^{-\frac{X_b}{X}} \left(\frac{x}{\frac{r_j}{X}}\right)^4 \mathrm{d}x + \int_{-\frac{X_b}{X}}^{\frac{X_b}{X}} \frac{\left(\frac{X_b}{X}\right)^2 x^2 \mathrm{d}x}{\left(\frac{r_j}{X}\right)^4} + \int_{\frac{X_b}{X}}^{1} \left(\frac{X}{\frac{r_j}{X}}\right)^4 \mathrm{d}x\right\} \tag{3.73}$$

and

$$\frac{\mathrm{E}[\Upsilon_j]}{S} = \frac{1}{2}\mathrm{E}\left[10^{\frac{\zeta}{10}}\right]\int_{-1}^{1} \frac{\left(\frac{X_b}{X}\right)^2 x^2 \mathrm{d}x}{\left(\frac{r_j}{X}\right)^4} \tag{3.74}$$

for $X_b \leq X$ and $2X > X_b > X$, respectively.

It will be recalled that for the cross-shaped microcells significant interference is generated by mobiles in the LOS streets but not in the OOS streets. Since mobiles are uniformly distributed over the microcells, the probability of a mobile located in the LOS street is 1/2, and so is the probability of a mobile in a cross-shaped microcell generating significant

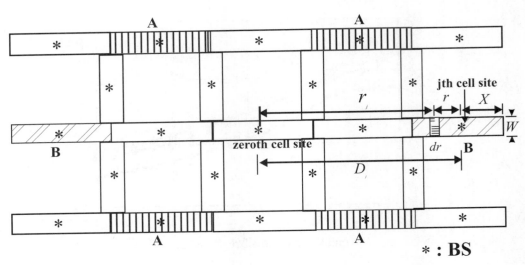

Figure 3.32: Up-link interfering microcells, shown shaded, for the zeroth rectangular-shaped microcell. There are four microcells per cluster.

interference. In contrast, for the rectangular-shaped microcells, all mobiles in the dominant interfering microcells B can generate interference as they are all in the LOS street. Hence the interference due to the mobiles in a rectangular-shaped microcellular system is twice that experienced in a cross-shaped microcellular arrangement. However, because the number of significant co-channel interfering microcells in the rectangular-shaped microcells is half that of the cross-shaped microcells, the average interference-to-signal power ratio is twice that experienced in a cross-shaped microcellular system. Consequently the SIR of the TDMA system in the rectangular-shaped microcellular network is the same as that of the cross-shaped microcellular network and the curves of Figure 3.31 apply for both microcellular arrangements.

Note that if VAF= 1, then all interfering transmitters are switched on to give the worst interference condition. When VAF= 3/8 it is assumed that for 3/8 of the time the interference is due to speech in progress, while for 5/8 of the time the interfering mobile transmitters are switched off. As can be seen in Figure 3.31, there is no point in using DTX to enhance the SIR because it is already more than adequate for typical GSM link requirements. However, DTX is used as to conserve battery power. Because the SIR in city street microcellular networks is much higher than the required SIR_{min}, the SIR decrease due to imperfect power control is negligible.

3.6.4 Microcellular GSM network capacity

The system capacity of the GSM in a microcellular environment is defined by Equation (3.50). When we ignore shadow fading, the *SIR* without power control, frequency hopping and DTX is given by [14]

$$\frac{S}{I} = \begin{cases} \frac{1}{J}\left(\frac{D}{X}\right)^4 & \text{for } X_b < X, \\ \frac{1}{J}\frac{D^4}{X_b^2 X^2} & \text{for } X_b \geq X, \end{cases} \tag{3.75}$$

where J is the number of first-tier co-channel cells, and is equal to four or two depending on whether the microcells are cross-shaped or rectangular shaped, respectively. D is the separation between two co-channel cells, X_b is the path loss break-distance, and X is the distance from the base station (BS) to the cell boundary. Since the number of first-tier co-channel cells is four and two for the cross-shaped and the rectangular-shaped microcells, respectively, the *SIR*s can be calculated using Equation (3.75) for a $X = 200$ m. The *SIR*s versus normalised path loss break-distance, X_b/X, are shown in Figure 3.33. Whereas a more rigorous analysis of the *SIR* showed that it was the same for both types of microcells, the simple Equation (3.75) gives the *SIR* of the rectangular-shaped microcells to be 3 dB higher than that of the cross-shaped microcells. This is because the number of co-channel microcells is only half that of the cross-shaped microcells. From the graphs, we observe that the optimal size of the microcell is $X_b \leq 1.25X$, where the *SIR*s are 15.1 and 18.1 dB for the cross-shaped and the rectangular-shaped microcells, respectively. For $X_b > X$, the path loss approaches a second-order path loss law. A simple TDMA system compared with the GSM system with power control, DTX (a VAF of 3/8), and frequency hopping provides a gain in the *SIR* of 9 dB and 6 dB for the cross-shaped and the rectangular-shaped microcells, respectively. Suppose the minimal required *SIR* is 12 dB, then irrespective of perfect power control, frequency hopping, and DTX, the system can operate in a two-cell cluster and four-cell cluster for cross-shaped and rectangular-shaped microcells, respectively, because of their very high *SIR*s.

For a carrier spacing of $B = 200$ kHz, 8 channels per carrier, $\chi = 0.1$, the capacity of the GSM having cross-shaped microcells with $M = 2$ is 18 channels per MHz per cell, and 9 channels per MHz per cell for the rectangular-shaped microcells, where $M = 4$. However, because the coverage area of a cross-shaped microcell is twice that of a rectangular-shaped microcell, the capacity in terms of channels per MHz per area, rather than channels per MHz per cell, is the same for both the cross-shaped and rectangular-shaped microcells.

While the GSM system using cross-shaped microcells has a higher capacity in terms of channels per MHz per cell than the rectangular-shaped microcellular system, it is difficult to deploy the cross-shaped microcells with their two-cell cluster in a street with irregular street patterns. On the other hand, rectangular-shaped microcells are merely segments of streets and more flexible in adapting to different street patterns. For streets that do not

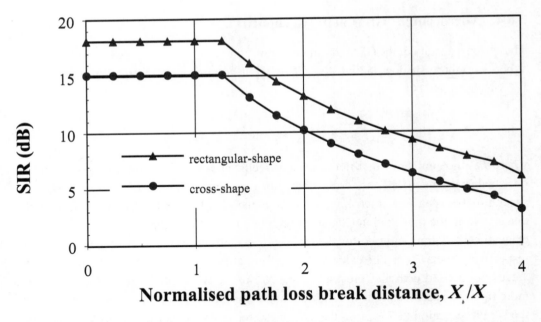

Figure 3.33: *SIR*s versus X_b/X for TDMA systems without power control, frequency hopping, and DTX in the cross-shaped and the rectangular-shaped microcellular arrangements.

have contiguous buildings along their sides, the TDMA system in the rectangular-shaped microcells is more robust than the TDMA system in the cross-shaped microcells. Frequency hopping also has the ability to reduce the effect of imperfect blocking of the buildings.

3.6.5 Irregular-shaped microcells

Microcells are essential if high capacity GSM networks are to be realised. By siting micro-cellular BS antennae below the urban skyline there is negligible diffraction over roof-tops into adjacent streets, although there is diffraction from roof-tops skywards to the upper floors of tall buildings. The signal emanating from the BS antenna propagates along the streets (and into buildings) and reflects and diffracts around corners. This process continues until the *SIR* along the streets goes below the minimum *SIR*, SIR_{min}, to support accept-able communications. Those streets, or parts of streets, where the $SIR > SIR_{min}$ define the street microcell. Tessellated street microcells are formed into tessellated clusters, but in most cities the microcells are not regular, as previously considered, but are shaped by the city buildings. Planning irregularly shaped microcells requires radio propagation prediction planning tools.

We define street microcells as being formed by BS antennae mounted below the skyline. Generally there will be other cells that are formed by BS antennae that are strategically placed at various heights within the city. By restricting the power from these antennae we can create *minicells* that form cells larger than microcells, although not necessarily so, but that operate in three dimensions. For example, they can be used to provide radio coverage in radio dead-spots within the microcells, and into multistorey buildings [29, 30]. Oversailing the microcells and minicells could be a sector of a macrocell. The macrocells, and the minicells, can also be used to support handovers between street microcells when microcellular channels are not available [22].

For GSM to continue on its successful commercial path it must have low cost microcellular BSs and their connecting and supporting infrastructure. Although current GSM BSs, BSCs and MSCs are relatively expensive, they are not inherently so, particularly for the microcellular environment. If they are connected by optical fibres, the innate cost of the fibres is also low. In the long term the cost of deploying microcellular networks will mainly be the cost of installation and site rental. It is not a question of whether GSM microcellular networks will be deployed in vast numbers, but when.

Currently many major cities have clusters of microcells, or single microcells, to relieve traffic hot-spots. In the present cellular radio culture some operators continue to use their existing macrocellular planning tools to plan microcells with limited success. So let us say a brief word about macrocellular planning tools. These tools are based on path loss calculations that include regression-line analysis of path loss with distance data, terrain information, edge-diffraction models, and clutter-loss models of buildings in different environments, such as urban and suburban areas. For urban street microcells, models that consider buildings as *clutter* are bound to give erroneous results. One single tall building can cast a large radio shadow. Indeed, the specific effect of *every* building and the influence of its local environment must be taken into consideration if the predictions of path loss, strongest server, overlap zones for hand overs, signal-to-interference ratios and signal-to-adjacent channel ratios are to be sufficiently accurate for planning a street microcellular network.

Conventional macrocellular planning tools use models that average the signal power losses (clutter losses) caused by buildings. Network operators tend to have little confidence in these predictions and frequently resort to measurements. If the measurements and the predictions show large discrepancies, then the operator *tunes* the planning tool. Surely the function of a predictor is to remove the need to make measurements. Since the clutter loss does not allow for individual buildings in the radio propagation path, operators must include shadow fading margins in their link budgets, i.e. allow for the shadows cast by buildings. As an example, shadow fading is generally modelled by a log-normal distribution with a standard deviation of 8 dB. If a cell were planned such that the average signal power at the cell edge was equal to the receiver sensitivity (i.e. the minimum received power for acceptable performance), then as a consequence of shadow fading the average received signal will be acceptable for 50% of the time at the cell boundary and unacceptable for the remaining 50% of the time. The levels of coverage may be improved by ensuring that the cell is planned such that the signal level is greater than the receiver sensitivity by an acceptable margin. Using the shadow fading model described above, the probability of coverage at the cell edge may be increased to 75% by including a shadow fading margin of around 5 dB.

A further drawback of conventional planning tools is their lack of resolution. The prediction information is usually presented by dividing the prediction area into a number of squares and generating a single signal strength value for each square. In some cases these squares may have sides from 50 m to 250 m. Tools of this nature are virtually useless when it comes to predicting the coverage of a street microcell that is only 200 m to 400 m across! Prediction tools that use accurate building information can predict with a much higher resolution and, in some cases, this can be a few metres. This shows that as cell sizes decrease, the radio prediction planning tools must include accurate building information in their prediction algorithms.

A number of small cell radio planning tools use a technique known as *ray tracing* to predict both the path loss and the propagation channel impulse response. In simple terms, ray tracing involves assuming that a number of rays are emitted from the transmitting antenna. The path of each of these rays is then traced as it is reflected and diffracted by the buildings. The characteristics of the signal at a receiving antenna are determined by examining the relative phase offsets and time delays of the arriving rays. Unfortunately, this approach has two major drawbacks. First, it requires substantial processing power to track each individual ray. This means that the predictions may take a long time. The second drawback is that the prediction accuracy of path loss and impulse response may be low because of the number of practical parameters and variables that cannot be included in the prediction.

A more appropriate approach to predicting radio propagation in small cells, e.g. street microcells, is to generate an average value of path loss at a given resolution, say, 5 m by 5 m, and make no attempt to predict the effects of fast fading of the signal characteristics, e.g. delay spread. The validity of this approach is strengthened by the fact that most modern cellular systems are designed to cope with delay spreads that are significantly larger than those found in microcells and so, in general, there is no need to predict fading phenomena in a microcellular network.

Figure 3.34 shows the plot of an arbitrary street microcell produced by the small cell radio planning tool, NPWorkplace.[1] The tool employs an empirical prediction model that was developed from a large database of radio propagation measurements. The bin size, i.e. the square area where the measurements were averaged, is 5 m by 5 m. The figure is taken

---

[1]NPWorkplace is a proprietary small cell radio planning tool from Multiple Access Communications Ltd. World wide web page: http://www.macltd.com

from a PC screen display. The buildings are represented by dark grey shapes, with roads, open spaces, and parks in black. Vegetation is quantised into three categories and would normally be displayed in different colours. Unfortunately we have had to adapt this figure for a monochrome presentation, and therefore these colours are not shown. Signal strength contours are usually displayed, where the distance between adjacent contours is represented by a colour corresponding to a band of signal strength values. So a street would appear on the PC screen as composed of coloured bands, with each band representing a range of signal strength values, e.g. -75 to -85 dBm. Here, with our monochrome displays we have defined a microcell as a zone where the signal strength is ≥ -85 dBm, and highlighted this zone in white. The position of the BS antenna is shown as a white arrow. A cross-polarised antenna having a gain of 7 dB is used at a height of 3 m, and the transmit power is 20 dBm. The predictions are made at a height of 2 m. The dimensions of the plot are 430 m by 420 m.

Figure 3.34 uses maps from a three-dimensional database. The resolution of the map data is 1 m $\times$ 1 m $\times$ 1 m which means that features such as elevator shafts are clearly discernible. Because of the three-dimensional nature of the building and terrain database we can predict the path loss contours at any height. So in Figure 3.35 we show the same microcellular BS as used in Figure 3.34, but at a height of 20 m. The coverage is seen to be much larger at 20 m than at 2 m, and while this may be used with advantage to provide coverage on higher floors in nearby buildings, care must be exercised to ensure that the microcell will not cause interference to minicells and the larger macrocells that may be overlaying the microcells.

At this point we observe that the shape of a street microcell is dependent upon the local buildings, and to a lesser extent on the terrain and vegetation. Although we may define a street microcell by carefully siting the BS antenna to be below the urban skyline, the radio propagation does not confine itself merely to the streets. The radiation will go skywards, reflecting and diffracting off buildings, and diffracting over roof edges. Larger microcells may be created in offices on higher floors in nearby buildings. There will also be penetration of electromagnetic energy into (and from) the buildings at the ground floor level.

For a given BS transmission power, the shape and size of a street microcell may be critically dependent upon the position of the BS antenna. For example, for a building located at a junction of two roads the choice of which side of the building to site the antenna may have a dramatic effect. The situation is illustrated in Figure 3.36. This sensitivity of street microcell size and shape to BS antenna location means that radio planners can site antennae to control the coverage and capacity they require in a given locality. The city is in essence a large electromagnetic mould in which electromagnetic energy is injected at carefully selected locations (BSs) in order to satisfy the network design plan.

An MS transmitting to its street microcellular BS may interfere with a minicellular BS or macrocellular BS because of its sky propagation which will suffer relatively small attenuation. The situation is made worse because a minicellular BS will usually employ a

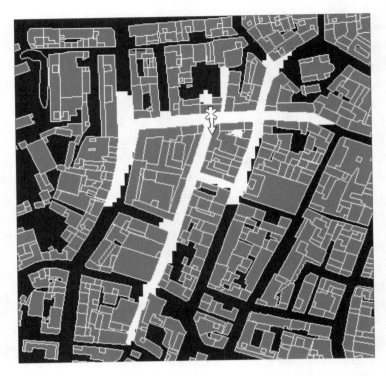

Figure 3.34: A street microcell is shown in white where the signal strength has been arbitrarily set
to ≥ -85 dBm. A cross-polarised BS antenna with a gain of 7 dB is used at a height
of 3 m, and transmit power is 20 dBm. The predictions are made at 2 m, and the plot
size is 430 m by 420 m.

high gain sectorised receiving antenna. When we consider an MS in a minicell communicating with its minicellular BS, and as power control of the MS transmit power is used,
this MS will cause little interference to the street microcellular BS because its transmissions
will be subjected to large attenuations as they diffract and reflect along the streets to reach
the microcellular BS which is only 6 m, say, above the street [30]. When we consider the
down-link transmissions there will be little interference at the MS in the minicell from the
microcellular BS, but the MS in the microcell may receive significant levels of interference
from the minicellular BS. This leads us to conclude that the stronger interference link in a
mixed cell environment are the unwanted transmissions between an MS in a microcell and
a minicellular BS [30].

In a dense urban environment there will be many cells. Low cost, small GSM BSs will
be housed in buildings; others in the streets, usually mounted on the sides of walls or lamp
posts; and small minicells will be mounted such that they have a partial view over the

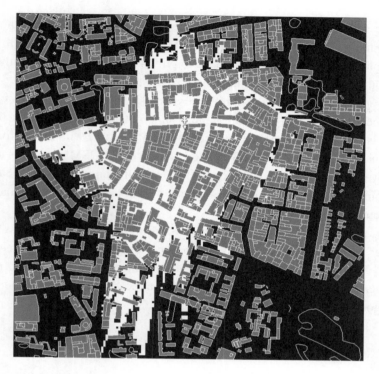

Figure 3.35: Coverage from the microcellular BS in Figure 3.34, but at a height of 20 m.

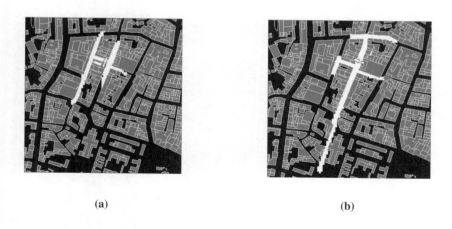

(a)   (b)

Figure 3.36: Sensitivity of cell size and shape to BS antenna position. The street microcell is shown
in white where the signal strength has been arbitrarily set to ≥ -85 dBm. A dipole
antenna at a height of 3 m is used. The transmit power is 20 dBm. The predictions are
at 2 m, and the size of each plot is 470 m by 460 m.

locality. We will not have many macrocells, i.e. large cells with antennas on the roofs of the tallest buildings. This is because large cells equates to low capacity. However, large cells are deployed in network start-up mode to provide coverage, and it is only as capacity increases that smaller cells, and eventually microcells, are introduced.

If the introduced microcells are significantly smaller than the other types of cells, then it is more spectrally efficient to partition the channel sets so that the microcells have their own channels. This means that the larger cells will not interfere with the microcells and the spectral efficiency increases. Figure 3.37 shows a network of GSM microcells using a reuse of four. However, when the minicells approach the size of microcells to assist handovers and to cover radio dead-spots in the microcell clusters, then partitioning of the radio channels may be spectrally inefficient.

To provide the high capacity multimedia networks of the future, GSM Phase 2+ networks (see Chapter 6) will require a myriad of small cells stacked in three dimensions. This situation may be a profound challenge for radio planners who will need to grapple with complex interference problems. Dynamic channel allocation of radio channels is likely to be one method of mitigating this problem.

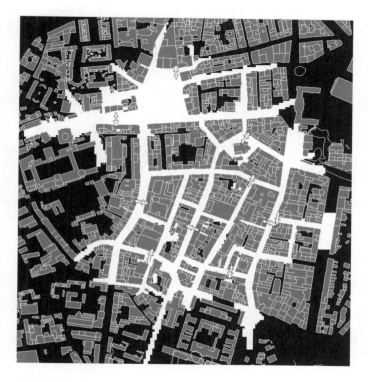

Figure 3.37: A network of tessellated street microcells. The predictions are at 2 m. The BS antennae are cross-polarised with a gain of 7 dB located at 3 m above street level. The transmit power is 20 dBm. The plot size is 840 m by 780 m.

Bibliography

[1] Goodman D.J., *Wireless Personal Communications Systems*, Addison-Wesley, 1997.

[2] Steele R. [Ed.], *Mobile Radio Communications*, Pentech Press, 1992.

[3] Eberspacher J. and H.J. Vogel, *GSM Switching Services and Protocols*. John Wiley, 1998.

[4] Gibson J.D. [Ed.], *The Mobile Communications Handbook*, CRC Press and IEEE Press, 2nd edn, 1999.

[5] Marcario R.C.V. [Ed.], *Modern Personal Radio Systems*, IEE, 1996.

[6] Rappaport T.S., *Wireless Communications*, Prentice-Hall, 1996.

[7] Redl S.M., M.K. Weber and M.W. Oliphant, *An Introduction to GSM*, Artech House, 1995.

[8] Heine G., *GSM Networks*, Artech House, 1999.

[9] Lee C.-C. and R. Steele, Signal-to-interference calculations for modern TDMA cellular communication systems, *IEE Proc. Commun.*, **142**(1), Feb. 1995, 21–30.

[10] Dornstetter J.-L. and D. Verhulst, Cellular efficiency with slow frequency hopping: Analysis of the digital SFH900 mobile system, *IEEE J. Sel. Area Commun.*, **5**(5), June 1987, 835–848.

[11] Chennakeshu S., A. Hassan and J. Anderson, Analysis of the capacity of a mixed-mode slow frequency hopping cellular system, *IEEE 43rd VTS Conf.*, Secaucus, New Jersey, May 1993, 540–543.

[12] ETSI, *Digital European Cordless Telecommunications—Common Interface*, Radio Equipment and Systems, Valbonne, France.

[13] Mewson P., An analysis of the effects of power control error within a direct sequence CDMA system for mobile radio, *Proc. IEEE*, Globecom, Houston, Texas, Dec 1991, 924–928.

[14] Lee W.C.Y., Elements of cellular mobile radio systems, *IEEE Trans. Veh. Technol.*, **VT-35**(2), May 1986, 48–56.

[15] Everitt D. and D. Manfield, Performance analysis of cellular mobile communication systems with dynamic channel assignment, *IEEE J. Sel. Area Commun.*, **7**(8), Oct. 1989, 1172–1180.

[16] Goodman D.J., Cellular packet communications, *IEEE Trans. Commun.*, **38**(8), August 1990, 1272–31280.

[17] Steele R., The cellular environment of lightweight handheld portables, *IEEE Commun. Mag.*, **27**(7), July 1989, 20–29.

[18] Steele R., Towards a high-capacity digital cellular mobile radio system, *IEE Proc.*, **132**, Pt F (5), Aug. 1985, 405–415.

[19] Greenstein L.J., N. Amitay, *et al*, Microcells in personal communication systems, *IEEE Commun. Mag.*, **30**(12), Dec. 1992, 76–78.

[20] Steele R., J. Widehead and W.C. Wong, System aspects of cellular radio. *IEEE Commun. Mag.*, **33**, Jan. 1995, 80–86.

[21] DaSilva J.S. and B.E. Fernandes, The European research program for advanced mobile systems, *IEEE Pers. Commun.*, Feb. 1995, 14–19.

[22] Steele R., Speech codecs for personal communications, *IEEE Commun. Mag.*, **31**, Nov. 1993, 76–83.

[23] Chia S.T.S., Radiowave propagation and handover criteria for microcells, *B. Telecom Tech. J.*, **8**, Oct. 1990, 50–61.

[24] Anderson J.B., T.S. Rappaport and S. Yoshida, Propagation measurements and models for wireless communications channels, *IEEE Commun. Mag.*, Jan. 1995, 42–49.

[25] Bultitude R.J.C. and K. Bedal, Propagation characteristics on Microcellular urban mobile radio channel at 910 MHz, *IEEE J. Sel. Area Commun.*, **7**(1), Jan. 1989, 31–39.

[26] Berg J.-E. *et al*, Path loss and fading models for microcells at 900 MHz, *IEEE 42nd VTS Conf.*, Denver, May 1992, 666–671.

[27] Simpson F. and J.M. Holtzman, Direct sequence CDMA power control, interleaving, and coding, *IEEE J. Sel. Area Commun.*, **11**(7), Sept. 1993.

[28] Erceg V., S. Ghssemzadeh, M. Taylor, D. Li and D.L. Schiling, Urban/suburban out-of-sight propagation modelling, *IEEE Commun. Mag.*, **30**(6), June 1992, 56–61.

[29] Webb W.T., Sizing up the microcell for mobile radio communications, *Electron. Commun. Eng. J.*, **5**(3), June 1993, 133–140.

[30] Dehghan S. and R. Steele, Small cell city, *IEEE Commun. Mag.*, **30**, Aug. 1997, 52–59.

[31] Steele R., J. Williams, D. Chandler, S. Dehghan and A. Collard, Teletraffic performance of GSM-900/DCS-1800 in street microcells, *IEEE Commun. Mag.*, **33**, Mar. 1995, 102–108.

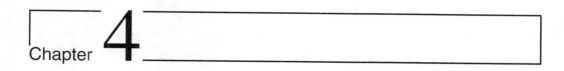

Chapter 4

The cdmaOne System

4.1 Introduction

In contrast to the GSM system, which was designed and developed by a number of different organisations working together, the cdmaOne technology was designed by a single company, Qualcomm Incorporated. The first commercial cdmaOne network was launched by Hutchison in Hong Kong on 28 September 1995 and since that time commercial networks based on the cdmaOne technology have been launched in many countries around the world including Korea and the United States.

Qualcomm's CDMA technology was 're-branded' as cdmaOne in 1997. Prior to this the technology was commonly referred to as 'IS-95', which is the name of the standard which describes the cdmaOne technology in the United States (i.e. Interim Standard number 95 [1]). The cdmaOne technology was originally designed to provide a high capacity overlay for the first generation analogue Advanced Mobile Phone System (AMPS) operating in the 800 MHz cellular band in the United States. This gave an AMPS operator the option of increasing its network capacity in specific areas by replacing a number of 30 kHz AMPS carriers with one or more 1.25 MHz cdmaOne carrier. Dual mode cdmaOne/AMPS mobile stations (MSs) are able to use the cdmaOne system, where available, and they will revert to the AMPS system in areas where there is no CDMA coverage.

With the introduction of personal communications systems (PCS) in the United States, the cdmaOne technology was modified to operate in the 1900 MHz PCS frequency band in a single mode configuration citecdma-pcs. This version of the cdmaOne technology was commonly referred to as 'CDMA-PCS' prior to the re-branding. In addition to the versions of cdmaOne described above, other variations exist which have been modified to operate in particular frequency bands in different countries throughout the world.

At this point it is important to clarify the terminology we shall be using in the remainder of this chapter. We shall use the term 'IS-95' to describe the CDMA system operating in the US cellular band (800 MHz) and we shall use the term 'CDMA-PCS' to describe the PCS system operating in the 1.9 GHz band. In many cases our discussion will relate to both systems and in this case we shall use the brand name cdmaOne to refer to both versions of the system simultaneously.

It is important to note that the cdmaOne system is basically an air-interface standard, in contrast to the GSM system which is specified up to the network gateway.

4.2   The cdmaOne Radio Interface

4.2.1   Operating frequencies

Before we proceed, we must make a point about terminology. In Europe the transmission path from the network towards the mobile station (MS) is known as the down-link and the transmission path from the MS to the network is known as the up-link. However, in North America the down-link and up-link are known as the forward and reverse links, respectively. Since IS-95 and CDMA-PCS are North American systems, we will use the North American terms throughout this chapter.

The IS-95 system operates in the US cellular frequency band. This band has been sub-divided into five blocks and distributed between two operators, A and B, thereby allowing two different cellular systems to be supported within the same geographical area. The US cellular spectrum allocations in the 800 MHz band are shown in Table 4.1

The IS-95 system uses frequency division duplex (FDD), i.e. the forward link and reverse link transmissions occur in different frequency bands. The duplex separation used in IS-95 (and AMPS) is 45 MHz and the carrier spacing is 1.25 MHz. We note that the IS-95 system has been conceived to operate in a dual mode configuration with the existing analogue AMPS systems in the United States and for the analogue carriers to be gradually replaced with CDMA carriers. In situations where a single CDMA carrier is placed in a

System	Frequencies (MHz)	
	Reverse link	Forward link
A″	824.040–825.000	869.040–870.000
A	825.030–834.990	870.030–879.990
B	835.020–844.980	880.020–889.980
A′	845.010–846.480	890.010–891.480
B′	846.510–848.970	891.510–893.970

Table 4.1: The US cellular bands.

band occupied by an analogue system, spectral guard bands must be provided between the CDMA service and the existing analogue service. Consequently, a single CDMA carrier operating within an analogue AMPS band will require around 1.8 MHz of spectrum.

The CDMA carrier numbering scheme for IS-95 is the same as that used for AMPS and is shown in Table 4.2, where N is the channel number, f_u is the reverse link frequency and f_d is the forward link frequency.

The table shows that the channel numbering is based on the AMPS carrier spacing of 30 kHz which allows the network operator to position a CDMA carrier at any point within the AMPS band with an accuracy of 30 kHz. It is important to note that a single 1.25 MHz CDMA carrier will occupy the same spectrum as around 40 AMPS carriers and, therefore, the channel numbers of adjacent CDMA carriers will differ by around 40. The CDMA carriers must be positioned in such a way as to allow sufficient guard bands between other services operating above and below the cellular band and between the A and B services. Consequently, the CDMA carriers are limited to using the channel numbers shown in Table 4.3.

The CDMA-PCS system has been designed to operate in the 1.9 GHz PCS band in the United States. This band is sub-divided into three 2×15 MHz blocks (i.e. 15 MHz for the reverse link and 15 MHz for the forward link) and three 2×5 MHz blocks. The PCS spectrum allocations are shown in Table 4.4.

The duplex spacing in the 1.9 GHz PCS band in the United States is 80 MHz and the channel numbering scheme is shown in Table 4.5, where N is the channel number, f_u is the reverse link frequency and f_d is the forward link frequency. This shows that the CDMA carriers may be placed anywhere within the PCS band in steps of 50 kHz. Each 1.25 MHz CDMA-PCS carrier will occupy 25 of these 50 kHz PCS channels and the channel numbers of adjacent CDMA-PCS carriers will differ by 25. The CDMA-PCS carriers must be positioned to ensure that there are sufficient spectral guard bands between the different operator frequency blocks (unless adjacent blocks are allocated to the same operator) and between the systems that occupy the frequency bands above and below the PCS band. For this reason a number of preferred CDMA channel numbers have been defined for each block, and these are shown in Table 4.6.

Having identified the operating frequencies of the IS-95 and CDMA-PCS systems we will

Table 4.2: IS-95 channel numbering.

Band	Frequency (MHz)	Channel numbers
Reverse link	$f_u = 0.030N + 825.000$	$1 \leq N \leq 777$
	$f_u = 0.030(N - 1023) + 825.000$	$1013 \leq N \leq 1023$
Forward link	$f_d = 0.030N + 870.000$	$1 \leq N \leq 777$
	$f_d = 0.030(N - 1023) + 870.000$	$1013 \leq N \leq 1023$

Table 4.3: Available channel numbers for IS-95 carriers.

System A	1	-	311
	689	-	694
	1013	-	1023

System B	356	-	644
	739	-	777

Table 4.4: PCS spectrum allocations.

Block	Frequency (MHz)	
	Reverse link	Forward link
A	1850–1865	1930–1945
D	1865–1870	1945–1950
B	1870–1885	1950–1965
E	1885–1890	1965–1970
F	1890–1895	1970–1975
C	1895–1910	1975–1990

Table 4.5: PCS channel numbers.

Band	Frequency (MHz)	Channel numbers
Reverse link	$1850.000 + 0.050N$	$0 \leq N \leq 1200$
Forward link	$1930.000 + 0.050N$	$0 \leq N \leq 1200$

Table 4.6: CDMA-PCS preferred channel numbers.

Block	Channel numbers
A	25, 50, 75, 100, 125, 150, 175, 200, 225, 250, 275
D	325, 350, 375
B	425, 450, 475, 500, 525, 550, 575, 600, 625, 650, 675
E	725, 750, 775
F	825, 850, 875
C	925, 950, 975, 1000, 1025, 1050, 1075, 1100, 1125, 1150, 1175

now examine the physical layer of the radio interface. In contrast to the GSM system, the coding systems employed on the reverse link and forward link are very different and, for this reason, we shall examine each link separately.

4.2.2 The cdmaOne Forward link

The forward link consists of the base station (BS) transmitter, the radio channel and the MS receiver. The cdmaOne system supports four different types of forward channels. The *pilot channel* is continuously transmitted by each CDMA carrier and is used by the MS to identify the BS. The pilot channel also acts as a cell beacon and is used by MSs in neighbouring cells to assess the suitability of the cell for handover. In this respect the pilot channel in the cdmaOne system may be likened to the BCCH carrier in the GSM system. The pilot carrier of the serving cells is also used by the MS as a coherent reference in the demodulation process and in the reverse link power control algorithm.

Another forward channel is the *synchronisation channel* which, as its name suggests, allows the MS to achieve time synchronisation with the BS and the network. The synchronisation channel also carries information relating to system time, and the contents of the BS's internal registers which are used in the coding, spreading and encryption processes. There are also a number of paging and traffic channels. The *paging channels* are used to page MSs to alert them to an incoming call. The paging channel is also used to carry general network information and channel assignment messages. The *traffic channels* are assigned to the users as required and they may carry speech or user data at bit rates of up to 9.6 kb/s for IS-95 and 14.4 kb/s for CDMA-PCS.

Each forward channel on a CDMA carrier is assigned a different 64-bit Walsh code, and these codes are shown in Figure 4.1. Each row of the table represents a different 64-bit Walsh code with the bit positions shown at the top of the table, and the index of the Walsh code shown in the left-hand column. We note that these codes are orthogonal, i.e. the value of any two codes, multiplied together and summed over a period of 64 chips, is zero, provided the '0' bits are replaced by a '−1' and the '1' bits are replaced by a '+1'. Multiplying a Walsh code by itself produces a constant level of +1 when the two codes are in time synchronisation. We note that although the codes shown in Figure 4.1 are true Walsh codes they are not indexed (or numbered) in the conventional manner. A Walsh code's index is normally given by the number of transitions that occur between the different levels during a code period (i.e. 64 chips). In the cdmaOne specifications, however, the Walsh codes have been numbered as shown in Figure 4.1. In this discussion we will always use the code index numbers shown in Figure 4.1 to avoid confusion and we will refer to the codes as Walsh Hadamard (WH) codes.

The full block diagram of the cdmaOne BS transmitter is shown in Figure 4.2. Each channel in the cdmaOne forward link uses a different coding scheme depending on the

Bit Position

Code Index

Figure 4.1: The Walsh Hadamard transform (WHT) matrix of order 64.

requirements of the channel. In the following sections we shall exam.
arately.

4.2.2.1 The pilot channel

The pilot channel is the simplest of all forward link channels, since it always carries an 'all zero' bit stream. Referring to Figure 4.2, this 'all zero' signal is EXORed (shown as a $\oplus$ in the figure) with the Walsh code with an index of 0 in Figure 4.1, i.e. a series of logical 0s. The result of this operation is another 'all zero' bit stream which is then divided into two and each part is EXORed with one of two different pseudo-random noise (PN) sequences, known as PNI, for the in-phase component, and PNQ, for the quadrature component. These two sequences are 2^{15} bits in length and they are based on the following characteristic polynomials:

$$PNI(x) = x^{15} + x^{13} + x^9 + x^8 + x^7 + x^5 + 1, \tag{4.1}$$
$$PNQ(x) = x^{15} + x^{12} + x^{11} + x^{10} + x^6 + x^5 + x^4 + x^3 + 1. \tag{4.2}$$

The sequences may be generated using a 15-bit feedback register. The maximal length sequences based on Equations (4.1) and (4.2) will be $2^{15} - 1$ bits in length. The sequences are extended to 2^{15} length sequences by inserting a '0' after 14 consecutive 0's, which will occur once for each repetition of the code.

The two PN sequences are generated at a chip rate of 1.2288 Mchips/s and the period will be

$$2^{15}/122880 = 32768/1228800 = 26.666 \text{ ms} \tag{4.3}$$

which results in exactly 75 PN sequence repetitions every 2 s.

EXORing the PN sequences with an all zeros data sequence will leave the PN sequences unchanged. The two sequences are then pulse shaped using low pass filters. The characteristics of the low pass filters are shown in Figure 4.3 in the form of a response mask taken from the specifications [1, 2]. In the diagram, $S(f)$ is the frequency response of the filter. The filter pass band extends from 0 to f_p and the stop band extends from f_s to ∞. Within the pass band the filter response is prescribed within the limits $\pm\delta_1$, and within the stop band the filter response shall be less than $-\delta_2$. The values for each of the parameters are $\delta_1=1.5$ dB, $\delta_2=40$ dB, $f_p=590$ kHz, and $f_s=740$ kHz.

The two data sequences are then multiplied by two quadrature carriers, PNI and PNQ, and the resulting signals are summed to produce a phase modulated carrier signal. The relationship between the input bit sequence and the resulting carrier phase is shown in Table 4.7. These phase transitions may be produced by translating the I and Q bit streams such that a 0 in the original bit stream is replaced by +1 level, and a 1 in the original bit stream is replaced by a -1 level. The constellation diagram is shown in Figure 4.4.

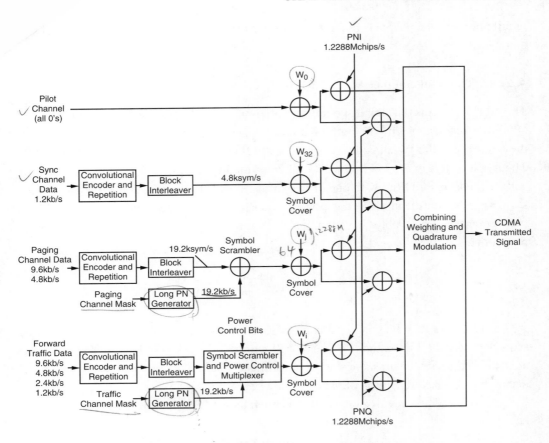

Figure 4.2: Block diagram of a cdmaOne BS transmitter (rate set 1).

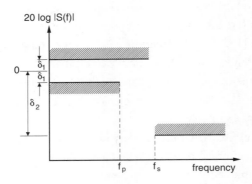

Figure 4.3: Pulse shaping filter requirements.

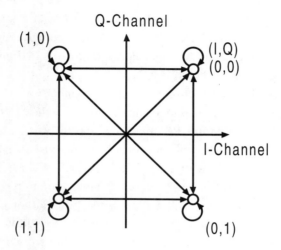

Figure 4.4: The phase constellation at the BS transmitter.

Table 4.7: I and Q data to phase transition mapping.

I	Q	Phase
0	0	$\frac{\pi}{4}$
1	0	$\frac{3\pi}{4}$
1	1	$\frac{-3\pi}{4}$
0	1	$\frac{-\pi}{4}$

We have described the construction of the pilot channel as it has been shown in the specifications. However, in practice the pilot channel is merely produced by modulating the
PNI and PNQ sequences onto two quadrature carriers; the use of Walsh code 0 and an 'all
zero' data sequence is irrelevant. We have also assumed that the translation from digital bits
(0 and 1) to logical levels (± 1) occurs just prior to quadrature modulation; however, this
translation may occur at an earlier stage. For example, the PNI and PNQ sequences could
be produced as logical levels directly.

Having described the construction of the pilot channel we now examine its functions.
One of the main functions of the pilot channel is to allow the MS to detect and identify the
BSs. Since all BSs use the same PN sequences and the same carrier frequency, the only
way in which the different pilot channels may be distinguished is by the phase of their PN
sequences. In IS-95, each BS within a geographical area will use a different time offset for
the PN sequence and this offset will be defined in integer multiples of 64 chips.

For the PN offset to have any meaning across the system it must be referenced to a common timing source. This requirement means that all BSs within a network must be time
synchronised. This is currently achieved using global positioning system (GPS) satellite
links as a source of universal coordinated time (UTC). The network *system time* is synchronised to UTC; however, it differs from UTC because the system time does not include the
leap seconds that are added to UTC. The *even seconds* of system time are also important
when we consider frame synchronisation. These represent points in system time when the
number of accumulated seconds is divisible by two, i.e. every other second.

The $2^{15} = 32768 = 512 \times 64$ length PN sequences allow 512 different offsets of 64 chips
from 0 (i.e. zero offset PN sequence) to 511. At switch-on, an MS will sweep a searcher
correlator over all possible pilot PN offsets to identify the different BSs within its local area.
The amplitude of the correlator output will indicate the strength of the BS using a given pilot
PN offset. An example of a searcher correlator output is shown in Figure 4.5, where both
the in-phase (I) and quadrature (Q) outputs are shown. The figure shows that the MS has
identified four strong BSs within the geographical area.

The pilot signal is also used by the MS to provide a coherent reference in the demodulation
of other signals transmitted on the same CDMA carrier. This is possible because the MS
is able to extract the RF carrier phase information from the pilot signal, and this will be
constant for all the channels on a single CDMA RF carrier.

The MS also uses the pilot signals to assess the suitability of neighbouring BSs for handover and, in this respect, the pilot signal is similar to the BCCH carrier in the GSM system.
The MS also uses the pilot channel to estimate what reverse transmitter power it should initially use. This estimate is known as the *open-loop* estimate, and once the MS is in a call, it
will continue to be used in conjunction with a closed-loop power control mechanism to allow more accurate control of the MS transmitted power to be made and over a wide dynamic

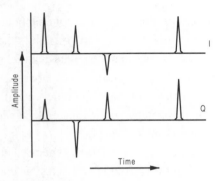

Figure 4.5: An example of a searcher correlator output.

range in the presence of fading.

4.2.2.2 The synchronisation channel

The synchronisation channel carries the information required to allow the MS to synchronise with a given BS. The channel data rate is 1.2 kb/s. The information data on the synchronisation channel, or sync channel, is one-half rate convolutionally encoded using a code with a constraint length of nine defined by the following generator polynomials:

$$g_0 = 1 + D + D^2 + D^3 + D^5 + D^7 + D^8,$$
$$g_1 = 1 + D^2 + D^3 + D^4 + D^8. \tag{4.4}$$

This code has a *minimum free distance* of 12; the interested reader is referred to Reference [3] for a detailed discussion of the operation and performance of this type of forward error correcting code.

This coding process results in a coded symbol rate of 2.4 ksymbols/s. Each symbol is repeated once to produce a symbol rate of 4.8 ksymbols/s. The symbols are then block interleaved over 128 symbols, i.e. over one period of the pilot code of 26.66 ms, and the resulting signal is then EXORed with the Walsh code that has an index of 32 (see Figure 4.1). The Walsh code is generated at a chip rate of 1.2288 Mchips/s and consists of 32 zero chips followed by 32 one chips.

We note that this Walsh code will not effectively spread the data signal over the full band of 1.25 MHz (i.e. 1.2288 Mchips/s) since its polarity changes only twice per 64 chip cycle. To achieve spectral spreading over the channel bandwidth of 1.25 MHz the synchronisation signal is EXORed with both the PNI and PNQ sequences, and the resulting signals are passed through two low pass pulse shaping filters that are identical to those used on the pilot channel. The filtered signals modulate two quadrature carriers following the same phase mapping conventions that were used on the pilot channel (see Table 4.7).

The sync channel will use the same PNI and PNQ offsets as the pilot channel on the same carrier. In this way the MS is able to associate the sync channel with the correct pilot channel, and in turn, with the correct cell. The sync channel carries a 15-bit system identification number (SID) and a 16-bit network identification number (NID). It also carries the pilot PN offset of the cell (PILOT_PN), the contents of the long code generator (LC_STATE, this will be described later) and the system time (SYS_TIME).

As we have already seen, the sync channel data is generated at a rate of 1.2 kb/s or, to be more specific, one frame of 32 bits every 26.66 ms. Each sync channel frame is aligned with the start of the PN sequences, and consequently the MS may acquire the sync channel frame timing information from the pilot channel. The interleaving on the sync channel is also performed over each 26.66 ms frame. Only one message is transmitted on the sync channel; the structure of the *sync channel message* is shown in Figure 4.6. The first eight bits of the message give the length of the message (MSG_LENGTH) in octets. This length will include the 8-bit MSG_LENGTH parameter itself, the message body and a 30-bit checksum. The message body contains the sync channel information (e.g. LC_STATE and SYS_TIME). The sync channel message is protected by a 30-bit cyclic redundancy checksum (CRC) which is appended at the end of the message and is defined by the following generator polynomial:

$$
\begin{aligned}
g(x) \;=\; & x^{30} + x^{29} + x^{21} + x^{20} + x^{15} \\
& + x^{13} + x^{12} + x^{11} + x^{8} + x^{7} + x^{6} + x^{2} + x + 1.
\end{aligned} \tag{4.5}
$$

The CRC is generated for both the 8 MSG_LENGTH bits and the message body. It is used by the MS to check for any errors in the sync channel message that remain uncorrected following the one-half rate convolutional forward error correction (FEC) decoding.

The sync channel message is mapped onto the sync channel frames as shown in Figure 4.6. Each frame consists of a single-bit start-of-message (SOM) flag followed by 31 information bits. The 31 information bits are used to carry the contents of the sync channel message, while the SOM flag is used to indicate the point at which a new message begins. Setting the SOM flag to a '1' indicates that the information contained in the remainder of the frame is the start of a new message. When the SOM flag is set to a '0' this indicates that the information contained in the frame is part of a message that began in an earlier frame.

The sync channel frames are formed into superframes, which consist of three consecutive frames. The superframe is 80 ms in length, as shown in Figure 4.6. A *sync channel message* will always be mapped onto an integer number of sync channel superframes, and consequently a certain amount of padding is added at the end of the message to fill-up the final superframe. This also means that a new sync channel message will only begin at the superframe boundaries. The sync channel superframes are aligned such that, for a zero off-

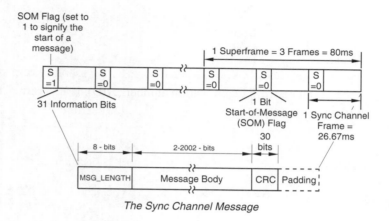

The Sync Channel Message

Figure 4.6: The sync channel structure.

set PN sequence, the start of a superframe will always coincide with the even seconds of system time. In the case of a non-zero PN offset, the start of a superframe will always align with a point equal to the PN offset after the even second marks of system time.

The even seconds of system time, which are called the even second marks in the specifications, are very important since they are used as a timing reference for the PN offsets. The PN sequences of a zero offset pilot will always start on the even second marks. For a pilot with a non-zero PN offset, the PN sequences will start at a time equal to the PN offset after the even second marks. This is shown in Figure 4.7.

The information in the sync channel message is time sensitive, i.e. it is only valid at a specific point in time, and it is important that the MS understands the precise time instant to which the information refers. In the case of a pilot with a zero PN offset, the information contained in the sync channel message will become valid 320 ms, which is equal to four superframe periods, after the end of the last superframe containing a part of the sync channel message. Alternatively, we can say that the LC_STATE and the SYS_TIME parameters contained in the sync channel message refer to a time 320 ms after the last message superframe. Where the pilot PN offset is not zero, the content of the message becomes valid at a time equal to 320 ms minus the PN offset after the last superframe carrying the message. This is shown in Figure 4.8.

4.2.2.3 The paging channel

The paging channel performs a number of different functions in addition to carrying paging messages between the network and an MS. It conveys general system information (e.g. the handover thresholds), access information (e.g. the maximum allowed number of unsuc-

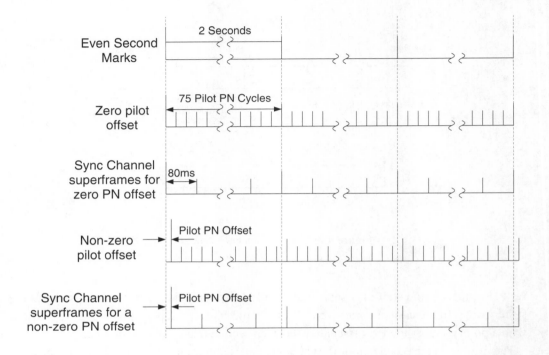

Figure 4.7: Pilot and sync channel timing.

cessful access attempts), a list of the surrounding cells and channel assignment messages. Referring to Figure 4.2, we note that the paging channel information is generated at a data rate of either 9.6 kb/s or 4.8 kb/s. All paging channels within a system use the same data rate. The paging channel data is one-half rate convolutionally encoded using the same code as that employed on the sync channel. This results in a symbol rate of either 19.2 ksymbols/s or 9.6 ksymbols/s, depending on the input data rate.

The code symbols for the lower data rate are repeated once to produce a constant symbol rate of 19.6 ksymbols/s regardless of the input data rate. Where the input data rate is 9.6 kb/s, the repetition process is not performed. The resulting modulation symbols are then block interleaved over a period of 20 ms, which is equivalent to 384 symbols at a rate of 19.2 ksymbols/s. We note that although the paging channel data is formed into 20 ms frames for the purposes of interleaving, the convolutional encoding process treats the data as a continuous bit stream. This means that no encoder tail bits are added between blocks prior to convolutional encoding in order to reset the encoder and the last bits of one block will influence the code symbols for the next block. This is in contrast to the convolutional encoding schemes used in GSM and on the cdmaOne traffic channel which use encoder tail

marks of system time, regardless of the pilot PN offset. This is achieved because the pilot PN offset is defined in units of 64 chips, or one Walsh code cycle. Following Walsh code spreading the data are quadrature spread, using the PNI and PNQ codes, baseband filtered and modulated onto two quadrature carriers using the phase mapping described in Table 4.7. The PNI and PNQ codes have the same offset as the pilot channel and the sync channel on the same CDMA carrier.

Having examined the construction of the paging channels, we now examine their frame structure. The paging channel may carry a number of different messages, e.g. *system parameters message, page message*; however, they all have the same basic format shown in Figure 4.11. The eight-bit MSG_LENGTH field defines the length of the paging channel message in octets, including the MSG_LENGTH field itself, the message body and the checksum. The maximum value of MSG_LENGTH is 148, which allows a maximum message size of 1184 bits. The message body contains the paging channel message information and the last 30 bits of the message are used to carry a cyclic redundancy checksum (CRC) which is generated for the MSG_LENGTH and message body fields. The CRC generator polynomial for the paging channel is the same as that used for the sync channel and is given by Equation (4.5).

When an MS is in the 'idle' mode, it constantly monitors one of the forward paging channels so that it can be alerted to the presence of an incoming call at any time. The paging channel is sub-divided into 80 ms slots and these are formed into maximum length cycles of 2048 slots, which corresponds to a cycle period of 163.84 s. The use of slots on the paging channel allows the system to support a 'slotted paging' mode of operation, whereby an MS is only required to monitor the paging channel within specific slots. This allows an MS to conserve power during periods where it is not required to monitor the paging channel, thereby prolonging battery life. This process is very similar to the discontinuous reception technique (DRX) employed by GSM. The system also supports a 'non-slotted paging' mode whereby an MS is required constantly to monitor a paging channel.

An MS may select its own paging channel slot cycle and this may range from 1.28 s (16 slots) up to the maximum cycle length of 163.84 s (2048 slots). The MS transmits its preferred slot cycle period to the network in the form of the three bit SLOT_CYCLE_INDEX

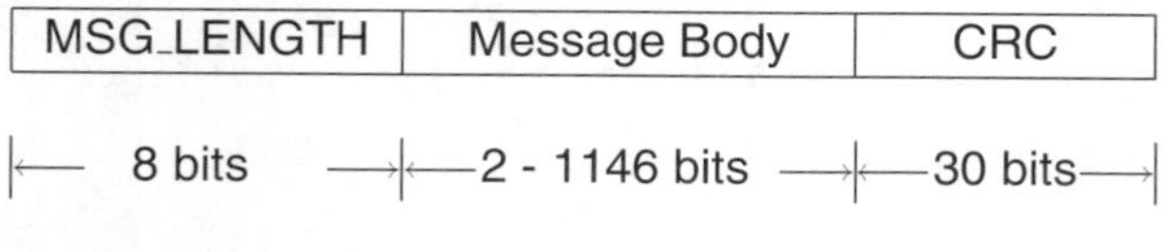

MSG_LENGTH	Message Body	CRC
⟵ 8 bits ⟶	⟵ 2 - 1146 bits ⟶	⟵ 30 bits ⟶

Figure 4.11: The paging channel message format.

parameter. The slot cycle period, T, is given by

$$T = 2^{\text{SLOT\_CYCLE\_INDEX}}, \tag{4.7}$$

where T is in units of 1.28 s, or 16 slots. For example, an MS with a SLOT_CYCLE_INDEX of 2 would monitor the paging channel once every 5.12 s or 64 slots. The MS chooses which slot to monitor within its paging channel cycle based on its mobile identification number (MIN). This is a 34-bit number which is a digital representation of the 10-digit telephone number that is assigned to a particular MS. In this way the MSs within a cell are pseudo-randomly distributed between the paging slots on the paging channel. The MS also uses its MIN to select the paging channel to be monitored in cases where a cell has more than one paging channel. Again the process is pseudo-random and it effectively distributes the MSs evenly between the available paging resources.

Each paging channel slot is composed of four 20 ms frames which are, in turn, composed of two 10 ms half frames, as shown in Figure 4.12. The half frame contains 96 bits, when the paging channel data rate is 9.6 kb/s, and 48 bits when the paging channel rate is 4.8 kb/s. The first bit of each half frame is used to indicate whether the paging messages are synchronised to the half frame boundaries and this bit is known as the synchronised capsule indicator (SCI). We note that the specifications use the word 'capsule' to refer to a message with its associated length indicator, CRC and bit padding. The remaining 95 or 47 bits of the half frames are used to carry the content of the paging channel message. Where the start of a paging message occurs directly after an SCI bit, the SCI bit is set to '1' and the paging message is deemed to be synchronous. In situations where the paging message directly follows the previous message it is deemed to be unsynchronised. Unsynchronised paging messages may only be transmitted when there are eight or more bits left in the half frame. If fewer than eight bits are left following the completion of a paging channel message, or the network decides not to transmit an unsynchronised message, then a number of '0' padding bits will be added to the message to extend it to the end of the half frame. In situations where the paging channel message does not begin on the second bit of a half frame, the SCI bit will be set to '0'. The first message of each paging channel slot will be transmitted in synchronised mode. The paging channel frame structure is shown in Figure 4.12.

The paging channel may carry two different types of paging message depending on the paging mode. The page message will only be transmitted when the system is operating in non-slotted mode, whereas the *slotted page message* is transmitted when the system is operating in the slotted paging mode. Each slotted page message contains a MORE_PAGES flag which, when set to '0', indicates that the particular paging slot contains no more valid paging messages. This allows the MS to stop monitoring the paging slot as soon as possible. In situations where an MS does not receive a slotted page message with a MORE_PAGES bit set to '0' within its chosen paging slot, the MS will continue to monitor the paging

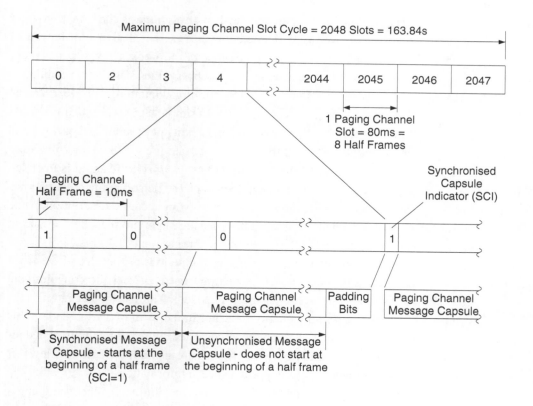

Figure 4.12: The paging channel frame structure.

channel for a further slot. This allows the network to extend its paging calls for a given MS into the slot after the chosen paging slot, when necessary.

4.2.2.4 The traffic channel

All the forward channels that have been considered so far are broadcast channels, conveying information that may be received by each MS within a cell. The traffic channels are used to carry both user traffic and control messages between the network and a specific MS. The traffic channels are termed *dedicated* channels, i.e. the traffic channel is used exclusively by a particular MS. The traffic channel format for the IS-95 and CDMA-PCS systems differ slightly, and we shall commence by describing the IS-95 forward traffic channel.

The IS-95 forward traffic channel conveys traffic data at rates of 1.2 kb/s, 2.4 kb/s, 4.8 kb/s and 9.6 kb/s. The traffic channel has been designed in this manner to support the code excited linear predictive (CELP) speech coder, whose output bit rate varies according to the speech activity. The IS-95 CELP coder is an analysis-by-synthesis coder that provides acceptable speech quality at an average data rate below 8 kb/s. The design and analysis of this type of coder is fairly complex and beyond the scope of this book. However, the reader is referred to Reference [4] for a full treatise on analysis-by-synthesis predictive coding.

The speech signal is encoded into frames of 20 ms duration containing 192, 96, 48 or 24 bits per frame, depending on the user's speech activity. In simple terms, speech information will be coded at the highest data rate (i.e. 9.6 kb/s or 192 bits per frame) and background noise in silence intervals will be coded at the lowest data rate (i.e. 1.2 kb/s or 24 bits per frame). The background noise is sometimes referred to as *comfort noise* and it is intended to reassure the listener that the link has not been lost when the person at the other end has stopped speaking. Since there is no requirement to convey an accurate description of the background noise, it can be encoded using fewer bits than the important speech information. The two intermediate data rates of 4.8 kb/s (96 bits per frame) and 2.4 kb/s (48 bits per frame) are used to provide a smooth transition between the periods of speech and silence.

Each frame contains eight encoder tail bits which are used to reset the convolutional encoder to a known state after the frame has been encoded. The two higher rate frames (9.6 kb/s and 4.8 kb/s) contain a frame quality indicator in the form of a CRC code. This code is used to detect bit errors that have not been corrected by the convolutional decoding process at the receiver. Only the higher rate frames are given this additional level of protection because these will be carrying important speech information.

Referring to Figure 4.2, the traffic channel frames are convolutionally encoded using a one-half rate coder. The code is exactly the same as that used on the synchronisation and paging channels, and the generator polynomials are given in Equation (4.4). The main difference between the convolutional coding on the traffic channels and that used on the paging and sync channels is that the convolutional coder is initialised to the 'all zero' state at the end of each frame in the case of the traffic channels.

The coder output rates will be 19.2 ksymbols/s, 9.6 ksymbols/s, 4.8 ksymbols/s and 2.4 ksymbols/s, depending on the input data rate. The symbols are then repeated to produce a constant symbol rate of 19.2 ksymbols/s, regardless of the input data rate. For example, an input data rate of 1.2 kb/s results in a symbol rate of 2.4 ksymbols/s after convolutional encoding. Each symbol is repeated seven times (i.e. eight copies of the symbol in total) to produce a symbol rate of 19.2 ksymbols/s. Table 4.8 gives the number of symbol repetitions that are required for each input data rate.

The symbols are interleaved over each 20 ms frame, which contains 384 code symbols at a rate of 19.2 ksymbols/s. Following interleaving, the code symbols are scrambled using a 19.2 ksymbols/s scrambling sequence derived from the long PN code using either the public or private long code mask. The construction of the 42-bit public long code mask is shown in Figure 4.13, where ESN signifies the electronic serial number of the MS. This is a 32-bit

Table 4.8: Symbol repetitions on the IS-95 traffic channel.

Input data rate (kb/s)	No. of symbol repetitions
9.6	0
4.8	1
2.4	3
1.2	7

number that is assigned to an MS by the manufacturer and is unique to each MS. The ESN is permuted, or rearranged, to prevent a high correlation between MSs with consecutive ESNs. This is important on the IS-95 reverse link where the long code is used to distinguish the different MSs. The public long code mask may be determined for any MS once the ESN is known, but this does not provide adequate protection against eavesdropping on the radio path. For this reason, although the private long code mask has been defined, its construction is not contained in the specifications and it is only available on a restricted basis. By this strategy the private long code mask provides protection against eavesdropping on the radio path. The 19.2 kb/s scrambling sequence is derived from the 1.2288 Mchips/s long PN code sequence by taking only one bit in each 64 and scrambling is achieved by EXORing the coded symbols with the scrambling sequence.

Following scrambling, the coded data are multiplexed with the power control sub-channel. This sub-channel data are used to adjust the transmitter power of an MS in an attempt to ensure that the received E_b/I_0 is within an acceptable range for good quality communications. The power control sub-channel consists of a single bit every 1.25 ms (i.e. 800 b/s) and this is used to instruct the MS either to increase its transmitter power by 1 dB, if the bit is a logical 0, or to decrease its power by 1 dB, if the bit is a logical 1.

Each 20 ms traffic channel frame on both the reverse link and the forward link is divided into 16×1.25 ms *power control groups*, which contain 24 modulation symbols at 19.2 ksymbol/s. The BS measures the reverse link received power over each power control group, decides whether the MS should increase or decrease its power, and transmits the corresponding power control bit two power control groups later on the forward link traffic channel. This is shown in Figure 4.14. We note that, at the BS, the reverse link and forward

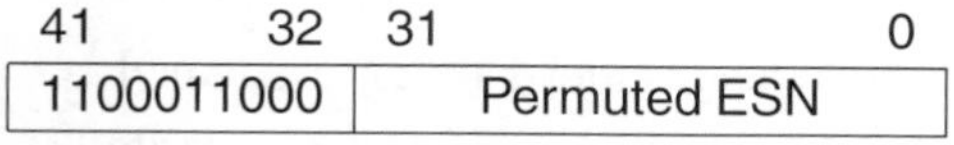

41	32	31		0
	1100011000		Permuted ESN	

Figure 4.13: Traffic channel public long code mask.

link traffic frames are displaced by the round trip propagation delay between the BS and the MS.

Each power control bit has the same duration as two code symbols (i.e. approximately $104\mu s$) and, consequently, it will be used to replace two symbols per power control group. The position of the power control bits is effectively 'randomised' within each power control group to prevent the generation of line spectra in the transmitted signal. Each power control group consists of 24 code symbols and the start of the power control bit may occur at any point within the first 16 symbols. The actual position is defined by the last four bits of the scrambling sequence that was used in the previous power control group. In the example in Figure 4.14, the last four bits of the scrambling sequence are 1101. Taking the last bit (i.e. bit 24) as the most significant, this gives the four-bit number $1011_2 = 11_{10}$, which means that the power control bit starts in position 11 in the power control group and it will replace code symbols 11 and 12.

The process of replacing the code symbols with the power control bits will introduce errors; however, these may be corrected by the powerful one-half rate code. We also note that the MS receiver will always know the position of the power control bits. This technique of removing code symbols is called *puncturing* and a similar technique has been used in GSM on the traffic channels to tailor a convolutional code to the channel bit rate.

Following the insertion of the power control sub-channel data, the traffic channel data are spread using a Walsh code generated at 1.2288 Mchips/s. In IS-95, the traffic channel may use any Walsh code with an index in the range 1 to 63 (see Figure 4.1), since Walsh code zero is reserved for the pilot channel. The spreading process consists of EXORing the traffic channel code symbols with the Walsh code to produce a spread signal at 1.2288 Mchips/s. The signal is then spread in quadrature by the PNI and PNQ sequences and the resulting bit streams are passed through the pulse shaping filters before being used to modulate the CDMA carrier.

In IS-95 the transmitted power of each traffic channel varies according to the channel data rate such that the transmitted energy per bit is always constant. This means that the transmitter power must be reduced in proportion to the number of times a symbol is repeated on each channel. We recall the example of a code symbol being repeated seven times when the channel bit rate is 1.2 kb/s so for this case. In order to maintain a transmitted energy per bit of E_b, each code symbol must be transmitted at a decreased power of $E_s = E_b/16$, as each data bit results in 16 codes symbols, i.e. two symbols from the one-half rate code plus seven repetitions. Table 4.9 shows the relative transmitter power that is used for each data rate. This technique allows the IS-95 system to reduce the interference caused by each user when it is operating below the maximum channel bit rate. Coupled with the variable rate CELP coder, this allows the IS-95 system to exploit the gaps in a user's speech to reduce interference and increase system capacity. However, the power control bits are

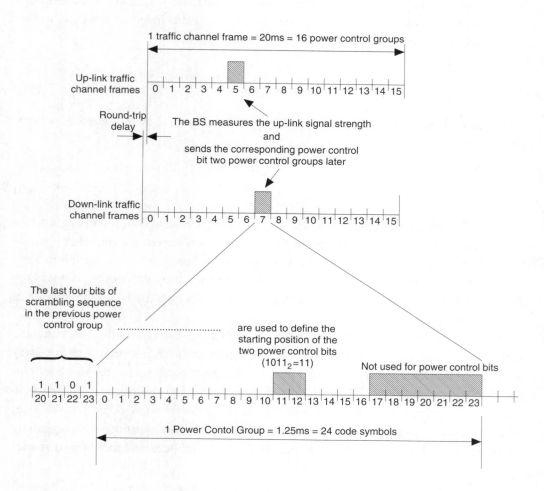

Figure 4.14: The position of the power control bits

always transmitted at full power regardless of the channel bit rate.

Having described the construction of the forward link IS-95 traffic channel on the radio path, we now examine its frame structure. The basic traffic channel frames are shown in Figure 4.15. As we have already discussed, each frame includes eight encoder tails bits. The 9.6 kb/s and the 4.8 kb/s frames also include a frame quality indicator in the form of a CRC code. In the case of the 9.6 kb/s frame, the CRC is 12 bits in length and is defined by the following generator polynomial:

$$g(x) = x^{12} + x^{11} + x^{10} + x^9 + x^8 + x^4 + x + 1. \tag{4.8}$$

The 4.8 kb/s frames carries an eight-bit CRC and this is defined by the generator polynomial

$$g(x) = x^8 + x^7 + x^4 + x^3 + x + 1. \tag{4.9}$$

The CRC performs two functions at the MS receiver. The first, and most obvious function is to allow the MS to check whether the frame has been received in error. However, the CRC also helps the MS to determine the data rate that has been used, since this information is not explicitly transmitted within the frame. The MS effectively decodes the frame at the four different data rates and determines the transmitted data rate based on the number of errors that are produced.

The IS-95 system supports the transmission of signalling data and traffic data within the same frame. It also supports the transmission of primary traffic (e.g. speech) and secondary traffic (e.g. fax data) within the same frame. This feature is referred to as *Multiplex Option 1* in the specifications. The term *dim and burst* is used to describe a situation where signalling (or secondary) traffic is transmitted in the same frame as primary traffic (e.g. speech), whereas the term *blank and burst* is used to described a situation where signalling (or secondary) traffic fills the entire frame. Both blank and burst and dim and burst may

Table 4.9: Relative forward link traffic channel transmitted power.

Data rate (kb/s)	Relative transmitted power
9.6	100%
4.8	50%
2.4	25%
1.2	12.5%

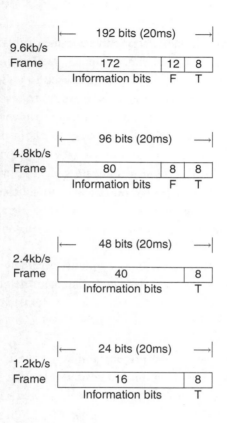

F = The Frame Quality Indicator (CRC)
T = Encoder Tail Bits

Figure 4.15: Forward link traffic channel frames.

only be used in a 9.6 kb/s frame. The 4.8 kb/s, 2.4 kb/s and 1.2 kb/s frames will only carry primary traffic (e.g. speech). Figure 4.16 [1, 2] shows the possible configurations of the 9.6 kb/s traffic frame, where MM is the mixed mode bit, TT is the traffic type bit and TM are the traffic mode bits.

We note that five different options exist whereby the primary traffic (e.g. speech) occupies either all, one-half, one-quarter, one-eighth, or none of the 9.6 kb/s frame. Table 4.10 shows the bit settings for the 9.6 kb/s frame. The MM is used to indicate whether the 9.6 kb/s frame carries both primary *and* signalling (or secondary) traffic. When this bit is set to a '0' it means that the frame carries primary traffic only. However, when the bit is set to a '1' it indicates that the frame carries mixed data, i.e. both primary and signalling (or secondary) traffic. The TT bit is used to show the type of traffic that is sharing the frame with the primary traffic. A '0' is used to indicate signalling traffic and a '1' is used to indicate secondary traffic. Finally, the TM bits are used to indicate the division between the primary traffic and the signalling (or secondary) traffic. For example, the TM bits are set to '00' to indicate that the frame contains 80 primary traffic bits and 88 bits of signalling (or secondary) traffic.

We note that the frames themselves do not explicitly carry any information to show their data rate (i.e. 9.6, 4.8, 2.4 or 1.2 kb/s). However, the rate is determined at the MS receiver by Viterbi decoding the received signal four times, once for each possible data rate. Using the decoder metrics and the CRC, the receiver is able to select the most probable transmitted bit rate [5].

During normal operation, when the primary traffic service is active and the secondary traffic service is not active, the signalling traffic is transmitted using one of the dim and burst formats. When neither the primary nor the secondary services are active, the signalling traffic is transmitted using the blank and burst format. The BS transmits traffic channel data frames with dummy data when there is no signalling traffic to be sent. The specifications state that the support of secondary traffic and other multiplex options are areas for further study.

Having discussed the format of the IS-95 traffic channels, we now consider the format of the CDMA-PCS traffic channels. The first point to note is that everything we have already said regarding the IS-95 traffic channels applies to the CDMA-PCS traffic channels. The main difference between the two systems is that the CDMA-PCS system has an additional mode of operation whereby the traffic channels can support bit rates of up to 14.4 kb/s. The CDMA-PCS specifications [2] refer to the 9.6 kb/s family of data rates as *Rate Set 1* and the new family of data rates as *Rate Set 2*. Rate Set 2 supports transmission at 1.8 kb/s, 3.6 kb/s, 7.2 kb/s and 14.4 kb/s and this allows the CDMA-PCS system to support a higher quality 13 kb/s speech codec.

The higher data rates of Rate Set 2 are achieved by lowering the effective power of the

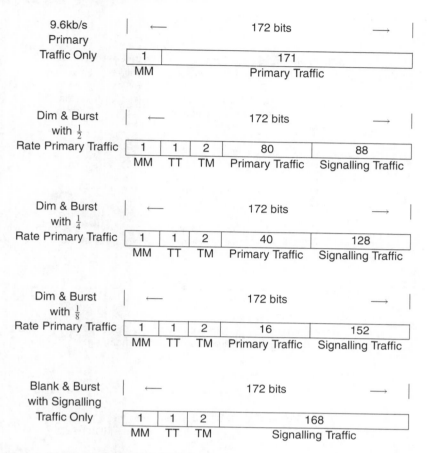

Figure 4.16: 9.6 kb/s traffic channel frame formats: MM=mixed mode, TT=traffic type and TM=traffic mode.

convolutional code from the one-half rate code used in IS-95 (and Rate Set 1) to a three-quarters rate code. This code rate reduction is produced by using the same one-half rate code described in Equation (4.4), but by removing, or 'puncturing', two out of every six code symbols. The puncturing process occurs after symbol repetition. The block diagram of the CDMA-PCS Rate Set 2 traffic channel is shown in Figure 4.17.

Figure 4.18 shows the traffic channel frame structures for Rate Set 2. This shows that every frame in Rate Set 2 carries a frame quality indicator in the form of a CRC. The size of the CRC varies from six bits, for the 1.8 kb/s frame, up to 12 bits, for the 14.4 kb/s frame. The generator polynomials for the frame quality indicators are given below:

$$g(x) = x^{12} + x^{11} + x^{10} + x^9 + x^8 + x^4 + x + 1, \quad \text{for the 12-bit CRC,} \quad (4.10)$$
$$g(x) = x^{10} + x^9 + x^8 + x^7 + x^6 + x^4 + x^3 + 1, \quad \text{for the 10-bit CRC,} \quad (4.11)$$

Table 4.10: Bit settings for the traffic channel frames.

Transmit rate (b/s)	Mixed mode (MM)	Traffic type (TT)	Traffic mode (TM)	Primary traffic (bits/ frame)	Signalling traffic (bits/ frame)	Secondary traffic (bits/ frame)
	0	-	-	171	0	0
	1	0	00	80	88	0
	1	0	01	40	128	0
9600	1	0	10	16	152	0
	1	0	11	0	168	0
	1	1	00	80	0	88
	1	1	01	40	0	128
	1	1	10	16	0	152
	1	1	11	0	0	168
4800	-	-	-	80	0	0
2400	-	-	-	40	0	0
1200	-	-	-	16	0	0

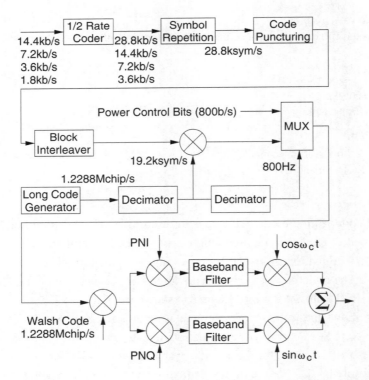

Figure 4.17: The Rate set 2 traffic channel configuration (CDMA-PCS).

$$g(x) = x^8 + x^7 + x^4 + x^3 + x + 1, \qquad \text{for the 8-bit CRC}, \qquad (4.12)$$

$$g(x) = x^6 + x^2 + x + 1, \qquad \text{for the 6-bit CRC}. \qquad (4.13)$$

The CDMA-PCS system supports the Multiplex Option 1 described above using Rate Set 1. However, it also supports a second service option, Multiplex Option 2, using Rate Set 2. Table 4.11 shows the manner in which the frame is allocated to primary, secondary or signalling traffic in Multiplex Option 2. We can see that the 14.4 kb/s frames consist of a reserved bit followed by an MM bit, which is used to indicate whether the frame is used to carry more than one type of traffic data (i.e. primary, secondary or signalling). In frames where the MM bit is set to a '1', the next four bits form the frame mode (FM) field and they are used to indicate the division between the different data types. The lower rate frames have a similar construction, however they do not include a reserved bit. Observe that Multiplex Option 2 supports the transmission of the three different data types within the same 14.4 kb/s or 7.2 kb/s frame.

4.2.2.5 The cdmaOne forward link channel arrangement

Having discussed the construction of each of the forward link IS-95 channels, we now examine the manner in which they are combined at the BS. As we have already seen, each forward link channel is assigned one of 64 different Walsh codes, giving a maximum of 64 forward link channels. WH code 0 (see Figure 4.1) is always used for the pilot channel on each CDMA carrier and every carrier always supports a pilot channel. The remaining 63 Walsh codes are distributed among the synchronisation (sync), paging and traffic channels. The sync channel, when present, always uses WH code 32 (see Figure 4.1). The specifications allow the possibility of a CDMA carrier without a sync channel; however, it is difficult to see where a carrier of this nature would be used. Each carrier may support up to a maximum of seven paging channels using WH codes one to seven. Each CDMA carrier may also support between 55 and 63 traffic channels, where a traffic channel can replace a paging channel or the sync channel on a one-for-one basis. Figure 4.19 summarises the possible forward link channel configuration for a single CDMA carrier. In situations where the allocated spectrum supports more than one cdmaOne carrier, the network operator can increase the system capacity by installing more than one CDMA carrier at each BS.

4.2.2.6 The MS receiver

The IS-95 and CDMA-PCS specifications [1,2] are essentially air-interface descriptions and give details of the formulation of the radio signals from the point of view of the transmitter. Unfortunately, the specifications provide scant details regarding the implementation of the cdmaOne receivers, except that they should perform the reverse operations to those at the

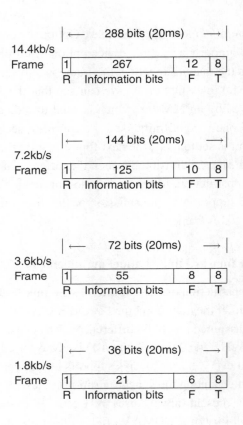

F = The Frame Quality Indicator (CRC)
T = Encoder Tail Bits
R = Reserved Bits

Figure 4.18: Forward link traffic channel frames (Rate Set 2).

Table 4.11: Bit settings for traffic channel frames (Rate Set 2).

| Transmit rate (bits/sec) | Format bits | | | Primary traffic (bits/frame) | Signalling traffic (bits/frame) | Secondary traffic (bits/frame) |
	Reserved (bits/frame)	Mixed mode (MM)	Frame mode (FM)			
	1	0	-	265	0	0
	1	1	0000	124	137	0
	1	1	0001	54	207	0
	1	1	0010	20	241	0
14400	1	1	0011	0	261	0
	1	1	0100	124	0	137
	1	1	0101	54	0	207
	1	1	0110	20	0	241
	1	1	0111	0	0	261
	1	1	1000	20	221	20
	0	0	-	124	0	0
	0	1	000	54	67	0
	0	1	001	20	101	0
	0	1	010	0	121	0
7200	0	1	011	54	0	67
	0	1	100	20	0	101
	0	1	101	0	0	121
	0	1	110	20	81	20
	0	0	-	54	0	0
	0	1	00	0	52	0
3200	0	1	01	0	0	52
	0	1	10	20	32	0
	0	1	11	20	0	32
1800	0	0	-	20	0	0
	0	1	-	0	0	20

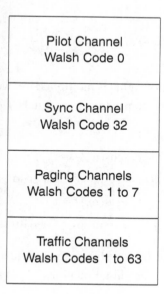

Figure 4.19: Forward link channel arrangement for IS-95.

transmitter. This allows the equipment manufacturers the freedom to use a range of different techniques and technologies at the receiver to improve the system performance. In this section we will describe some of the main features of the cdmaOne mobile station receiver; however, we stress that this discussion is very much based on the opinions and views of the authors and not the specifications themselves.

The MS searcher   The MS receiver must have some means of detecting and identifying the BSs within its locality and this is achieved using the forward link pilot channel which, as we have already seen, must be transmitted at a constant power by every cdmaOne BS. The sub-system within the MS which is used to detect BSs is called a searcher and it is shown in block diagram form in Figure 4.20. The pilot channel consists of an RF carrier whose in-phase and quadrature components have been BPSK modulated by the 2^{15} length PNI and PNQ sequences, respectively. At the transmitter the pilot signal, $p(t)$, is given by

$$p(t) = i(t)\cos(\omega_c t) + q(t)\sin(\omega_c t), \tag{4.14}$$

where $i(t)$ and $q(t)$ are the pulse shaped PNI and PNQ sequences, respectively, and ω_c is the CDMA angular carrier frequency. Suppose a pilot signal experiences a propagation delay of τ_p seconds and an attenuation, α, as it travels to the MS. The pilot signal that arrives at the MS is given by

$$r(t) = \alpha[i(t - \tau_p)\cos(\omega_c(t - \tau_p)) + q(t - \tau_p)\sin(\omega_c(t - \tau_p))] \tag{4.15}$$

assuming a *single* path channel.

At the receiver, the received pilot signal is multiplied by two locally generated quadrature carriers with angular frequency ω_c to produce two signals, $r_I(t)$ and $r_Q(t)$, given by

$$
\begin{aligned}
r_I(t) &= [i(t - \tau_p)\cos(\omega_c(t - \tau_p)) \\
&\quad + q(t - \tau_p)\sin(\omega_c(t - \tau_p))]\alpha\cos(\omega_c t + \phi), \tag{4.16} \\
r_Q(t) &= [i(t - \tau_p)\cos(\omega_c(t - \tau_p)) \\
&\quad + q(t - \tau_p)\sin(\omega_c(t - \tau_p))]\alpha\sin(\omega_c t + \phi), \tag{4.17}
\end{aligned}
$$

where ϕ represents the phase of the locally generated quadrature carriers relative to the carrier phase at the transmitter. Expanding Equations (4.16) and (4.17) gives

$$
\begin{aligned}
r_I(t) &= \frac{\alpha i(t - \tau_p)}{2}[\cos(2\omega_c t + \phi - \omega_c \tau_p) + \cos(\phi + \omega_c \tau_p)] \\
&\quad + \frac{\alpha q(t - \tau_p)}{2}[\sin(2\omega_c t + \phi - \omega_c \tau_p) - \sin(\phi + \omega_c \tau_p)],
\end{aligned}
$$

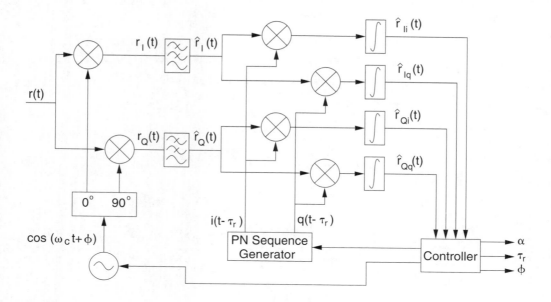

Figure 4.20: Block diagram of an MS searcher.

$$(4.18)$$

$$r_Q(t) = \frac{\alpha i(t - \tau_p)}{2}[\sin(2\omega_c t + \phi - \omega_c \tau_p) + \sin(\phi + \omega_c \tau_p)]$$
$$+ \frac{\alpha q(t - \tau_p)}{2}[-\cos(2\omega_c t + \phi - \omega_c \tau_p) + \cos(\phi + \omega_c \tau_p)],$$

$$(4.19)$$

and low pass filtering $r_I(t)$ and $r_Q(t)$ to remove the '$2\omega_c t$' component gives

$$\hat{r}_I(t) = \frac{\alpha i(t - \tau_p)}{2}\cos(\phi + \omega_c \tau_p)$$
$$- \frac{\alpha q(t - \tau_p)}{2}\sin(\phi + \omega_c \tau_p), \qquad (4.20)$$

$$\hat{r}_Q(t) = \frac{\alpha i(t - \tau_p)}{2}\sin(\phi + \omega_c \tau_p)$$
$$+ \frac{\alpha q(t - \tau_p)}{2}\cos(\phi + \omega_c \tau_p). \qquad (4.21)$$

The two signals $\hat{r}_I(t)$ and $\hat{r}_Q(t)$ are then correlated with locally generated and delayed versions of the PNI and PNQ codes to produce four signals, $\hat{r}_{Ii}(t)$, $\hat{r}_{Iq}(t)$, $\hat{r}_{Qi}(t)$ and $\hat{r}_{Qq}(t)$, given by

$$
\hat{r}_{Ii}(t) = \frac{\alpha\cos(\phi + \omega_c\tau_p)}{2}\int_0^T i(t-\tau_p)i(t-\tau_r)dt
$$
$$
-\frac{\alpha\sin(\phi + \omega_c\tau_p)}{2}\int_0^T q(t-\tau_p)i(t-\tau_r)dt, \tag{4.22}
$$

$$
\hat{r}_{Qi}(t) = \frac{\alpha\sin(\phi + \omega_c\tau_p)}{2}\int_0^T i(t-\tau_p)i(t-\tau_r)dt
$$
$$
+\frac{\alpha\cos(\phi + \omega_c\tau_p)}{2}\int_0^T q(t-\tau_p)i(t-\tau_r)dt, \tag{4.23}
$$

$$
\hat{r}_{Iq}(t) = \frac{\alpha\cos(\phi + \omega_c\tau_p)}{2}\int_0^T i(t-\tau_p)q(t-\tau_r)dt
$$
$$
-\frac{\alpha\sin(\phi + \omega_c\tau_p)}{2}\int_0^T q(t-\tau_p)q(t-\tau_r)dt, \tag{4.24}
$$

$$
\hat{r}_{Qq}(t) = \frac{\alpha\sin(\phi + \omega_c\tau_p)}{2}\int_0^T i(t-\tau_p)q(t-\tau_r)dt
$$
$$
+\frac{\alpha\cos(\phi + \omega_c\tau_p)}{2}\int_0^T q(t-\tau_p)q(t-\tau_r)dt, \tag{4.25}
$$

where T represents the period of the correlation and τ_r represents the delay in the locally generated versions of $i(t)$ and $q(t)$. Assuming the perfect case where the cross correlation of $i(t)$ and $q(t)$ is zero, i.e.

$$
\int_0^T i(i-\tau_i)q(t-\tau_k)dt = 0, \quad \text{for any } \tau_i, \tau_k, \tag{4.26}
$$

Equations (4.22)–(4.25) become

$$
\hat{r}_{Ii}(t) = \frac{\alpha\cos(\phi + \omega_c\tau_p)}{2}\int_0^T i(t-\tau_p)i(t-\tau_r)dt, \tag{4.27}
$$

$$
\hat{r}_{Qi}(t) = \frac{\alpha\sin(\phi + \omega_c\tau_p)}{2}\int_0^T i(t-\tau_p)i(t-\tau_r)dt, \tag{4.28}
$$

$$
\hat{r}_{Iq}(t) = \frac{-\alpha\sin(\phi + \omega_c\tau_p)}{2}\int_0^T q(t-\tau_p)q(t-\tau_r)dt, \tag{4.29}
$$

$$
\hat{r}_{Qq}(t) = \frac{\alpha\cos(\phi + \omega_c\tau_p)}{2}\int_0^T q(t-\tau_p)q(t-\tau_r)dt. \tag{4.30}
$$

We shall also assume that the autocorrelation functions of both $i(t)$ and $q(t)$, R_i and R_q, are perfect such that

$$
\int_0^T i(t-\tau_i)i(t-\tau_k) = 0, \quad \text{for } \tau_i \neq \tau_k, \tag{4.31}
$$

$$\int_0^T q(t - \tau_i)q(t - \tau_k) = 0, \quad \text{for } \tau_i \neq \tau_k, \tag{4.32}$$

$$\frac{1}{T}\int_0^T i(t - \tau_i)i(t - \tau_i) = 1, \tag{4.33}$$

$$\frac{1}{T}\int_0^T q(t - \tau_i)q(t - \tau_i) = 1, \tag{4.34}$$

and therefore

$$\hat{r}_{Ii}(t) = \hat{r}_{Qi}(t) = \hat{r}_{Iq}(t) = \hat{r}_{Qq}(t) = 0, \quad \text{for } \tau_r \neq \tau_p, \tag{4.35}$$

and

$$\hat{r}_{Ii}(t) = \frac{\alpha\cos(\phi + \omega_c\tau_p)T}{2}, \tag{4.36}$$

$$\hat{r}_{Qi}(t) = \frac{\alpha\sin(\phi + \omega_c\tau_p)T}{2}, \tag{4.37}$$

$$\hat{r}_{Iq}(t) = \frac{-\alpha\sin(\phi + \omega_c\tau_p)T}{2}, \tag{4.38}$$

$$\hat{r}_{Qq}(t) = \frac{\alpha\cos(\phi + \omega_c\tau_p)T}{2}, \tag{4.39}$$

when $\tau_r = \tau_p$.

The receiver is able to determine the propagation delay, the attenuation and the phase shift in the channel from Equations (4.37)–(4.39). With the perfect cross-correlation condition of Equation (4.26) and perfect autocorrelation conditions of Equations (4.31)–(4.34), the values of $\hat{r}_{Ii}(t)$, $\hat{r}_{Qi}(t)$, $\hat{r}_{Iq}(t)$ and $\hat{r}_{Qq}(t)$ will only be non-zero when the value of τ_r equals the propagation delay τ_p. Therefore the receiver is able to detect the BSs by sweeping τ_r over a range of possible values and noting the values of τ_r that cause $\hat{r}_{Ii}(t)$, $\hat{r}_{Qi}(t)$, $\hat{r}_{Iq}(t)$ and $\hat{r}_{Qq}(t)$ to become non-zero.

The receiver can determine the phase shift introduced by the channel by adjusting the phase of its locally generated carrier signal, ϕ, such that $\hat{r}_{Qi}(t) = \hat{r}_{Iq}(t) = 0$. This means that ϕ will be equal to the phase difference introduced by the channel but with opposite polarity (i.e. $-\omega_c\tau_p$) and

$$\hat{r}_{Ii}(t) = \hat{r}_{Qq}(t) = \frac{\alpha}{2}. \tag{4.40}$$

By measuring the amplitude of $\hat{r}_{Ii}(t)$ and $\hat{r}_{Qq}(t)$, the MS is able to determine the attenuation, α, caused by the channel.

This analysis is a gross over-simplification of the situation in an actual cdmaOne system. It has been introduced to demonstrate the basic principles of pilot detection. In a full cdmaOne system the MS will be surrounded by a number of BSs, each transmitting its own pilot signal and these signals will all arrive at the input to the searcher. We have already seen that the BSs in the cdmaOne system are distinguished by the offset of their pilot PN sequences (i.e. PNI and PNQ) and therefore we must add a second factor, τ_{PNn}, into the description of the pilot channel to represent the PN offset of the nth BS. Therefore Equation (4.15) becomes

$$
r(t) = \sum_{j=1}^{B} \alpha_j i(t - \tau_{PNj} - \tau_{pj}) \cos(\omega_c(t - \tau_{pj}))
$$

$$
+ \sum_{j=1}^{B} \alpha_j q(t - \tau_{PNj} - \tau_{pj}) \sin(\omega_c(t - \tau_{pj})), \tag{4.41}
$$

where τ_{pj} represents the propagation delay from the jth BS to the MS, α_j is the attenuation experienced by the signal from the jth BS and B is the total number of BSs in the vicinity of the MS.

With multiple pilot signals arriving at the MS receiver, the values of $\hat{r}_{Ii}(t)$, $\hat{r}_{Qi}(t)$, $\hat{r}_{Iq}(t)$ and $\hat{r}_{Qq}(t)$ will become non-zero each time the value of τ_r becomes equal to $\tau_{PNj} + \tau_{pj}$, assuming perfect cross-correlation (see Equation (4.26)) and autocorrelation (see Equations (4.31)–(4.34)). In this way the MS is able to identify the pilot signals from a number of different BSs, to calculate the phase shift and attenuation introduced by the channel, and to form the sum of the pilot PN offset (τ_{PNj}) and the propagation delay (τ_{pj}) for each pilot signal.

Up to this point we have assumed that the pilot signal arriving from the jth BS is composed of a single path with attenuation α_j and phase shift $\omega_c \tau_{pj}$. Generally, there will be a number of paths between the jth BS and the MS each with an attenuation of α_{jk} and phase offset θ_{jk} for the kth path. Therefore Equation (4.15) becomes

$$
r(t) = \sum_{j=1}^{B} \sum_{k=1}^{K_j} \alpha_{jk} i(t - \tau_{PNj} - \tau_{pjk}) \cos(\omega_c t + \theta_{jk})
$$

$$
+ \sum_{j=1}^{B} \sum_{k=1}^{K_j} \alpha_{jk} q(t - \tau_{PNj} - \tau_{pjk}) \sin(\omega_c t + \theta_{jk}), \tag{4.42}
$$

where τ_{pjk} represents the propagation delay for the pilot signal arriving on the kth path from the jth BS, and K_j is the total number of paths arriving from the jth BS. When this signal is applied to the input of the searcher, the values of $\hat{r}_{Ii}(t)$, $\hat{r}_{Qi}(t)$, $\hat{r}_{Iq}(t)$ and $\hat{r}_{Qq}(t)$ will become non-zero whenever τ_r is equal to $\tau_{PNj} + \tau_{pjk}$. In this way the MS is able to identify the individual paths arriving from each BS and it also determines the attenuation and phase shift of each path.

Our analysis has been simplified by the assumption that the PNI and PNQ sequences display perfect cross-correlation (see Equation (4.26)) and autocorrelation (see Equations (4.31)–(4.34)) properties. This is not so in the practical cdmaOne systems and a certain amount of cross-correlation and autocorrelation noise will be present in the values $\hat{r}_{Ii}(t)$, $\hat{r}_{Qi}(t)$, $\hat{r}_{Iq}(t)$ and $\hat{r}_{Qq}(t)$. We also note that the searcher will also receive interference from the traffic channels that are in use in the surrounding cells and this will also manifest itself as noise in the values of $\hat{r}_{Ii}(t)$, $\hat{r}_{Qi}(t)$, $\hat{r}_{Iq}(t)$ and $\hat{r}_{Qq}(t)$. This noise will lead to errors in the calculation of α_{jk}, $\tau_{PNj} + \tau_{pj}$ and θ_{jk}.

The magnitude of the noise contaminating $\hat{r}_{Ii}(t)$, $\hat{r}_{Qi}(t)$, $\hat{r}_{Iq}(t)$ and $\hat{r}_{Qq}(t)$ depends to a certain degree on the value of the integration period of the correlation, T. Integrating over the entire 2^{15} length PN sequence (i.e. $T = 26.666$ ms) tends to produce less noise than performing a partial correlation over, say, 64 chips. However, there is a trade-off in the size of T. In addition to detecting the surrounding BSs, the searcher is also used to identify the best forward link paths while the MS is using a traffic channel. As the MS moves, the received paths will fade and the MS must constantly ensure that it is decoding the best paths. The value of T is the period during which the MS is able to update the parameters it has calculated for each path, and a smaller T allows the MS to react faster to changes in the channel. The equipment manufacturer must ensure that T is large enough to keep the errors in $\hat{r}_{Ii}(t)$, $\hat{r}_{Qi}(t)$, $\hat{r}_{Iq}(t)$ and $\hat{r}_{Qq}(t)$ below acceptable levels, but small enough to allow the MS to respond to changes in the channel.

Once the MS has detected a BS it attempts to decode the information contained on the corresponding synchronisation (sync) channel. Since the sync channel and the pilot channel signals pass through the same radio channel, the channel coefficients calculated for the pilot channel will also be valid for the sync channel. This means that the MS knows the value of τ_r for each path in the received sync channel signal. The MS receiver is equipped with at least three receiving elements and these will be locked to the three best (i.e. strongest) paths in the received signal, as determined by the searcher.

Each receiving element recovers the sync channel data from the chosen path by correlating the received signal not only with PNI and PNQ, as described above, but also WH code 32 (see Figure 4.1). In the absence of noise, the result of this correlation will be positive when a logical '0' is transmitted and negative when a logical '1' is transmitted. Since the MS has determined the attenuation and phase shift imposed by each path it is able to perform maximal ratio combining on the signals received from each path. Having recovered the coded symbols, the MS performs de-interleaving and convolutional decoding to recover the sync channel information.

The MS receiver functions in a similar manner when it is used to recover a paging or traffic channel; however, the Walsh code in the correlation process will be selected according to the forward link channel to be received. The device used at the MS receiver to extract the information from each path is called a RAKE receiver and we examine the basic theory behind its operation in the next section.

The RAKE receiver   The RAKE receiver was first described by Price and Green [6] in 1958 and the reader is referred to Proakis [7] for a detailed description and analysis of the RAKE receiver. The RAKE receiver is used to achieve path diversity in a multipath fading channel by effectively collecting the signal energy from each received path using a delay line at the receiver. (This has been likened to the garden rake, hence the name.) The different signal paths are combined using a maximal ratio combining scheme whereby the signal on each path is weighted according to its received power prior to combining.

Considering a basic, single user CDMA system where the data is transmitted in the form of a spreading code $c(t)$, such that a logical '1' is represented by the code $c_1(t) = c(t)$ and a logical '0' is represented by the inverse of the code, $c_2(t) = -c_1(t) = \bar{c}(t)$, we will assume that the codes have a highly peaked autocorrelation function such that

$$\int_0^T c_m\left(t - \frac{k}{W}\right) c_m\left(t - \frac{l}{W}\right) dt \approx 0, \qquad m = 1, 2 \text{ and } k \neq l, \tag{4.43}$$

and

$$\int_0^T c_m\left(t - \frac{k}{W}\right) c_m\left(t - \frac{k}{W}\right) dt = T, \qquad m = 1, 2, \tag{4.44}$$

where T is the period of a bit (i.e. the length of the code) and W is the bandwidth of the transmitted signal (i.e. $1/W = T_c$, the chip period). The codes $c_1(t)$ and $c_2(t)$ are also antipodal such that

$$\int_0^T c_1\left(t - \frac{k}{W}\right) c_2\left(t - \frac{k}{W}\right) dt = -T. \tag{4.45}$$

The multipath channel may be modelled by a delay line with delay blocks of $T_c = 1/W$ and a number of complex taps, $a_k(t)$, for each signal path, as shown in Figure 4.21. Suppose the excess path delay in the radio channel is T_m, then the channel model has L taps equal to

$$L = T_m W + 1. \tag{4.46}$$

The channel tap coefficients, $a_k(t)$, are time variant; however, we may assume that the taps are constant over a bit period, T. Therefore $a_k(t)$ may be represented as a_k where we consider each bit period individually.

The received baseband signal, $r_b(t)$, is given by

$$r_b(t) = \sum_{k=1}^{L} a_k c_m\left(t - \frac{k}{W}\right) + z(t), \qquad 0 \leq t \leq T, \tag{4.47}$$

where $z(t)$ is a complex-valued zero mean white Gaussian noise process. At the receiver the objective is to extract the signal energy in each path, weight it according to its received

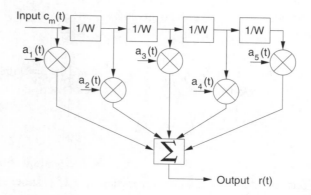

Figure 4.21: Wideband baseband channel model.

power and sum the result to produce a decision parameter to determine the transmitted data bit. Now each path introduces a different phase shift which must be removed to ensure that the various paths are co-phased and thereby summed constructively. Both the signal weighting and the phase shift normalisation may be achieved by multiplying each path by the complex conjugate of its channel tap, i.e. a_k^*.

To demonstrate this we express the channel taps, a_k, in polar form such that

$$a_k = \alpha_k e^{-j\theta_k}, \tag{4.48}$$

where α_k is the amplitude of the kth tap and θ_k is the phase shift of the kth tap. The received baseband signal from the kth tap, $r_k(t)$, will be given by

$$r_k(t) = c_m(t) a_k e^{-j\theta_k}, \qquad m = 1,2. \tag{4.49}$$

Multiplying the received signal by $a_k^*(t)$ gives

$$
\begin{aligned}
s_k(t) &= c_m(t)\alpha_k e^{-j\theta_k}\alpha_k e^{j\theta_k}, & m &= 1,2, \\
&= c_m(t)\alpha_k^2, & m &= 1,2.
\end{aligned}
\tag{4.50}
$$

This shows that the received signal in the kth path has been weighted by the amplitude of the channel tap, α_k, and the phase change, θ_k, has been removed.

The individual received signal paths are recovered by passing the received signal through a delay line and multiplying each delayed replica of the signal by the complex conjugate of the code $c_1(t)$, i.e. $c_1^*(t)$. Alternatively we could use the code $c_2^*(t) = -c_1^*(t)$; however, we would have to invert the decision variable. In our example both $c_1(t)$ and $c_2(t)$ are real, and consequently $c_1(t) = c_1{}^*(t)$ and $c_2(t) = c_2{}^*(t)$.

At the receiver, the decision variable, U, is given by

$$
U = \text{Re}\left[\sum_{k=1}^{L} a_k^* \sum_{n=1}^{L} a_n \int_0^T c_m\left(t - \frac{n}{W}\right) c_1^*\left(t - \frac{k}{W}\right) dt\right]
$$
$$
+ \text{Re}\left[\sum_{k=1}^{L} a_k^* \int_0^T z(t) c_1^*\left(t - \frac{k}{W}\right) dt\right], \qquad m = 1, 2.
$$

(4.51)

The first term in Equation (4.51) is only non-zero where $n = k$, as shown in Equation (4.43), and so the equation may be simplified to give

$$
U = \text{Re}\left[\sum_{k=1}^{L} |a_k|^2 \int_0^T c_m\left(t - \frac{k}{W}\right) c_1^*\left(t - \frac{k}{W}\right) dt\right]
$$
$$
+ \text{Re}\left[\sum_{k=1}^{L} a_k^* \int_0^T z(t) c_1^*\left(t - \frac{k}{W}\right) dt\right], \qquad m = 1, 2. \qquad (4.52)
$$

In the case of antipodal signals which satisfy the property given in Equation (4.45), the first term in the right-hand side of Equation (4.52) will always be ± 1, depending on the transmitted signal, and Equation (4.52) may be simplified to

$$
U = \begin{cases} \text{Re}(2\varepsilon \sum_{k=1}^{L} |a_k|^2 + \sum_{k=1}^{L} |a_k| N_k), & m = 1, \\[2mm] -\text{Re}(2\varepsilon \sum_{k=1}^{L} |a_k|^2 + \sum_{k=1}^{L} |a_k| N_k), & m = 2, \end{cases}
$$

(4.53)

where ε is defined by

$$
|c_1(t)| = |c_2(t)| = \sqrt{\frac{2\varepsilon}{T}}, \qquad (4.54)
$$

N_k is given by

$$
N_k = e^{j\phi_k} \int_0^T z(t) c_1^*\left(t - \frac{k}{W}\right) dt, \qquad (4.55)
$$

and ϕ_k is the phase shift imposed by a_k^*. The bit decision is made on the basis of the polarity of U such that where $U > 0$, the transmitted data bit is a logical '1' and where $U \leq 0$, the transmitted data bit is a logical '0'. If we had used the code $c_2(t)$ the decision variable U would have been inverted, i.e. positive = '0' and negative = '1'.

We may explain the operation of the RAKE receiver with reference to Figure 4.22 and Table 4.12. In Figure 4.22 the received baseband signal, $r(t)$, is passed through a tapped delay line. In our example we have chosen five taps which would allow us to extract a maximum of five paths. Each delay block represents a single chip delay, $T_c = 1/W$, and the signals from each tap are multiplied by a locally generated version of the code sequence, $c_1(t)$. For exemplary purposes we have assumed that the code sequence is eight chips in length (bits 0 to 7) and the input signal, $r(t)$, contains three distinct paths, with path 1 arriving first, path 2 arriving two chip periods later, and path 3 arriving a further two chip periods later. This is shown in Table 4.12 where point A represents the input to the RAKE receiver (see Figure 4.22). The table shows that, at time $= 0$, the zeroth chip of the sequence is received from path 1, the sixth chip of the sequence is received from path 2 and the fourth chip of the sequence is received from path 3. The table also shows the phasing of the sequences on each path at different points within the delay line (i.e. points B to E).

If we ignore the use of the channel taps, $a_k^*(t)$, for the moment, the multiplication by $c_1(t)$ and integration over a bit period, T, represents a correlation process. The result of the correlation (i.e. the output of the integrator) will be zero when the locally generated code sequence, $c_1(t)$, does not align with the code sequences in any path, due to the highly peaked autocorrelation properties of the code $c_1(t)$. However, in the case where $c_1(t)$ does align with the code sequence in a particular path, the output of the integrator will be non-zero. Referring to Table 4.12 we can see that the locally generated version of $c_1(t)$ (point F) will align with the code sequence in path 1 at point E, the code sequence in path 2 at point C and the code sequence in path 3 at point A.

This shows that the RAKE receiver has effectively separated the received signal, $r(t)$, into its component paths by correlating with delayed versions of $r(t)$. The factors $a_k^*(t)$ are used to provide maximal ratio combining of the signals in each path. Multiplying each path by the complex conjugate of its corresponding channel model tap weight has the effect of removing any phase differences between the different paths and also weighting each path according to its received power, as discussed previously. This allows the outputs of each integrator to be summed to produce a single decision variable, U. The received bit is considered to be a logical '1' if U is positive and a logical '0' if U is negative. Observe that the position of the summation block and the integrator blocks may be reversed such that the signals from each tap are summed and then integrated over the bit period, T, allowing the use of a single integrator.

The system shown in Figure 4.22 represents one possible implementation of the RAKE receiver, where a number of delayed replicas of the received signal are produced. The integration period will commence at exactly the same time for each tap and, consequently, the results of the correlation of each path will be available simultaneously, removing the requirement to store the results of some correlations until every correlation is complete. Another implementation of the RAKE receiver is where a single version of the received signal is correlated with a number of different delayed replicas of the locally generated

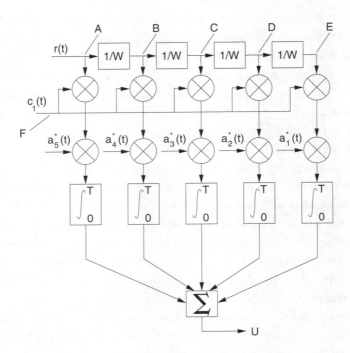

Figure 4.22: The RAKE receiver (with delayed input signal).

Table 4.12: Signals for the RAKE receiver (delayed input version), where points A, B, C, D and E correspond to those in Figure 4.22.

Time →

		0	1	2	3	4	5	6	7	8	9	10	11	12	13	14	15	16	17	18	19	20	21	22	23
	Path 1	0	1	2	3	4	5	6	7	0	1	2	3	4	5	6	7	0	1	2	3	4	5	6	7
Point A	Path 2	6	7	0	1	2	3	4	5	6	7	0	1	2	3	4	5	6	7	0	1	2	3	4	5
	Path 3	4	5	6	7	0	1	2	3	4	5	6	7	0	1	2	3	4	5	6	7	0	1	2	3
	Path 1	7	0	1	2	3	4	5	6	7	0	1	2	3	4	5	6	7	0	1	2	3	4	5	6
Point B	Path 2	5	6	7	0	1	2	3	4	5	6	7	0	1	2	3	4	5	6	7	0	1	2	3	4
	Path 3	3	4	5	6	7	0	1	2	3	4	5	6	7	0	1	2	3	4	5	6	7	0	1	2
	Path 1	6	7	0	1	2	3	4	5	6	7	0	1	2	3	4	5	6	7	0	1	2	3	4	5
Point C	Path 2	4	5	6	7	0	1	2	3	4	5	6	7	0	1	2	3	4	5	6	7	0	1	2	3
	Path 3	2	3	4	5	6	7	0	1	2	3	4	5	6	7	0	1	2	3	4	5	6	7	0	1
	Path 1	5	6	7	0	1	2	3	4	5	6	7	0	1	2	3	4	5	6	7	0	1	2	3	4
Point D	Path 2	3	4	5	6	7	0	1	2	3	4	5	6	7	0	1	2	3	4	5	6	7	0	1	2
	Path 3	1	2	3	4	5	6	7	0	1	2	3	4	5	6	7	0	1	2	3	4	5	6	7	0
	Path 1	4	5	6	7	0	1	2	3	4	5	6	7	0	1	2	3	4	5	6	7	0	1	2	3
Point E	Path 2	2	3	4	5	6	7	0	1	2	3	4	5	6	7	0	1	2	3	4	5	6	7	0	1
	Path 3	0	1	2	3	4	5	6	7	0	1	2	3	4	5	6	7	0	1	2	3	4	5	6	7
Point F					4	5	6	7	0	1	2	3	4	5	6	7	0	1	2	3	4	5	6	7	0	1	2	3

code sequence, $c_1(t)$. The alternative system is shown in Figure 4.23 where $c_1(t)$ is passed through a delay line to produce a number of delayed replicas. Ignoring the weighting factors, $a_k^*(t)$, for the moment, each version of $c_1(t)$ is correlated with the received baseband signal, $r(t)$, to extract each path signal. The integration (correlation) occurs over the entire length of the respective delayed replica of $c_1(t)$ and the results of the correlation become available at different times. This means that the individual results must be stored until they all become available. The channel coefficients, $a_k(t)$, are used to remove the phase differences caused by the channel and weight each path according to its received power to allow maximal ratio combining.

Table 4.13 shows the phasing of the codes at each point within the system in Figure 4.23. We note that a delayed version of $c_1(t)$ aligns with path 1 at point A, with path 2 at point C and with path 3 at point E. Each integration occurs over the entire length of the delayed version of $c_1(t)$, from bit 0 to bit 7. Referring to Table 4.13 we see that the result of the correlation of the version of $c_1(t)$ at point A and path 1 will become available at time=7, whereas the result of the correlation between the delayed version of $c_1(t)$ at point E and path 3 does not become available until time=11.

In the cdmaOne system the code word $c_1(t)$ is the product of the channel Walsh code and the PNI and PNQ sequences which adds to the complexity of the description provided above. We have also assumed that $c_1(t)$ has perfect autocorrelation properties; however, in the practical cdmaOne system this will not be the case. Each path will experience interference from the other paths as a result of an imperfect autocorrelation. The RAKE receiver will also experience interference from other users within the same cell and from users in neighbouring cells. The interference will tend to manifest itself as noise on the decision variable, U.

The cdmaOne MS receiver is equipped with at least three receiving elements and these are locked on to the three best received signal paths. This means that, whereas we have shown the delays in the RAKE receiver to be fixed at $1/W$, in the cdmaOne system they

Table 4.13: The signals for the RAKE receiver (delayed reference version).

Time $\longrightarrow$

		0	1	2	3	4	5	6	7	8	9	10	11	12	13	14	15	16	17	18	19	20	21	22	23
	Point A	0	1	2	3	4	5	6	7	0	1	2	3	4	5	6	7	0	1	2	3	4	5	6	7
Delayed	Point B	7	0	1	2	3	4	5	6	7	0	1	2	3	4	5	6	7	0	1	2	3	4	5	6
Versions	Point C	6	7	0	1	2	3	4	5	6	7	0	1	2	3	4	5	6	7	0	1	2	3	4	5
of $c_1(t)$	Point D	5	6	7	0	1	2	3	4	5	6	7	0	1	2	3	4	5	6	7	0	1	2	3	4
	Point E	4	5	6	7	0	1	2	3	4	5	6	7	0	1	2	3	4	5	6	7	0	1	2	3
Received	Path 1	0	1	2	3	4	5	6	7	0	1	2	3	4	5	6	7	0	1	2	3	4	5	6	7
Signal	Path 2	6	7	0	1	2	3	4	5	6	7	0	1	2	3	4	5	6	7	0	1	2	3	4	5
$r(t)$	Path 3	4	5	6	7	0	1	2	3	4	5	6	7	0	1	2	3	4	5	6	7	0	1	2	3

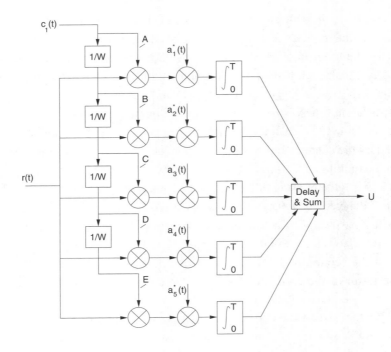

Figure 4.23: The RAKE receiver (with delayed reference signal).

are programmable and they may be varied to ensure that the receiver always uses the three best paths to form the decision variable. The RAKE receiver channel coefficients, a_k^*, are determined by the searcher from the received pilot signal.

4.2.3 The cdmaOne reverse link

The reverse link radio path consists of the mobile station (MS) transmitter, the radio channel and the base station (BS) receiver. The cdmaOne system supports two different types of radio channel on the reverse link. The *access channel*, as its name would suggest, is used by the MS initially to access the network, e.g. at call initiation or in response to a paging message. Each BS may support up to 32 access channels per forward link paging channel and the MSs within a cell are pseudo-randomly distributed between the access channels. In cells with a large number of users this process ensures that the access channel resources are approximately equally loaded.

The second type of channel supported on the cdmaOne reverse link is the *traffic channel* and reverse link channels are assigned to individual users as required. A reverse link traffic channel may carry speech or user data at bit rates up to 9.6 kb/s for the IS-95 system, and up to 14.4 kb/s for the CDMA-PCS system. The cdmaOne MS does not transmit a pilot signal as this would significantly decrease reverse link capacity. For this reason cdmaOne reverse link signals are very different from the forward link signals. We will now describe each of the reverse link channels in detail.

4.2.3.1 The access channel

A block diagram of the cdmaOne transmitter supporting the access channel is shown in Figure 4.24. The access channel data are presented in the form of one 96-bit frame every 20 ms. Of these 96 bits, 88 are used to carry information and the remaining eight are encoder tail bits. The resulting input data rate is 4.8 kb/s. The data are passed through a one-third rate convolutional encoder with a constraint length of nine defined by the following generator functions:

$$
\begin{aligned}
g_0 &= 1 + D^2 + D^3 + D^5 + D^6 + D^7 + D^8, \\
g_1 &= 1 + D + D^3 + D^4 + D^7 + D^8, \\
g_2 &= 1 + D + D^2 + D^5 + D^8.
\end{aligned}
\tag{4.56}
$$

This code has a minimum free distance of 18 [3].

This coding process generates three code symbols for each input bit and the resulting symbol rate is 14.4 ksymbols/s. Following convolutional encoding, the code symbols are repeated once to produce a symbol rate of 28.8 ksymbols/s and they are then block interleaved over the 20 ms frames (i.e. 576 code symbols at 28.8 ksymbols/s). The resulting code symbols are formed into six-bit words and these are used to select one of the 64-chip Walsh codes (see Figure 4.1) in a '64-ary orthogonal modulator'. The index of the Walsh code, W_x, is chosen according to the formula

$$
W_x = c_0 + 2c_1 + 4c_2 + 8c_3 + 16c_4 + 32c_5,
\tag{4.57}
$$

where c_0 to c_5 represent the six bits of the code word, c_5 being the most recent bit and c_0 being the oldest bit. For example, a six-bit code word of 110011 ($c_5 \ldots c_0$) would be used to select the Walsh code with an index of 51, i.e. W_{51}.

The effect of this 64-ary modulation process is to convert the 28.8 ksymbols/s code symbols into orthogonal Walsh codes with a chip rate of 307.2 kchips/s. *Note that the Walsh codes are used in an entirely different manner on the reverse link compared to the forward link.* On the forward link the Walsh codes are used to identify the different channels; however, on the reverse link the Walsh codes are used to carry information. The use of the 64-ary orthogonal modulation scheme enables the BS receiver to perform non-coherent detection in the absence of a pilot signal.

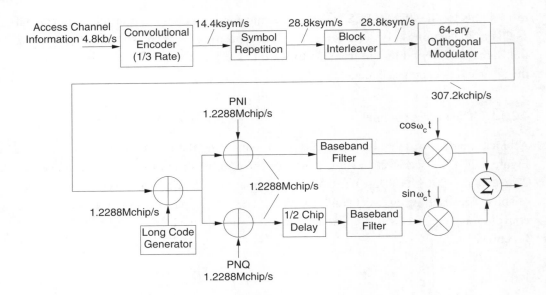

Figure 4.24: A block diagram of the cdmaOne access channel.

Following the 64-ary orthogonal modulation, the modulation symbols are spread by the $2^{42} - 1$ length PN sequence (i.e. the long code) which is generated at a chip rate of 1.2288 Mchips/s. This involves EXORing the modulation symbols with the output signal from the long code generator. The format of the long code generator has already been described in the previous section which dealt with the forward link channel. The contents of the long code generator are determined from the information carried by the sync channel, and the same generator is used for the reverse and forward links. For the access channel, the MS sets the 42-bit long code mask according to the format shown in Figure 4.25. Bits 28 to 32 give the access channel number (ACN) and this defines one of the 32 possible access channels associated with each paging channel. The number of access channels associated with a particular paging channel is contained within an *access parameters message* which is transmitted on the forward link paging channel. The MS randomly selects one of the access channels associated with its paging channel by setting the ACN field in the long code mask according to the output of a pseudo-random number generator. The BS is required to continuously monitor every reverse link access channel.

The paging channel number (PCN) in the long code mask gives the identity of the forward link paging channels used by the MS. Together, the PCN and the ACN uniquely identify one of the possible 7×32 reverse link access channels. The 16-bit BASE_ID field is used to carry the base station identification code and the last nine bits of the mask carry the pilot

41	33	32	28	27	25	24	9	8	0
110001111		ACN		PCN		BASE_ID		PILOT_PN	

ACN - Access Channel Number
PCN - Paging Channel Number
BASE_ID - Base Station Identification
PILOT_PN - PN Offset for the Forward CDMA Channel

Figure 4.25: The access channel long code mask format.

PN offset (PN_OFFSET) for a particular BS.

Following the long code spreading, the resulting 1.2288 Mchips/s signal is spread in quadrature by EXORing the data with the PNI and PNQ codes. We note that the PNI and PNQ codes used by the MS on the reverse link always have a zero offset, regardless of the PN offset used by the BS on the forward link.

Following this process, the data that have been spread by the PNQ code are delayed by half a chip period before the two bit streams are passed through baseband filters. The outputs of the filters are used to modulate two quadrature carriers and the resulting signals are summed to produce a phase modulated carrier signal. The specifications of the baseband filters are identical to those used on the forward link. The relationship between the input bit streams and the carrier phase is given in Table 4.14 and this is achieved by translating the I and Q bit streams such that a logical '0' in the original bit stream is replaced by a +1 level, and a logical '1' in the original bit stream is replaced by a -1 level.

The half chip delay in the Q channel ensures that only one chip is changing at any one time. This means that the amplitude of the carrier signal does not pass through zero during phase transitions; the resulting phase transition state diagram is shown in Figure 4.26. This simplifies the design of the MS transmitter output stages.

Having discussed the construction of the reverse link access channel we now examine the frame structure used on this type of channel. The access channel message structure

Table 4.14: Reverse link phase mapping.

I	Q	Phase
0	0	$\pi/4$
1	0	$3\pi/4$
1	1	$-3\pi/4$
0	1	$-\pi/4$

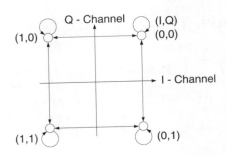

Figure 4.26: The reverse link signal constellation.

is shown in Figure 4.27. The eight-bit MSG_LENGTH field gives the message length in octets, including the MSG_LENGTH field itself, the message body and the 30-bit checksum. Padding bits (set to 0) are added to the end of the access channel message to produce an access channel message *capsule* that always occupies an integer number of 20 ms access channel frames. The CRC code generator polynomial for the access channel message is the same as that used for the sync channel message on the forward link (see Equation (4.5)). The maximum length of the access channel message is 880 bits (i.e. MSG_LENGTH $\leq$ 110).

The contents of the access channel message are partitioned into 88-bit sections which form the body of consecutive access channel frames, as shown in Figure 4.28. The remaining eight bits of the 96-bit access channel frame are encoder tail bits. The maximum size of the access channel message is limited by the parameter MAX_CAP_SZ which is transmitted to the MS on the paging channel as part of the access parameters message. The access channel message may occupy up to 3 + MAX_CAP_SZ frames. Each access channel message is always preceded by a number of preamble frames, as shown in Figure 4.28. The preamble frames consist of 96 zeros which are used by the BS receiver to aid the decoding of the access channel message. The MS will transmit 1 + PAM_SZ preamble frames before each access channel message, where the parameter PAM_SZ is contained in the access parameters message. The access channel message and the associated preamble frames occupy an

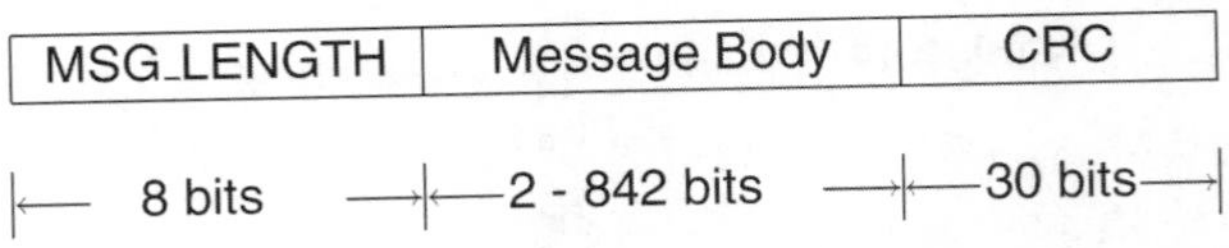

MSG_LENGTH	Message Body	CRC
8 bits	2 - 842 bits	30 bits

Figure 4.27: The access channel message format.

access channel slot which consists of

$$(3 + \text{MAX\_CAP\_SZ}) + (1 + \text{PAM\_SZ}) \text{ frames} \qquad (4.58)$$

and each access channel slot will begin and end on a 20 ms access channel frame boundary.

Having examined the frame structure used on the access channel, we now consider the behaviour of the MS during an *access attempt*, which is the term used to describe the entire process of sending an access request and gaining (or failing to gain) access to the system. Each access channel message may be transmitted a number of times before a valid acknowledgement is received (or the MS gives up) and each individual transmission is called an *access probe*. As we have already seen, the access probe consists of an access channel preamble followed by the access channel message, and each access probe in the same access attempt will contain exactly the same access channel message. Within each access attempt the access probes are grouped into *access probe sequences*, each of which will consist of up to (1+NUM_STEP) access probes. The NUM_STEP parameter is transmitted as part of the access parameters message on the paging channel. Each access probe within a particular sequence will use the same access channel (i.e. the same ACN). However, the access channel used for each access probe sequence will be chosen pseudo-randomly by the MS from among all the access channels associated with the current paging channel.

The cdmaOne access procedure is designed such that the MS uses the minimum transmitted power level to reach the BS and, in this way, the interference generated by each MS during access is minimised. Since the MS is not engaged in a two-way connection with the BS during the access procedure, the closed-loop power control algorithm cannot be used to determine the required MS transmitter power. For this reason, the MS commences by transmitting access probes at a relatively low transmission power and then gradually increases its transmitted power on subsequent access probes until an acknowledgement message is received from the BS.

The transmitted power of the first access probe in an access probe sequence is calculated using the pilot power received from the BS, i.e. the open-loop estimate. The average path loss, PL, between the BS and the MS is given by

$$PL = BS_{TX} - MS_{RX}, \qquad (4.59)$$

where BS_{TX} is the BS forward link transmitted power in dBm and MS_{RX} is the forward link power received at the MS in dBm. Assuming the path loss is the same on the reverse and the forward links, the transmitted power of the MS, MS_{TX}, for a given received reverse link power at the BS, BS_{RX}, is given by

$$\begin{aligned} MS_{TX} &= BS_{RX} + PL, \\ MS_{TX} &= BS_{RX} + BS_{TX} - MS_{RX}. \end{aligned} \qquad (4.60)$$

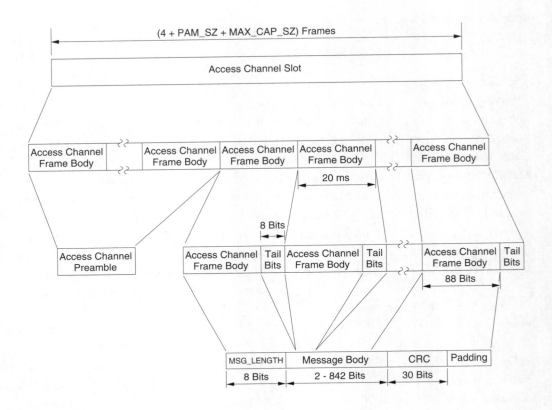

Figure 4.28: The access channel frame structure.

In the specifications [1, 2], the expression for the transmitted power of the first access probe in the access probe sequence, MS_{TX0}, includes a number of correction parameters as shown below:

$$
\begin{aligned}
MS_{TX0} \quad = \quad & -MS_{RX} \quad \text{(dBm)} \\
& -73 \\
& +\text{NOM\_PWR} \quad \text{(dB)} \\
& +\text{INIT\_PWR} \quad \text{(dB)}.
\end{aligned} \tag{4.61}
$$

We notice that the '$BS_{RX} + BS_{TX}$' term in Equation (4.60) has been replaced by the '$-73+$ NOM_PWR + INIT_PWR' term in Equation (4.61). The -73 is a fixed system parameter and the NOM_PWR and INIT_PWR are system variables, which are transmitted to the MS in the access parameters message.

Although NOM_PWR and INIT_PWR could have been combined into a single parameter, they have remained separate to provide a distinction between the different functions they perform. The NOM_PWR (nominal power) parameter is used to adjust the MS transmitted power such that the required nominal power is received at the BS, assuming the path loss on the reverse link and forward link is the same. This parameter must be adjusted according to the transmitted power of the BS and the required power at the BS receiver. The NOM_PWR parameter is in the range -8 to 7 dB and its nominal value is 0 dB.

The INIT_PWR parameter is used to ensure that the first access probe is received at a level below the required nominal received power. This parameter is included to allow for the difference between the fast fading characteristics on the reverse link and forward link. Setting INIT_PWR to a value of less than 0 dB ensures that the initial MS power is below that calculated using the open-loop estimate. This ensures that the MS does not use excessive transmitted power when the forward link signal has faded, but the reverse link signal has not faded. Setting INIT_PWR such that the initial MS transmitted power is very low ensures that the MS will generate the minimum amount of interference during its access attempt. However, a lower initial power is likely to result in the transmission of more access probes because the MS transmitted power will take longer to reach the required power to establish communications with a given BS. The value of INIT_PWR is in the range -16 to 15 dB and the nominal value is 0 dB.

Following the first access probe, transmitted at a power MS_{TX0}, the MS increases its transmitted power in subsequent access probes in steps of PWR_STEP dB, where PWR_STEP is a system parameter transmitted as part of the access parameters message. The transmitted power of the nth access probe in the access probe sequence, MS_{TXn}, is given by

$$
\begin{aligned}
MS_{TXn} \quad = \quad & -MS_{RX} \quad \text{(dBm)} \\
& -73 \\
& +\text{NOM\_PWR} \quad \text{(dB)} \\
& +\text{INIT\_PWR} \quad \text{(dB)} \\
& +n \times \text{PWR\_STEP} \quad \text{(dB)}.
\end{aligned}
\tag{4.62}
$$

The value of PWR_STEP is in the range 0 to 7 dB. Careful adjustment of the NOM_PWR, INIT_PWR and PWR_STEP parameters allows the network operator to ensure that the MS uses just sufficient transmitted power to satisfy the acceptable minimum received power at the BS.

Control of the MS transmitter power during an access attempt is important in ensuring that the interference generated by the MS during this phase of its operation is kept to a minimum. The timing of the access probe transmissions is also important to maximise the throughput of each access channel. As we have already seen, the MS chooses its access channel using a pseudo-random process and consequently there is a finite probability that two (or more) MSs will attempt to access the system simultaneously using the same access channel. This could result in neither reverse link access message being decoded correctly at the BS nor the MS receiving a valid acknowledgement. In this situation both MSs must attempt to access the system again. Should both MSs wait exactly the same time before re-transmitting their next access message, they will collide again and neither MS would be able to gain access to the system. For this reason, the timing of the access probes and access probe sequences is effectively randomised to ensure that MSs do not continue to collide during an access attempt.

We now examine the access attempt timing in more detail. The reverse link access channel supports two different types of message, the response message and the request message, and the timing is different for each. The response message, as its name would suggest, is transmitted in response to an earlier message from the BS (e.g. a paging call). The request message is not triggered by a message from the BS but is sent autonomously by the MS (e.g. call origination). This type of message determines the timing between the access probe sequences of an access attempt. For a response message, the time between access probe sequences is determined pseudo-randomly. This time is known as the *back-off delay* and it is given the symbol RS in the specifications. RS will take a value in the range 0 to $1 + BKOFF$ slots, where BKOFF is a system parameter transmitted as part of the access parameters message. The timing for the access probe sequences in an access attempt for a response message is given in Figure 4.29. The maximum number of access probe sequences in a response message access attempt is limited by the MAX_RSP_SEQ parameter, which is transmitted as part of the access parameters message. MAX_RSP_SEQ has a maximum value of 15.

The system has the ability to limit the number of response messages generated by the MSs within a cell by limiting the rate at which it transmits forward link messages that require a response (e.g. paging messages). This level of control does not occur with request messages which are transmitted autonomously by the MS. For this reason an additional delay is added between the access probe sequences and this is used to control the rate at which the request messages are generated by each MS. This additional delay is known as the *persistence delay* (*PD*) in the specifications. Following each access probe sequence the MS waits RS slots. For each subsequent access channel slot the MS performs a persistence test which is used to determine whether the next access probe sequence should commence. If the persistence test is passed, the MS begins to transmit the next access probe sequence. In the event that the test fails, the MS delays the transmission of the access probe sequence and proceeds to test the next slot. The persistence delay is shown in Figure 4.30.

The persistence test is a pseudo-random process whereby a number, RP, is generated between 0 and 1. RP is compared with a value P and the persistence test is passed when $RP < P$. The persistence test fails if $RP \geq P$ or $P = 0$. This means that as P decreases, the persistence delay increases and the rate at which the request messages are generated

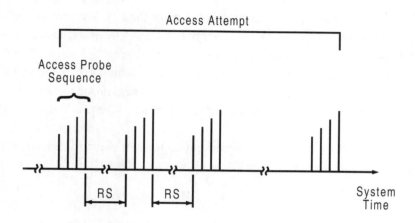

Figure 4.29: Access attempt timing (response message).

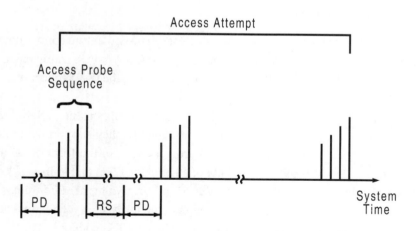

Figure 4.30: Access attempt timing (request message).

decreases. Consequently the system may use the value of P to control the request message rate on each access channel. The value of P depends on the type of request message (e.g. registration, call origination) and the *overload class* of the MS. The overload class is used to distinguish between different MSs (e.g. normal MS, test MS, emergency MS) and assign them different priorities in terms of the access channel resources. The actual calculation of P is quite complex and the reader is referred to the specifications for a full description. In the case of a request message, the maximum number of access probe sequences in an access attempt is given by the parameter MAX_REQ_SEQ which is transmitted to the MS as part of an access parameters message.

Following the transmission of each access probe the MS waits for a response from the BS. The length of this waiting period, TA, is given by

$$TA = (2 + ACC\_TMO) \times 80 \text{ ms}, \tag{4.63}$$

where ACC_TMO is a system parameter transmitted to the MS as part of the access parameters message. If the MS receives a valid acknowledgement during this waiting period, then the access attempt ends. If the MS does not receive a valid acknowledgement during this waiting period, then the MS transmits the next access probe in the access probe sequence after a further delay of RT slots, where RT is an additional back-off delay. RT is a pseudo-random number in the range 0 to 1+ PROBE_BKOFF, where the PROBE_BKOFF parameter is transmitted as part of the access parameters message.

4.2.3.2   The traffic channel

The reverse link traffic channels are used to carry speech, user data and control data between the MS and the BS in a dedicated mode. The IS-95 and CDMA-PCS traffic channels differ slightly in their format and we shall commence by describing the IS-95 transmitter, as shown in Figure 4.31.

The reverse link traffic channels have been designed to support the variable rate CELP speech coder used in the IS-95 systems and, consequently, they support data rates of 1.2, 2.4, 4.8 and 9.6 kb/s. For each rate, the data are generated in the form of 20 ms frames containing 192 bits (9.6 kb/s), 96 bits (4.8 kb/s), 48 bits (2.4 kb/s) or 24 bits (1.2 kb/s). Each frame includes eight encoder tail bits, set to zero, and the two higher rate frames (9.6 kb/s and 4.8 kb/s) also include frame quality indicators in the form of a cyclic redundancy check (CRC) code. The 192-bit frame (9.6 kb/s) has a 12-bit CRC code and the 96-bit frame (4.8 kb/s) has an eight-bit CRC code. These frame formats and the CRC codes are identical to those used on the forward link traffic channel for the same data rates.

Each frame is one-third rate convolutionally encoded using the same code as that used on the access channel (see Equation (4.56)) and this results in a code symbol rate of 3.6, 7.2, 14.4 or 28.8 ksymbols/s, depending on the input data rate. Following convolutional

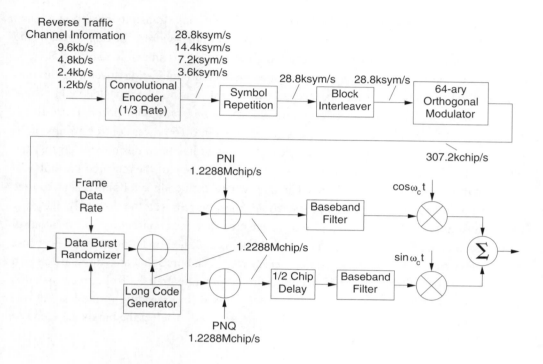

Figure 4.31: Block diagram of the reverse link traffic channel used in IS-95.

encoding, the code symbols are repeated to maintain a constant symbol rate of 28.8 ksymbols/s when the input data rate is below 9.6 kb/s. For example, when the input data rate is 2.4 kb/s, the code symbols are repeated three times (i.e. four copies of the same data) to produce a data rate of 28.8 ksymbols/s. We note that the repetition process is included to simplify the description of the following interleaver and burst randomiser and an equipment manufacturer is entitled to use alternative methods, provided the same result is achieved.

The repeated code symbols are interleaved over one 20 ms frame, which consists of 576 code symbols at 28.8 ksymbols/s. Following interleaving, the code symbols are formed into six-bit words and the binary value of these words is used to select a 64-chip WH code having the same index, e.g. the six bit word 000010 will select WH code 2. The index of the modulation symbol is selected according to Equation (4.57), in the same manner as that used on the access channel. In common with the reverse link access channel, the Walsh codes are used to carry the information, which is in contrast to the forward link channels where the Walsh codes are used to identify the individual channels.

The reverse link also differs from the forward link in the way it deals with the repeated symbols. On the forward link, the BS reduces its transmitted power to maintain a constant energy per data bit; however, on the reverse link, the MS only transmits one copy of each code symbol and discards the remaining repetitions. In other words, the duty cycle of the MS transmissions is adjusted according to the input data rate as shown in Table 4.15.

Removing the repeated symbols in a regular manner produces line spectra in the output signal which may affect the speech quality. For this reason, the repeated symbols are removed in a 'random' manner in a process known as *burst randomisation*. Each 20 ms frame, consisting of 576 code symbols, is sub-divided into 16 power control groups of 36 code symbols lasting 1.25 ms. The interleaving process has been designed such that the removal of a single power control group will remove one copy of the repeated symbols. To illustrate this we consider the 2.4 kb/s data rate, where each code symbol is repeated three times. Prior to transmission three of these copies must be removed to allow only one copy to be transmitted. The interleaving process ensures that four consecutive power control groups each contain copies of the same code symbols and the burst randomisation process must remove three power control groups in every four. The bursts that are erased are chosen based on the state of the long code generator in the previous frame. To be more specific, the last 14 bits of the long code generator (b_0–b_{13}) used for spreading in the last but one power control group of the previous speech frame are used to select the bursts to be erased according to the following rules.

Data rate = 9.6 kb/s
All power control groups are transmitted

Data rate = 4.8 kb/s
The following power control groups are transmitted:
$b_0, 2 + b_1, 4 + b_2, 6 + b_3, 8 + b_4, 10 + b_5, 12 + b_6, 14 + b_7$.

Data rate = 2.4 kb/s
The following power control groups are transmitted:
b_0 if $b_8 = 0$ or $2 + b_1$ if $b_8 = 1$,
$4 + b_2$ if $b_9 = 0$ or $6 + b_3$ if $b_9 = 1$,

Table 4.15: Reverse link traffic channel transmitted duty cycle.

Data rate	Duty cycle
9.6 kb/s	100%
4.8 kb/s	50%
2.4 kb/s	25%
1.2 kb/s	12.5%

$8 + b_4$ if $b_{10} = 0$ or $10 + b_5$ if $b_{10} = 1$,
$12 + b_6$ if $b_{11} = 0$ or $14 + b_7$ if $b_{11} = 1$,

Data rate = 1.2 kb/s
The following power control groups are transmitted:
b_0 if ($b_8 = 0$ and $b_{12} = 0$), or $2 + b_1$ if ($b_8 = 1$ and $b_{12} = 0$),
or $4 + b_2$ if ($b_9 = 0$ and $b_{12} = 1$), or $6 + b_3$ if ($b_9 = 1$ and $b_{12} = 1$),
$8 + b_4$ if ($b_{10} = 0$ and $b_{13} = 0$), or $10 + b_5$ if ($b_{10} = 1$ and $b_{13} = 0$),
or $12 + b_6$ if ($b_{11} = 0$ and $b_{13} = 1$), or $14 + b_7$ if ($b_{11} = 1$ and $b_{13} = 1$).

Figure 4.32 shows an example of the burst randomisation process for each of the four data rates taken from the specifications. In this example (b_0, b_1, ..., b_{13}) are set to 0 0 1 0 1 1 0 1 1 0 0 1 0 0. Using the rules above for the 2.4 kb/s data rate, b_8 is set to '1' and so power control group $2 + b_1 = 2$ is transmitted. b_9 is set to '0' and so power control group $4 + b_2 = 5$ is transmitted. b_{10} is set to '0' and so power control group $8 + b_4 = 9$ is transmitted. Finally, b_{11} is set to '1' and so power control group $14 + b_7 = 15$ is transmitted. We note that the burst randomisation algorithm ensures that the bursts transmitted at the 1.2 kb/s data rate are a subset of the bursts transmitted at the 2.4 kb/s data rate, which are, in turn, a subset of the bursts transmitted at the 4.8 kb/s data rate.

Following the burst randomiser the data are spread (i.e. EXored) by the output of the long code generator. The long code mask is identical to that used for the forward link traffic channel (see Figure 4.13) with the private long code mask being used for security. This process allows the BS to identify the different MSs on the reverse link. The resulting 1.2288 Mchips/s signal is then spread using the two, zero offset PNI and PNQ sequences. We note that the MS will always use the zero offset versions of the PNI and PNQ codes. The resulting data streams are then used to modulate the phase of a radio carrier in the same manner as employed in the access channel data described in the previous section.

Having discussed the construction of the IS-95 reverse link traffic channel we now consider its frame structure. The formation of the traffic channel frames used on the IS-95 reverse link are identical to those used on the forward link; Figures 4.15 and 4.16, and Table 4.10, also apply to the reverse link traffic channel. The reverse link traffic channels support Multiplex Option 1 which allows primary, secondary and signalling traffic to share the same channel using 'dim and burst' and 'blank and burst' techniques.

Our discussion of the reverse link traffic channel has been limited to the IS-95 system, but it also applies equally to the CDMA-PCS system. However, the CDMA-PCS system also supports an additional rate set, Rate Set 2, on its reverse link providing data rates of up to 14.4 kb/s. In common with the forward link traffic channel, these higher data rates are supported by reducing the power of the channel coding and, in the case of the reverse link,

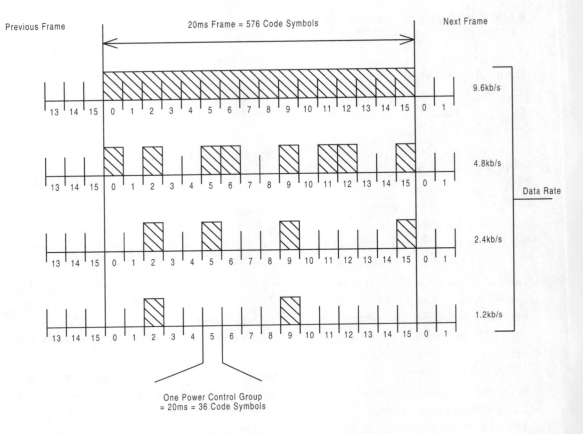

Figure 4.32: The burst randomisation process.

the one-third rate code is reduced to a one-half rate code.

For Rate Set 2, the data on the reverse link traffic channel are presented in the form of 20 ms frames, which are identical to those used on the forward link (see Figure 4.18). The frames are one-half rate convolutionally encoded using the same code as that used on the sync, paging and Rate Set 1 forward link traffic channels (see Equation (4.4)). The resulting code symbols are repeated to produce a constant symbol rate of 28.8 ksymbols/s prior to interleaving. From this point, the remainder of the Rate Set 2 traffic channel (i.e. 64-ary modulation, burst randomisation, spreading and modulation) is identical to the Rate Set 1 reverse link traffic channel. We note that the burst randomisation algorithm given above applies to Rate Set 2 with the 14.4 kb/s, 7.2 kb/s, 3.6 kb/s and 1.8 kb/s rates replacing the 9.6 kb/s, 4.8 kb/s, 2.4 kb/s and 1.2 kb/s rates, respectively.

The reverse link traffic channel supports Multiplex Option 2 using Rate Set 2, and Table 4.11 also applies equally to the reverse link.

4.2.3.3 The cdmaOne reverse link channel arrangement

Each cdmaOne cell will support a number of reverse link access channels and reverse link traffic channels. The number of reverse link access channels supported is determined by the number of forward link paging channels in use, with each paging channel having up to 32 associated access channels. The actual number of access channels associated with each particular paging channel, ACC_CHAN, is transmitted to the MS as part of the access parameters message carried on each individual paging channel. This value can be dynamically varied with varying teletraffic load within the cell.

The specifications do not define a maximum number of reverse link traffic channels. We note that, since the reverse link traffic channels are defined by the $2^{42} - 1$ bit long code mask, the number of channels is not limited to the number of Walsh codes as in the case of the forward link. However, it seems reasonable to assume that the reverse link will not support more traffic channels than the forward link and, consequently, the maximum number of reverse link traffic channels is likely to be 62. Figure 4.33 summarises the reverse link channel configuration for a single carrier BS. Where an operator is allocated sufficient spectrum to support more than one CDMA carrier, system capacity may be increased by installing additional carriers at each cell site.

4.2.3.4 The BS receiver

As we have already noted in our discussion of the MS receiver, the IS-95 and CDMA-PCS systems are essentially air-interface specifications, and although the construction of the MS transmitter is well defined, there are very few details relating to the organisation of the BS receiver. For this reason, the discussion contained in this section is, to a large extent, based on the views and opinions of the authors rather than the specifications themselves.

The BS receiver will use a RAKE receiver to demodulate the reverse link signal from a particular MS and the basic discussion of the RAKE receiver contained in the previous

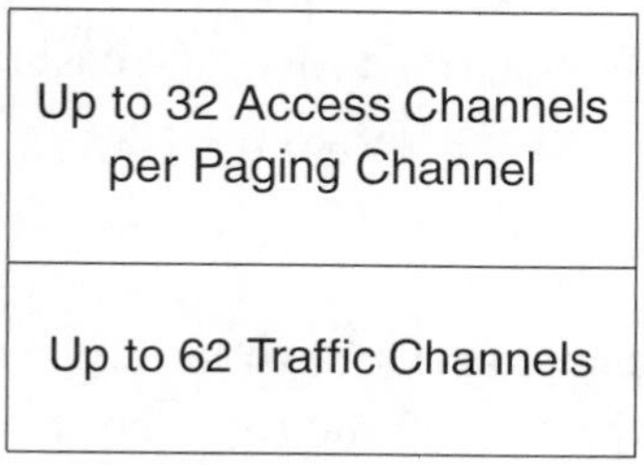

```
Up to 32 Access Channels
per Paging Channel


Up to 62 Traffic Channels
```

Figure 4.33: Reverse link channel arrangement for cdmaOne.

section may also be applied to the BS receiver. The main difference between the reverse
link and forward link, in terms of the receivers, is a result of the absence of a pilot channel
for each MS on the reverse link. This means that the BS receiver must identify and track the
significant multipath components from each MS using the data signal itself and this will tend
to complicate the searching process. Despreading is achieved by multiplying the received
signal by appropriately delayed versions of the pilot PN codes, PNI and PNQ, and the user's
long code, $c_l(t)$, to recover the transmitted Walsh code. The signal is then passed through a
bank of 64 correlators to determine which Walsh code has been sent. Once this decision has
been made, the Walsh code is converted into its six-bit address and the resulting data are
de-interleaved and convolutionally decoded to recover the original data frame. Figure 4.34
gives a possible block diagram of the BS receiver for a single reverse link channel. This
process would appear to be reasonably straightforward; however, the process of identifying
and tracking the multipath components using the data signal itself is complex. We will now
examine this searching process in more detail and the manner in which the BS determines
the channel estimate for use in the RAKE receiver.

Ignoring the processes prior to the 64-ary modulator, the transmitted reverse link signal
from the mobile is given by

$$
\begin{aligned}
s(t) \; = \; & w_x(t)c_l(t)i(t)\cos\omega_c(t) \\
& + w_x(t-\tau_c/2)c_l(t-\tau_c/2)q(t-\tau_c/2)\sin\omega_c(t),
\end{aligned}
\tag{4.64}
$$

where $w_x(t)$ is the Walsh code produced by the 64-ary modulator, $c_l(t)$ is the long code
of the user, ω_c is the RF carrier frequency, $\tau_c/2$ represents the 1/2 chip delay between the
in-phase and quadrature components of the data, and $i(t)$ and $q(t)$ are the PNI and PNQ
codes, respectively.

For a single path, the transmitted signal will experience a delay of τ_p seconds and an
attenuation of α as it passes through the radio channel. The received signal at the BS is
given by

$$
\begin{aligned}
r(t) \; = \; & \alpha[w_x(t-\tau_p)c_l(t-\tau_p)i(t-\tau_p)\cos\omega_c(t-\tau_p) \\
& + w_x(t-\tau_p-\tau_c/2)c_l(t-\tau_p-\tau_c/2)q(t-\tau_p-\tau_c/2) \\
& \times \sin\omega_c(t-\tau_p)].
\end{aligned}
\tag{4.65}
$$

At the receiver this signal is multiplied by two locally generated quadrature carriers of
frequency ω_c to produce two signals, $r_I(t)$ and $r_Q(t)$, given by

$$
\begin{aligned}
r_I(t) \; = \; & [w_x(t-\tau_p)c_l(t-\tau_p)i(t-\tau_p)\cos\omega_c(t-\tau_p) \\
& + w_x(t-\tau_p-\tau_c/2)c_l(t-\tau_p-\tau_c/2)q(t-\tau_p-\tau_c/2) \\
& \times \sin\omega_c(t-\tau_p)]\alpha\cos(\omega_c t + \phi),
\end{aligned}
\tag{4.66}
$$

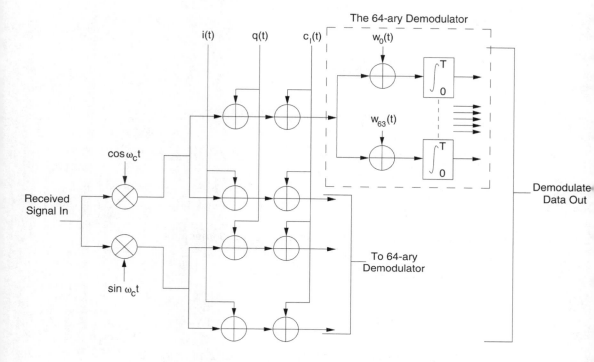

Figure 4.34: Block diagram of the BS receiver.

$$r_Q(t) = [w_x(t - \tau_p)c_I(t - \tau_p)i(t - \tau_p)\cos\omega_c(t - \tau_p)$$
$$+ w_x(t - \tau_p - \tau_c/2)c_I(t - \tau_p - \tau_c/2)q(t - \tau_p - \tau_c/2)$$
$$\times \sin\omega_c(t - \tau_p)]\alpha\sin(\omega_c t + \phi), \tag{4.67}$$

where ϕ represents the phase of the locally generated quadrature carriers relative to the carrier phase at the transmitter.

Expanding Equations (4.66) and (4.67) gives

$$r_I(t) = \frac{\alpha w_x(t - \tau_p)c_I(t - \tau_p)i(t - \tau_p)}{2}$$

$$\times [\cos(2\omega_c t + \phi - \omega_c \tau_p) + \cos(\phi + \omega_c \tau_p)]$$

$$+ \frac{\alpha w_x(t - \tau_p - \tau_c/2)c_I(t - \tau_p - \tau_c/2)q(t - \tau_p - \tau_c/2)}{2}$$

$$\times [\sin(2\omega_c t + \phi - \omega_c \tau_p) - \sin(\phi + \omega_c \tau_p)], \tag{4.68}$$

$$r_Q(t) = \frac{\alpha w_x(t - \tau_p)c_l(t - \tau_p)i(t - \tau_p)}{2}$$

$$\times [\sin(2\omega_c t + \phi - \omega_c \tau_p) + \sin(\phi + \omega_c \tau_p)]$$

$$+ \frac{\alpha w_x(t - \tau_p - \tau_c/2)c_l(t - \tau_p - \tau_c/2)q(t - \tau_p - \tau_c/2)}{2}$$

$$\times [-\cos(2\omega_c t + \phi - \omega_c \tau_p) + \cos(\phi + \omega_c \tau_p)], \tag{4.69}$$

and low pass filtering $r_I(t)$ and $r_Q(t)$ to remove the '$2\omega_c t$' component gives

$$\hat{r}_I(t) = \frac{\alpha w_x(t - \tau_p)c_l(t - \tau_p)i(t - \tau_p)}{2}$$

$$\times \cos(\phi + \omega_c \tau_p)$$

$$- \frac{\alpha w_x(t - \tau_p - \tau_c/2)c_l(t - \tau_p - \tau_c/2)q(t - \tau_p - \tau_c/2)}{2}$$

$$\times \sin(\phi + \omega_c \tau_p), \tag{4.70}$$

$$\hat{r}_Q(t) = \frac{\alpha w_x(t - \tau_p)c_l(t - \tau_p)i(t - \tau_p)}{2}$$

$$\times \sin(\phi + \omega_c \tau_p)$$

$$+ \frac{\alpha w_x(t - \tau_p - \tau_c/2)c_l(t - \tau_p - \tau_c/2)q(t - \tau_p - \tau_c/2)}{2}$$

$$\times \cos(\phi + \omega_c \tau_p). \tag{4.71}$$

The two signals $\hat{r}_I(t)$ and $\hat{r}_Q(t)$ are then despread by correlating them with locally generated and delayed combinations of the two pilot sequences, PNI and PNQ, the user's long code, $c_l(t)$, and one of the 64 possible Walsh codes, $w_x(t)$. Since the BS does not know which Walsh code is being sent, it must correlate $\hat{r}_I(t)$ and $\hat{r}_Q(t)$ with every possible Walsh code, i.e. the MS receiver must perform 256 correlations (two different PN codes [$i(t)$ and

$q(t)] \times 2$ quadrature received signals $[\hat{r}_I(t)$ and $\hat{r}_Q(t)] \times 64$ different Walsh codes $[w_0(t) \ldots w_{63}(t)] = 256$ different combinations).

We use the symbols $c_{ix}(t)$ and $c_{qx}(t)$ to define the combinations of $i(t)$ or $q(t)$, $c_l(t)$ and $w_x(t)$ such that

$$
\begin{aligned}
c_{ix}(t) &= w_x(t)c_l(t)i(t), \\
c_{qx}(t) &= w_x(t)c_l(t)q(t),
\end{aligned} \tag{4.72}
$$

where $w_x(t)$ is any one of the 64 possible Walsh codes.

Using the definitions given in Equation (4.72) above, Equations (4.70) and (4.71) may be re-written as

$$
\begin{aligned}
\hat{r}_I(t) &= \frac{\alpha c_{ix}(t - \tau_p)}{2} \cos(\phi + \omega_c \tau_p) \\
&\quad - \frac{\alpha c_{qx}(t - \tau_p - \tau_c/2)}{2} \sin(\phi + \omega_c \tau_p),
\end{aligned} \tag{4.73}
$$

$$
\begin{aligned}
\hat{r}_Q(t) &= \frac{\alpha c_{ix}(t - \tau_p)}{2} \sin(\phi + \omega_c \tau_p) \\
&\quad + \frac{\alpha c_{qx}(t - \tau_p - \tau_c/2)}{2} \cos(\phi + \omega_c \tau_p).
\end{aligned} \tag{4.74}
$$

These two signals, $\hat{r}_I(t)$ and $\hat{r}_Q(t)$, are then correlated with each of the 64 possible $c_{ix}(t)$ and $c_{qx}(t)$ sequences. This results in 256 signals given by

$$
\begin{aligned}
r_{Iiy}(t) &= \frac{\alpha \cos(\phi + \omega_c \tau_p)}{2} \int_0^T c_{ix}(t - \tau_p)c_{iy}(t - \tau_r)dt \\
&\quad - \frac{\alpha \sin(\phi + \omega_c \tau_p)}{2} \int_0^T c_{qx}(t - \tau_p - \tau_c/2)c_{iy}(t - \tau_r)dt,
\end{aligned} \tag{4.75}
$$

$$
\begin{aligned}
r_{Qiy}(t) &= \frac{\alpha \sin(\phi + \omega_c \tau_p)}{2} \int_0^T c_{ix}(t - \tau_p)c_{iy}(t - \tau_r)dt \\
&\quad + \frac{\alpha \cos(\phi + \omega_c \tau_p)}{2} \int_0^T c_{qx}(t - \tau_p - \tau_c/2)c_{iy}(t - \tau_r)dt,
\end{aligned} \tag{4.76}
$$

$$
\begin{aligned}
r_{Iqy}(t) &= \frac{\alpha \cos(\phi + \omega_c \tau_p)}{2} \int_0^T c_{ix}(t - \tau_p)c_{qy}(t - \tau_r - \tau_c/2)dt \\
&\quad - \frac{\alpha \sin(\phi + \omega_c \tau_p)}{2} \int_0^T c_{qx}(t - \tau_p - \tau_c/2)c_{qy}(t - \tau_r - \tau_c/2)dt,
\end{aligned} \tag{4.77}
$$

$$r_{Qqy}(t) = \frac{\alpha \sin(\phi + \omega_c \tau_p)}{2} \int_0^T c_{ix}(t - \tau_p)c_{qy}(t - \tau_r - \tau_c/2)dt$$

$$+ \frac{\alpha \cos(\phi + \omega_c \tau_p)}{2} \int_0^T c_{qx}(t - \tau_p - \tau_c/2)c_{qy}(t - \tau_r - \tau_c/2)dt, \quad (4.78)$$

where T represents the correlation period, τ_r represents the delay in the locally generated versions of $c_{ix}(t)$ and $c_{qx}(t)$ and y represents the order of the locally generated Walsh code used to generate c_{iy} and c_{qy}. We note that 64 versions of $\hat{r}_{Iiy}(t)$, $\hat{r}_{Iqy}(t)$, $\hat{r}_{Qiy}(t)$ and $\hat{r}_{Qqy}(t)$ will be generated, one for each of the 64 possible transmitted Walsh codes, i.e. $y = 0$ to 63. These will appear at each one of the 64 outputs of the four correlator banks (see Figure 4.34, where only one of the four correlator banks has been shown).

We now examine the properties of $c_{ix}(t)$ and $c_{qx}(t)$. Both $c_{ix}(t)$ and $c_{qx}(t)$ are formed by multiplying two PN sequences and a Walsh code together. We will assume that the resulting sequence may be approximated by another PN sequence and the usual cross correlation and autocorrelation properties apply. Assuming the cross correlation of $c_{ix}(t)$ and $c_{qx}(t)$ is zero, i.e.

$$\int_0^T c_{ix}(t - \tau_i)c_{qy}(t - \tau_k)dt = 0, \quad \text{for any } \tau_i, \tau_k, x \text{ and } y, \quad (4.79)$$

Equations (4.75)–(4.78) become

$$\hat{r}_{Iiy}(t) = \frac{\alpha \cos(\phi + \omega_c \tau_p)}{2} \int_0^T c_{ix}(t - \tau_p)c_{iy}(t - \tau_r)dt, \quad (4.80)$$

$$\hat{r}_{Qiy}(t) = \frac{\alpha \sin(\phi + \omega_c \tau_p)}{2} \int_0^T c_{ix}(t - \tau_p)c_{iy}(t - \tau_r)dt, \quad (4.81)$$

$$\hat{r}_{Iqy}(t) = \frac{\alpha \sin(\phi + \omega_c \tau_p)}{2}$$
$$\times \int_0^T c_{qx}(t - \tau_p - \tau_c/2)c_{qy}(t - \tau_r - \tau_c/2)dt, \quad (4.82)$$

$$\hat{r}_{Qqy}(t) = \frac{\alpha \cos(\phi + \omega_c \tau_p)}{2}$$
$$\times \int_0^T c_{qx}(t - \tau_p - \tau_c/2)c_{qy}(t - \tau_r - \tau_c/2)dt. \quad (4.83)$$

The cross correlation of two sequences formed using different Walsh codes (i.e. c_{iy} and c_{ix} where $x \neq y$) is also zero, namely

$$\int_0^T c_{ix}(t - \tau_i)c_{iy}(t - \tau_k)dt = 0, \quad \text{for any } \tau_i, \tau_k \text{ and } x \neq y,$$

$$\int_0^T c_{qx}(t - \tau_i)c_{qy}(t - \tau_k)dt = 0, \quad \text{for any } \tau_i, \tau_k \text{ and } x \neq y. \quad (4.84)$$

This property means that $\hat{r}_{Iiy}$, $\hat{r}_{Iqy}$, $\hat{r}_{Qiy}$ and $\hat{r}_{Qqy}$ will only be non-zero when $y = x$, i.e. the Walsh code used to form $\hat{r}_{Iiy}$, $\hat{r}_{Iqy}$, $\hat{r}_{Qiy}$ and $\hat{r}_{Qqy}$ is the same as the transmitted Walsh code.

We will represent these non-zero values of $\hat{r}_{Iiy}$, $\hat{r}_{Iqy}$, $\hat{r}_{Qiy}$ and $\hat{r}_{Qqy}$ using the symbols $\hat{r}_{Iix}$, $\hat{r}_{Iqx}$, $\hat{r}_{Qix}$ and $\hat{r}_{Qqx}$, respectively.

If the autocorrelation functions of both $c_{ix}(t)$ and $c_{qx}(t)$ are perfect such that

$$\int_0^T c_{ix}(t - \tau_i)c_{ix}(t - \tau_k) = 0, \quad \text{for } \tau_i \neq \tau_k, \tag{4.85}$$

$$\int_0^T c_{qx}(t - \tau_i)c_{qx}(t - \tau_k) = 0, \quad \text{for } \tau_i \neq \tau_k, \tag{4.86}$$

$$\frac{1}{T}\int_0^T c_{ix}(t - \tau_i)c_{ix}(t - \tau_i) = 1, \tag{4.87}$$

$$\frac{1}{T}\int_0^T c_{qx}(t - \tau_i)c_{qx}(t - \tau_i) = 1, \tag{4.88}$$

then

$$\hat{r}_{Iix}(t) = \hat{r}_{Qix}(t) = \hat{r}_{Iqx}(t) = \hat{r}_{Qqx}(t) = 0, \quad \text{for } \tau_r \neq \tau_p, \tag{4.89}$$

and

$$\hat{r}_{Iix}(t) = \frac{\alpha\cos(\phi + \omega_c\tau_p)T}{2}, \tag{4.90}$$

$$\hat{r}_{Qix}(t) = \frac{\alpha\sin(\phi + \omega_c\tau_p)T}{2}, \tag{4.91}$$

$$\hat{r}_{Iqx}(t) = \frac{\alpha\sin(\phi + \omega_c\tau_p)T}{2}, \tag{4.92}$$

$$\hat{r}_{Qqx}(t) = \frac{\alpha\cos(\phi + \omega_c\tau_p)T}{2}, \tag{4.93}$$

when $\tau_r = \tau_p$.

This shows that the receiver is able to determine the propagation delay, the attenuation and the phase shift in the channel. With the perfect cross correlation and autocorrelation conditions described in Equations (4.79) and (4.85)–(4.88), respectively, the values of $\hat{r}_{Iix}(t)$, $\hat{r}_{Qix}(t)$, $\hat{r}_{Iqx}(t)$ and $\hat{r}_{Qqx}(t)$ will only be non-zero when the value of τ_r equals the propagation delay τ_p. Therefore the search receiver is able to 'find' the MS by sweeping τ_r over a range of possible values and noting the values of τ_r that cause the output of one of its bank of 64 correlators, i.e. $\hat{r}_{Iix}(t)$, $\hat{r}_{Qix}(t)$, $\hat{r}_{Iqx}(t)$ and $\hat{r}_{Qqx}(t)$, to become non-zero.

The receiver can determine the phase shift introduced by the channel by adjusting the phase of its locally generated carrier signal, ϕ, such that $\hat{r}_{Qix}(t) = \hat{r}_{Iqx}(t) = 0$. This means that ϕ will be equal to the phase difference introduced by the channel but with opposite polarity (i.e. $-\omega_c\tau_p$), and

$$\hat{r}_{Iix}(t) = \hat{r}_{Qqx}(t) = \frac{\alpha}{2}. \qquad (4.94)$$

Finally, by measuring the amplitude of $\hat{r}_{Iix}(t)$ and $\hat{r}_{Qqx}(t)$, the MS is able to determine the attenuation, α, caused by the channel.

We note that this analysis is a gross over-simplification of the situation in the full cdmaOne system; however, it has been used to demonstrate the basic principles of the BS search receiver. In a full IS-95 system the BS will receive reverse link signals from a number of different MSs and these signals will all arrive at the input to the searcher. If α_l and τ_{pl} are the attenuation and propagation delay experienced by the signal from the lth MS, then the received signal at the BS is given by

$$
\begin{aligned}
r(t) \quad = \quad & \sum_{l=1}^{N} \{\alpha_l[w_x(t - \tau_{pl})c_l(t - \tau_{pl})i(t - \tau_{pl})\cos\omega_c(t - \tau_{pl}) \\
& + w_x(t - \tau_{pl} - \tau_c/2)c_l(t - \tau_{pl} - \tau_c/2)q(t - \tau_{pl} - \tau_c/2) \\
& \times \sin\omega_c(t - \tau_{pl})]\},
\end{aligned}
\qquad (4.95)
$$

where N is the total number of MSs received at the BS. The reverse link signals from the $N-1$ other MSs produce intracellular interference. However, the BS is able to select the required MS by relying on the cross correlation properties of the long code sequence, $c_l(t)$. Assuming these sequences exhibit perfect cross correlation, i.e.

$$\int_0^T c_l(t)c_k(t)\mathrm{d}t \quad = \quad 0, \quad \text{for } l \neq k, \qquad (4.96)$$

and

$$\frac{1}{T}\int_0^T c_l(t)c_l(t)\mathrm{d}t \quad = \quad 1, \qquad (4.97)$$

the intracellular interference is removed in the despreading process.

Up to this point we have assumed that the reverse link signal arriving from the lth MS is composed of a single path with attenuation α_l and phase shift $\omega_c\tau_{pl}$. In a practical system, the channel between the lth MS and the BS will consist of a number of paths each with an attenuation of α_{lk} and propagation delay of τ_{plk} for the kth path. Therefore Equation (4.65) becomes

$$
\begin{aligned}
r(t) \quad = \quad & \sum_{l=1}^{N}\sum_{k=1}^{K_l} \{\alpha_{lk}[w_x(t - \tau_{plk})c_l(t - \tau_{plk})i(t - \tau_{plk})\cos\omega_c(t - \tau_{plk}) \\
& + w_x(t - \tau_{plk} - \tau_c/2)c_l(t - \tau_{plk} - \tau_c/2)q(t - \tau_{plk} - \tau_c/2) \\
& \times \sin\omega_c(t - \tau_{plk})]\},
\end{aligned}
\qquad (4.98)
$$

where K_l is the number of paths arriving from the lth MS. When this signal is applied to the input of the searcher, the values of $\hat{r}_{Iix}(t)$, $\hat{r}_{Qix}(t)$, $\hat{r}_{Iqx}(t)$ and $\hat{r}_{Qqx}(t)$ will become non-zero whenever τ_r is equal to τ_{pjk}. In this way the MS is able to identify the individual paths arriving from each BS and also determine the attenuation and phase shift of each path.

Thus far our analysis has been simplified by the assumption that the $c_{ix}(t)$ and $c_{qx}(t)$ sequences display perfect cross correlation and autocorrelation properties (see Equations (4.79), (4.85)–(4.88) and Equation (4.98)). This will not be so in a practical system and a certain amount of cross correlation and autocorrelation noise will be present in the 64 versions of $\hat{r}_{Iiy}(t)$, $\hat{r}_{Qiy}(t)$, $\hat{r}_{Iqy}(t)$ and $\hat{r}_{Qqy}(t)$. The searcher will also receive interference from the traffic channels that are in use in the surrounding cells and this will also manifest itself as noise in the values of $\hat{r}_{Iiy}(t)$, $\hat{r}_{Qiy}(t)$, $\hat{r}_{Iqy}(t)$ and $\hat{r}_{Qqy}(t)$. This noise may lead to errors in the calculation of α_{lk} and τ_{plk}.

A searcher is used to calculate the amplitude and delay of the most significant multipath components of the reverse link signal. These values are then used to lock the demodulating fingers of the RAKE receiver onto each of these components. Since the phase of the locally generated quadrature carriers can be adjusted to make $\hat{r}_{Iqx}(t)$ and $\hat{r}_{Qix}(t)$ go to zero, the demodulator needs only to correlate each of the 64 Walsh codes with $\hat{r}_{Iix}(t)$ and $\hat{r}_{Qqx}(t)$ to determine the identity of the transmitted Walsh code, i.e. 128 correlations. Once the identity of the transmitted Walsh code has been determined, by detecting a non-zero output from each of the bank of 64 correlators, the data are recovered by converting the Walsh code into its six-bit address. The resulting data are then de-interleaved and convolutionally decoded to remove any errors that have been introduced as a result of the demodulation process.

4.3 Control of the Radio Resources

In the previous sections we have examined the radio interface in some detail. In this section we turn our attention to the manner in which the network manages the available radio resources. We will examine the cell selection procedure performed by the MS at switch-on, and the behaviour of the MS whilst in idle mode. The operation of the network and the MS during an access attempt will be described. Finally, we will examine the cdmaOne handover process.

4.3.1 Cell selection

The cell selection procedure is performed by the MS with the aim of detecting and identifying a BS belonging to an appropriate network. The first task of the cell selection procedure is to determine the most appropriate network and this choice will be based on a number of user-defined preferences. In the case of the IS-95 system, which is a dual mode standard, the user may choose from analogue and CDMA networks, whereas in the CDMA-PCS system, the user may only choose from a number of different CDMA networks. Assuming that a CDMA network is chosen in each case, the MS tunes to the primary CDMA carrier frequency for the chosen network. This frequency is used at every BS within the network and the MS must store details of the primary CDMA carrier frequency associated with each network. Each network also has a secondary CDMA carrier frequency, and although the specifications do not specifically state the purpose of this channel, it seems reasonable to assume that it may be used as an alternative to the primary channel. Once the CDMA carrier frequency has been selected for the required network, the MS begins to search for pilot channels on that frequency. This is achieved by sweeping the search receiver over a range of phase offsets as described in Section 4.2.2.6 and noting its output, with a non-zero output indicating the presence of a pilot signal. If the MS does not 'find' a suitable pilot signal within 15 s, it returns to the network selection procedure and allows the user to select an alternative network.

Once the MS identifies a suitable pilot signal it attempts to demodulate the corresponding sync channel. We note that a pilot channel and its associated sync channel are transmitted with the same PN offset, and consequently once the MS has identified a suitable pilot signal it automatically knows the PN offset of the sync channel. The MS decodes the sync channel information by setting its code to WH code 32 (see Figure 4.1). The sync channel carries a number of different system parameters which are used by the MS to identify and synchronise with the network. These parameters include:

- The 15-bit system identification (SID) and 16-bit network identification (NID). These are used by the MS to identify the different CDMA networks.

- The Pilot PN Sequence Offset Index (PILOT_PN) applie for a particular BS. This parameter gives the phase of the PNI and PNQ sequences relative to the zero offset PNI and PNQ sequences which always start on the even second marks of system time. Using its pilot searcher the MS has determined the arrival time of the pilot signal, and knowing the PN offset of this signal, the MS may identify the position of the even second marks to the nearest chip.

- System Time (SYS_TIME). This parameter gives the system time, in units of 80 ms, for a time of 320 ms minus the PN offset after the end of the last sync channel superframe (see Figure 4.8). Using the SYS_TIME and PILOT_PN parameters the MS is able fully to synchronise with the network.

- The Long Code State (LC_STATE). As we have already seen, the long code generator plays an important part in both the reverse link and forward link data paths and the

MS uses the LC_STATE parameter to synchronise its long code generator with the network.

- The Paging Channel Data Rate (PRAT). The paging channel data rate may be set to 4.8 kb/s or 9.6 kb/s and this parameter is used to indicate the rate used.

Once the MS has recovered the information from the sync channel and fully synchronised with the network, it enters the idle mode and awaits paging messages from the network.

4.3.2 The idle mode and paging

In the idle mode, the MS must monitor the paging channel of the current BS to detect incoming calls and it must also monitor the pilot channels of surrounding cells to ensure that the MS is always 'camped-on' the most appropriate BS. We also note that, in addition to paging messages, the paging channels also contain network information which the MS must decode and store. This information is carried in the form of a number of messages which include the following:

- The System Parameters Message. This contains general system information, including the number of paging channels available at the BS, the handover thresholds and a number of registration parameters which allow the MS to identify the location area and determine whether a location update is required.

- The Access Parameters Message. This carries a number of different parameters relating to system access (e.g. NOM_PWR, INIT_PWR and PWR_STEP). We have already discussed the function of a number of these parameters in Section 4.2.3.1 .

- Neighbour List Message. This message gives details (i.e. the Pilot PN offset) of the neighbouring BSs. The MS uses this information to determine the cells that are possible handover candidates.

- The CDMA Channel List Message. This message lists all the CDMA frequencies available at a BS which support at least one paging channel.

Prior to receiving a System Parameters Message and CDMA Channel List Message, an MS is not aware of the paging channel arrangement at the particular BS and is therefore unable to determine its own allocated paging channel. For this reason, the MS initially tunes to the primary paging channel which is always present on Channel 1 (i.e. Walsh code W1, see Figure 4.1). Once the MS has received the information carried by this channel it is able to determine its allocated paging channel and slot as described in Section 4.2.2.3 dealing with the forward link paging channel.

Whilst in the idle mode, the MS continually monitors the received signal strength of the pilot channel in neighbouring cells. If the MS detects that a neighbouring BS is stronger than the current BS, it performs an idle mode handover whereby it will camp-on the new BS.

4.3.3   The access procedure

The access procedure is used by the MS to access the network. It may be triggered by the network (e.g. by paging the MS) or by the user (e.g. by originating a call). We have already examined the timing of access messages on the access channel. We now discuss the contents of these messages and the data flow between the MS and the network.

The access channel may carry a number of different messages which include the following:

- The Registration Message. This is used to perform a location update which allows the network to track the MS. The message contains the identity of the MS and a field to indicate the reason for the message (e.g. the MS has moved between location areas, or a preset period of time has elapsed since the previous location update).

- The Order Message. A degree of signalling information may be carried between the MS and the network without the need to set up a dedicated signalling link. This type of information is known as orders and they are carried in the form of order messages on both the forward link paging channel and the reverse link access channel.

- The Origination Message. This message is used to indicate that the user has originated a call. The message contains the identity of the MS and its class mark, details of the service that has been requested by the user, and as many of the dialled digits as possible without exceeding the message capsule size.

- The Page Response Message. This message is used by the MS to respond to a paging message on the forward link paging channel. It contains the MS identity and the MS class mark.

Once the MS has transmitted a message on the access channel it monitors its assigned paging channel for a valid response. In general, the MS's access request will either be accepted or denied and the network's response will reflect this decision. In the situation where the access attempt is accepted the BS will respond with a *channel assignment message* on the paging channel. Where access is denied the BS responds with a release order and the MS returns to the idle mode.

The *channel assignment message* contains full details of the channel that has been allocated by the network including the CDMA frequency and the channel number (i.e. W1 to

W63). On receiving this message the MS retunes to the new channel and attempts to receive the null traffic channel frames transmitted by the BS. These frames are transmitted at the 1.2 kb/s rate and consist of 16 1s followed by eight 0s (i.e. the encoder tail bits). The MS must receive at least two consecutive 'good' frames within 0.2 s of tuning to the forward link traffic channel. If this is achieved, then the MS begins to transmit preamble traffic channel frames on the corresponding reverse link traffic channel. These frames are transmitted at the 9.6 kb/s data rate and they consist of 192 0s. (We note that this applies to the IS-95 system and the preamble frame construction may vary for the CDMA-PCS system.)

The BS attempts to acquire the reverse link signal transmitted by the MS and when this has been achieved it transmits a *base station acknowledgement order* on the forward link traffic channel. When this message has been received at the MS, the link has been successfully established and the call may be setup. The access procedure is summarised in Figure 4.35 for a mobile terminated call.

4.3.4 Handover

The cdmaOne system supports three main types of handover: idle mode handover, soft handover, and hard handover. An *idle mode handover* occurs when an MS moves from the coverage area of one BS into the coverage area of a second BS while an MS is in idle mode. The MS determines that an idle mode handover should occur when it detects a sufficiently strong pilot channel signal from a BS other than from its current BS. A *soft handover* is used between BSs having CDMA carriers with identical frequency assignments, and the mobile communicates simultaneously with these BSs until it is evident that only one BS is required. *Hard handovers* occur when a mobile is switched between two BSs using different carriers. There is also handover that occurs when a mobile is switched between the IS-95 BS and an AMPS BS. We will focus here on the commonly used soft handover process.

To expedite the handover process, four types of channel sets are defined. The *active set* contains those channels that are carrying forward traffic to the MS. During normal operation there is usually only one channel in the active set. However, during a soft handover when the MS is communicating with more than one BS, there will be at least two channels in the active set. Note that a forward CDMA pilot channel is fully defined by its PN offset index and its frequency assignment.

The channels that are likely candidates for soft handover are held in the *candidate set*. A channel is added to this set if its received pilot power exceeds a threshold value T_ADD and it is removed if the power falls below a threshold value T_DROP for a fixed period of time, or if the channel is transferred to the active set during the handover process.

The *neighbour set* contains pilot channel offsets for BSs that are close to the mobile. Initially this set is composed of channels received in a neighbour list message from the current BS. Members are added to the set in a number of circumstances. For example, a

Mobile Station	Channel	Base Station
	← Paging Channel ←	Send Paging Message
Send Page Response Message	→ Access Channel →	Activates Traffic Channel *and*
	← Forward link Traffic ← Channel	Begin sending null Traffic Channel Data
Retune to assigned Traffic Channel and Receive two consecutive 'good' frames	← Paging Channel ←	Send Channel Assignment Message
Begin sending the Traffic Channel Preamble	→ Reverse link Traffic → Channel	Acquire Reverse link Traffic Channel
Base Station Acknowledgement Order Received Successfully	← Forward link Traffic ← Channel	Send Base Station Acknowledgement Order

Figure 4.35: Simplified data flow for the access procedure of a mobile terminated call.

neighbour list update message may be received by the MS. Also, if an active set member, i.e. the current BS, becomes inactive due to a handover, then the current BS is now a neighbour and is added to the list. Finally, if the received signal strength of a pilot in the candidate set falls below T_DROP for a given period of time, then it is also added to the neighbour list. All members of the neighbour set are assigned a variable that relates to their duration in the set. The variable is incremented when a handover occurs, or a neighbour list update message is received. Members are dropped from the set when their age has exceeded a threshold, or when they become a member of the candidate set as a result of their pilot power exceeding T_ADD.

The final type of channel set is the *remaining set*. This contains all possible pilot offsets in use on the current CDMA frequency, excluding those contained in other sets.

Each of the channel sets above has an associated 'search window' in which the MS must search for usable multipath components. Details of window size are transmitted as part of the system parameters message on the forward link paging channel. The search window size may vary from 4 to 452 chips (i.e. 3.3 μs to 368 μs). In the case of the active and candidate sets, the search window is centred on the earliest arriving usable multipath for each pilot. For example, if the search window is 40 chips wide, the MS will search for multipath components ± 20 chips either side of the earliest usable multipath for each pilot in the set. In the case of the neighbour and remaining sets, the search window is centred on the PN offset of each pilot in the set.

The MS assists in the handover process by measuring the strength of the pilot channels in the active and candidate sets and reporting these measurements to the network in the form of a *pilot strength measurement message* on the reverse link traffic channel. The signal strength of each pilot is computed as the ratio of the received energy per chip, E_c, to the total received spectral density (noise and signals), I_0. Where a base station has a number of usable multipath components, the E_c/I_0 values are summed for the k best paths, where k represents the number of demodulating elements at the MS. This value is sent as a six-bit number which is given by the integer value of

$$-2 \times 10 \times \log_{10} \frac{E_c}{I_0}. \tag{4.99}$$

Note that the energy per chip, E_c, is used instead of the energy per bit, E_b, because the pilot channel does not carry any information bits. The sending of a *pilot strength measurement message* may be triggered by any one of a number of different events including

- if a *Pilot Measurement Request Order* is received from the BS;

- if the strength of a pilot in either the neighbour set or in the remaining set exceeds T_ADD;

- if the strength of a pilot in the candidate set exceeds the strength of a pilot in the active set by more than T_COMP $\times 0.5$ dB, where T_COMP is a system parameter which is transmitted as part of the system parameters message on the forward link paging channel;

- if the strength of a pilot in the active set falls below T_DROP for a predetermined period of time, measured by the handoff drop timer, T_TDROP.

4.3.4.1 Soft handover

During a soft handover, the MS maintains simultaneous connections with more than one BS. The MS is allocated a forward link channel at each of the BSs, and the information transmitted on each channel is the same (except for the power control bits). The MS performs diversity combining of the forward link signals by directing its demodulating elements to the three best forward link paths, regardless of their origin. This is shown graphically in Figure 4.36 where we have shown that the received signal at a particular MS consists of a number of paths from two different BSs (cell A and cell B). The MS continually scans the pilot signal of each BS to search for usable multipath components and it always assigns its three demodulating elements to the three best, i.e. strongest, paths. This is because the MS has a RAKE receiver with an ability to operate on three paths. These paths may be combined coherently using maximal ratio combining, since the MS is able to establish the amplitude, phase and time delay of each path from the pilot signal transmitted by each BS. The forward link channels involved in a soft handover must use the same RF frequency at each BS, but the channels may use different Walsh codes.

On the network side, each BS receives and decodes the MS s reverse link frames independently. These frames are passed to the base station controller (BSC), along with a quality indicator, which is determined from the convolutional decoding metrics and the CRC protection within each frame. Given this information, the BSC selects the best frame and passes this to the speech decoder. This means that the network is able to select the best BS on a frame-by-frame basis. We note that the signals arriving at the two BSs cannot be combined in a coherent manner because the amount of information that would have to be passed between the BSs, via the BSC, is too large. For this reason, the switched diversity approach has been adopted, whereby each BS independently decodes the speech using the four best paths from the MS and then the best frame is chosen at the BSC. This ability to switch rapidly between two (or more) different BSs, sometimes called *macrodiversity*, improves the reverse link performance when an MS is in soft handover.

Reverse link performance may also be improved by using antenna diversity at each individual BS. In this case, each BS or sector will be fitted with two (or more) antennas and the demodulating fingers are assigned to the four best received paths, regardless of the antenna at which they arrive. In this case, the paths from the different antennas all arrive at the same BS and maximal ratio combining may be used to decode the information. The soft handover process from the reverse link point of view is summarised in Figure 4.37.

So far we have only considered a soft handover between two different BSs located at two geographically separated sites. A second form of soft handover may occur between two

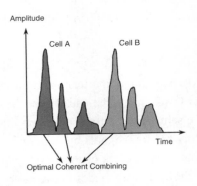

Figure 4.36: Soft handover combining at the MS.

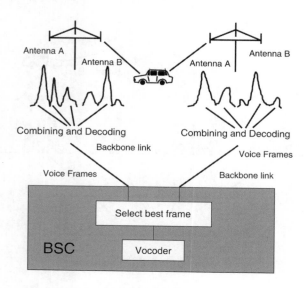

Figure 4.37: Soft handover combining at the BS.

sectors at the same cell site. In this case the signals that arrive at the two different sectors from the same MS will be available at the same BS, and therefore maximal ratio combining may be performed. In practice, each demodulating element (i.e. RAKE finger) will be able to 'see' the signals arriving at all the antennae in each of the sectors and the receiver will always select the four best paths, regardless of the antenna on which they arrive. This ability means that a soft handover between different sectors of the same BS is more effective, from an reverse link view point, than a soft handover between different BSs. To make the distinction between these two types of handovers, a soft handover between different sectors of the same cell is called a 'softer' handover. The softer handover is shown in Figure 4.38.

On the forward link, the overall effect of a soft handover will be to increase the number of available signal paths at the MS and this can only enhance the quality of the forward link as the MS moves towards the cell edge. Reverse link power control continues to operate during soft handover; however, we note that the power control bits on the forward links from the two BSs may be contradictory. This problem is resolved by increasing the MS transmitted power only when both bits indicate that an increase is required. For any other combination of power control bits the reverse link power is decreased. This means that the MS is always power controlled by the BS that is receiving the best signal and, as a result, an MS in soft handover does, on average, have a lower transmitted power than an MS that is not in soft handover.

The system controller begins the handover process by assigning a CDMA channel at the new BS. The new BS searches for, and acquires, the signal from the MS, and then it commences transmission of the appropriate forward link traffic to the MS. The active

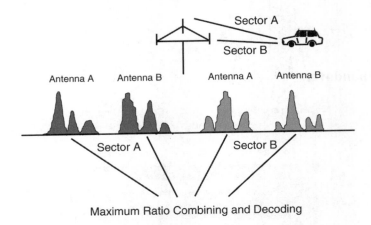

Figure 4.38: Softer handover.

set at the MS is now updated to include both the new and the old BS channels using a *handoff direction message* transmitted by the network. Both BSs are able to demodulate the reverse link traffic channel data from the MS. This enables the system controller to perform handover diversity combining thereby improving the reverse link traffic channel performance.

If, while a soft handover is in progress, a BS becomes fully loaded, then the system controller can direct an MS to discontinue using all channels, except for one, by sending a handover direction message. In other words, the MS is directed to use a particular channel at a particular BS.

Figure 4.39 gives an example of the message exchanges between the BS and MS during a typical soft handover. The key points highlighted on the diagram are as follows.

1. The pilot strength exceeds T_ADD. Mobile sends a pilot strength measurement message and transfers the pilot to the candidate set.

2. The BS sends a handoff direction message.

3. The MS transfers the pilot to its active set and sends a handoff completion message.

4. The pilot strength drops below T_DROP and the MS starts the handoff drop timer.

5. The handoff drop timer expires. The MS sends a pilot strength measurement message.

6. The BS sends a handoff direction message.

7. The MS moves the pilot from the active set to the neighbour set and sends a handoff completion message.

4.3.5 Hard handover

Handovers between CDMA channels having different frequency assignments are implemented as hard handovers. They are initiated by a BS sending a message to an MS directing it to the new BS channel. The MS acknowledges receipt of this message using the old CDMA channel. The MS then disables its transmitter and attempts acquisition of the newly assigned channel. Following successful acquisition of this channel, the MS re-enables its transmitter. No diversity advantage is possible when executing a hard handover; hence soft handovers are preferred.

The need for a hard handover may arise where an operator has two different frequency allocations in different areas. This is a situation which may be common in the United States where a PCS operator may have acquired the 'A' spectrum block in one area and the 'B' block in a neighbouring area. We also note that a hard handover may be required between BSs that are connected to different BSCs, since there may be no facility to transfer the additional signalling information that occurs during a soft handover between two different BSCs.

4.3.6 Power control

4.3.6.1 Reverse link power control

In the previous sections we have described the mechanisms that the cdmaOne system uses to accurately control the received power of each MS within a cell. The closed-loop mechanism allows the MS output power to be adjusted by ± 1 dB every 1.25 ms (i.e. 800 times per second). The range of the closed-loop power control must be ± 24 dB and this is perfectly adequate to cope with the fast fading that is imposed by a radio channel where the larger fades are typically around 30 dB. However, it is not sufficient to cope with the large variations in the path loss that will occur as the MS moves relative to the BS which could be greater than 120 dB. These variations will have two components: the attenuation due to the distance between the MS and BS, and the shadow fading. For this reason the MS continuously moves the centre of its closed-loop power control window so that it aligns with the open-loop estimate. In practice, the MS generates this open-loop estimate by low-pass filtering the received power from the BS to remove the fast fading components. In this way the MS tracks both the rapid fades and the slower variations in the path loss.

During our explanations we have stated that MSs adjust their transmitted power such that the received power at their BS is the same irrespective of the location of the MSs. In an

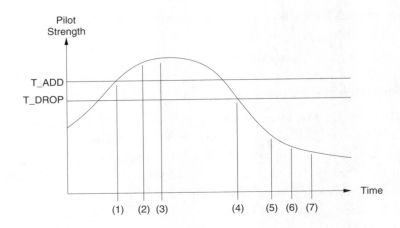

Figure 4.39: Soft handover example.

operational situation the main objective of the reverse link power control system is to deliver a constant reverse link quality for each MS within a cell, and this quality is measured in terms of the frame erasure rate (FER). A frame will be 'erased', or discarded, when the CRC shows that it still contains errors that have not been corrected by the convolutional decoding process. For speech transmissions, the erased frame is interpolated from the speech frames on either side of it. The FER measures the ratio of erased frames to the total number of frames transmitted, and typically a FER of 1% is acceptable.

The FER is measured at the BSC, and during a soft handover it is the combined value for all the BSs involved in the soft handover process. Using this measurement, the BSC adjusts the E_b/I_0 for each MS to achieve the target FER (e.g. 1%). When the FER is lower than the target it reduces the E_b/I_0, and when it is higher than the target value it increases the E_b/I_0. The E_b/I_0 value is sent to the BS and the power control sub-channel is used to maintain the received E_b/I_0 for each MS at this target value. The BS receiver measures the I_0 value by measuring the total received power at its input and dividing this value by the spread bandwidth, W. Although the receiver knows approximately both the wanted signal and the unwanted interference levels at its input, the wanted signal level is very much smaller than the interference, and it is fair to assume that I_0 is the power spectral density (PSD) of the signal at the input of the receiver. E_b is calculated by summing the signal power that is received on each finger of the RAKE receiver and dividing this by the data rate.

The reverse link power control mechanism involves two loops. The outer loop is closed at the BSC and it attempts to maintain a constant FER by adjusting the E_b/I_0 set point. The inner loop is closed at the BS and it attempts to maintain a constant received E_b/I_0 for each MS by adjusting the MS transmitter power using the power control sub-channel. The outer loop will have a much larger time constant than the inner loop.

Within a cdmaOne system, each MS has its own target E_b/I_0 which will vary according to its location and speed. For example, a stationary or slow moving MS is likely to have a very low target E_b/I_0, since its channel will be changing very slowly and the search receiver and power control system are able to track these changes. As the MS speed increases, the target E_b/I_0 increases as the receiver and power control mechanism are less able to track the more rapid changes in the channel, and this leads to bursts of errors as the MS moves through fades. At slow MS speeds these error bursts will be relatively long, i.e. longer than the interleaving depth of 20 ms and they are likely to overload the error correction processes at the receiver. As the MS speed increases further the length of the error bursts reduces, since the MS moves through the fades more rapidly. Where the error bursts are relatively short, compared with the interleaving depth, the error correction processes at the receiver will be more effective and this will result in a reduction in the target E_b/I_0. Reference [5] includes a graph of the target E_b/I_0 versus MS speed for an operating frequency of 850 MHz and an FER of 1%. This shows that the E_b/I_0 value reaches a peak of around 6dB at around 60 km/h. It also shows that the required E_b/I_0 is around 3 dB for stationary MSs and around 5 dB for a MS speed of 200 km/h. The power control mechanism will also tend to reduce the target E_b/I_0 for an MS in a soft handover, since the use of two BSs will result in the system being able to achieve the required FER at a lower E_b/I_0.

4.3.6.2   Forward link power control

In a single cell CDMA system there would be no requirement for forward link power control. However, in a practical system consisting of a number of cells, the forward link performance is improved by transmitting more power to MSs at a cell edge, since these are experiencing more interference from the neighbouring cell sites. The forward link power for a particular MS is adjusted in steps of around 0.5 dB once every 15–20 ms and the total dynamic range is limited to around ±6 dB [8]. The forward link power control mechanism is a closed-loop process whereby the BS periodically reduces the power that is transmitted to each subscriber. This process continues until the FER at the MS exceeds an acceptable level and triggers a request for an increase in the forward link power from the MS. At this point the BS increases the power that is allocated to the particular MS until the requests for additional power cease.

Bibliography

[1] TIA/EIA/IS-95, *Mobile Station – Base Station Compatibility Standard for Dual-Mode Wideband Spread Spectrum Cellular System*, July 1993.

[2] JTC(AIR)/94.08.01-022R2, *PN-3384, Personal Station – Base Station Compatibility Requirements for 1.8 to 2.0 GHz Code Division Multiple Access (CDMA) Personal Communications Systems*, 1 Aug. 1994.

[3] Viterbi A.J., *CDMA – Principles of Spread Spectrum Communication*, Addison-Wesley, 1995.

[4] Steele R. (Ed.), *Mobile Radio Communications*, Pentech Press, 1992.

[5] Padovani R., Reverse link performance of IS-95 based cellular systems, *IEEE Pers. Commun.*, Third Quarter 1994, 28–34.

[6] Price R. and P.E. Green Jr, A communication technique for multipath channels, *Proc. IRE*, **46**, Mar. 1958, 555–570.

[7] Proakis J.G., *Digital Communications*, McGraw-Hill, 1989.

[8] Qualcomm Inc., *An Overview of the Application of Code Division Multiple Access (CDMA) to Digital Cellular Systems and Personal Cellular Networks*, Qualcomm Inc. Doc. No. EX60-10010, 21 May 1992.

Chapter 5

Analysis of IS-95

5.1 List of Mathematical Symbols

a_{ij}	path loss and shadow fading between the zeroth BS and the ith MS in the jth microcell
$a_0(t)$	path loss and shadow fading multiplicative factor
$b_i(t)$	data sequence for the ith user
b_0	value of $b(t)$ over the zeroth symbol period $(= \pm 1)$
$C_{PN}(t)$	down-link pilot codes
$C_{W(i+1)}(t)$	$(i+1)$th Walsh code
c_{ij}	spreading code of of ith MS in the jth cell
$c_i(t)$	code for the ith user
$cc(n,k,K)$	convolutional code
D_j	distance between the zeroth BS and the adjacent jth cell site
da	area housing an MS
d_f	minimal free distance
$(E_b/I_0)_{im}$	SIR in the presence of power control errors
$E[(\cdot)]$	expectation of $(\cdot)$
$erfc\,[(\cdot)]$	complementary error function of $(\cdot)$
e_s	sectorisation efficiency
F	a factor used in estimating E_b/I_0
$f(r)$	PDF of an MS being in a ring of area $2\pi r dr$
$f(r_j/r,\sigma)$	expectation of $10^{\zeta/10}$ with the constraint function $\phi(\zeta,r/r_j)$

G_c	asymptotic coding gain of convolutional code
G_p	processing gain $= T_b/T_c$
g	controls the rate of increase in the step size by the adptive power control algorithm
$g(r_j/\gamma,\sigma)$	variance of $10^{\zeta/10}$ with the constraint function $\phi(\zeta, r/r_j)$
$h_j(t - \tau_j)$	jth component of the impulse response
I	total interference power at the output of the matched filter
I_{ext}	intercellular interference power at the output of the matched filter
I'_{ext}	I_{ext} in the presence of power control errors
I_{int}	intracellular interference power
I'_{int}	I_{int} in the presence of power control errors
I_j	interference power from all MSs in the jth cell to the zeroth BS
I_{MAI}	multiple access interference
I_0	interference PSD
$I(r_j,r)$	interference power at the zeroth BS from the active MS in the jth cell
J	number of interfering cells
K	constraint length of the convolutional code
k	number of message bits in the convolutional code
k	number of active users within a cell
k_{CDMA}	capacity as channels per cell per MHz
L	number of match filters, or number of resolvable paths by the RAKE receiver
M	number of chips in a spreading code $(= G_p)$
M'	spreading factor in the presence of convolutional coding
N	number of mobile users
N_0	single-sided PSD of the AWGN
$n_{I,m}(t); n_{Q,m}(t)$	inphase and quadrature components of the multipath interference
n	convolutional code length
$n_{ext}(t)$	equivalent baseband intercellular interference
$n_{I,ext}(t); n_{Q,ext}(t)$	inphase and quadrature componenets of $n_{ext}(t)$
$n_{int}(t)$	equivalent baseband intracellular interference
$n_{I,total}(t); n_{Q,total}(t)$	inphase and quadrature components of the total interference noise
N_s	number of sectors per cell

$n(t)$	receiver noise
P_i	transmitted power of the ith MS, or from the BS for the ith MS
P_{ij}	transmitted power from the ith MS in the jth cell (up-link), or the transmitted power allocated for the ith channel at the jth BS (down-link)
P_m	power allocated to each MS by a BS
P_o	outage probability, i.e. probability of the BER $> 10^{-3}$
P_p	transmitted pilot power
P_R	received wanted signal power
P_T	transmitted power of an MS in a power control system
P_{tar}	target received signal power at a BS
p_b	bit error probability
P_p	transmitted power of the pilot signal
$\Pr(\cdot)$	probability of $(\cdot)$
$Q(\cdot)$	Gaussian Q-function
R	cell radius
R_b	bit rate of the message sequence
R_c	chip rate
$R_{dn}(t)$	received signal at an MS
$R_I(t); R_Q(t)$	inphase and quadrature components of $R(t)$, the received baseband signal at the BS
R_0	distance from a BS where 'near-in' MSs are present
$R_{up}(t)$	received signal at the BS from an MS
r	distance of an MS from a BS
S	desired received power at a BS
S'	signal power at the output of the matched filter in the presence of imperfect power control
$S_I(t); S_q(t)$	inphase and quadrature components of the wanted signal
S_p	received pilot power compoment for the zeroth MS on the down-link
$s_{dn}(t)$	signal transmitted from a BS
$s_{ij}(t)$	transmitted signal from the ith MS to the jth BS
$s_{dn}^j(t)$	signal transmitted from the jth neighbouring BS
$s_0(t)$	spread BPSK signal for zeroth MS
T_b	bit duration
T_c	chip duration

W	chip rate and bandwidth of the CDMA signal, also the width of a street in a street microcell
X	distance from a street microcellualr BS to the end of the microcell
X_b	a break distance in a street microcell where the propagation path loss exponent changes
$x_{ij}(t)$	transmitted baseband signal from the ith MS to the jth BS
$Z_{ext}(T_b)$	intercellular interference component of $Z(T_b)$
$Z_{int}(T_b)$	intracellular interference at the output of the matched filter
$Z_n(T_b)$	receiver noise at the output of the matched filter
$Z(T_b)$	output of the matched filter at time $t = T_b$
$Z_w(T_b)$	wanted component of $Z(T_b)$
β_d	coefficient of the transfer function $T(D,H)$ of the convolutional code
β_l	magnitude of the lth path of the fast fading channel impulse response
$\oplus$	exclusive-OR operation
$\wedge$	system parameter in the power control algorithm
δ_i	normally distributed received error power random variable at a BS for MS_i
δ_{ij}	power control error for the ith MS in the jth cell
$\delta(t-u)$	delta function at time u
η	AWGN power at the output of the matched filter
γ_b	E_b/I_0, or energy per bit per interference PSD
γ_c	E_c/I_0, or energy per symbol per interference PSD
γ_{req}	required (E_b/I_o) for BER $< 10^{-3}$
λ_{ij}	normally distributed random variable with standard deviation σ and zero mean
μ	voice activity factor (VAF)
ν_i	voice activity variable of the ith user
ν_{ij}	voice activity variable of the ith user in the jth cell
ω_2	down-link angular carrier frequency
ω_1	up-link angular carrier frequency
Φ	fixed step size used in power control algorithm
ϕ_{ij}	carrier phase between the interference signal from the ith MS in the jth cell and the zeroth MS in the zeroth cell
ϕ_0	carrier phase difference $\hat{\theta}_0 - \hat{\theta}$
$\phi(\zeta, \frac{r}{r_j})$	constraint function

ρ	density of MSs in a cell
σ_ε	standard deviation of ε
σ_e	standard deviation of δ_i
τ_i	random delay of the ith user signal at the BS on the up-link, or the random offset at the BS on the down-link
τ_{ij}	relative propagation delays of the ith MS in the jth cell with respect to the zeroth MS in the zeroth cell
τ_p	time offset of the pilot signal at the BS
θ	received carrier phase angle at the BS, or the transmitted phase angle of the carrier at the BS
θ_i	random phase angle of the trnsmitted ith mobile carrier
$\hat{\theta}_i$	change in the phase angle of the ith MS $= \omega_1(\tau_0 - \tau_i) + \theta_i$
θ_{ij}	carrier phase of ith MS in the jth cell
θ_0	overlapping angle of adjacent sectors
$\triangle_i$	adaptive step size used in the power control algorithm
$\text{var}[(\cdot)]$	variance of $(\cdot)$
ε	$\delta_j - \delta_0$ random variable having a normal distribution
ξ	error in estimating P_R in the power control
ζ	$\lambda_{ij} - \lambda_0$

5.2   Introduction

In CDMA many mobiles use the same RF bandwidth at the same time, and a CDMA receiver is able to separate the wanted signal from the other mobile signals if it knows the spreading code used in the generation of the wanted CDMA signal. This demodulation process occurs in the presence of interference generated by other mobile users. This interference is a major limitation on the capacity of a CDMA system.

In this chapter the capacity of a CDMA system in tessellated hexagonal cells and city street microcells is investigated. The system performance in terms of outage probability for a bit error rate (BER) larger than a minimal required level is analysed. The number of users that can be supported by a cell for a given outage probability is evaluated. The corresponding capacity in terms of channels per cell per MHz is calculated according to this number of users per cell. Our discussion concentrates on the capacity evaluation rather than on other issues, such as code synchronisation. We begin by examining a single cell CDMA system before moving on to a multiple cell CDMA system. Since the arrangement of the up-link, or forward link, is different from the down-link, or reverse link, the performances of both the up-link and down-link are considered. The effect of sectorisation and channel coding on CDMA systems is also discussed.

5.3   CDMA in a Single Macrocell

Consider a single cell CDMA communication system using binary phase shift keying (BPSK) spread spectrum modulation. As shown in Figure 5.1, the BS uses the angular carrier frequency ω_2 on the down-link to communicate with all its mobiles, while mobiles transmit to their base stations (BSs) via the angular carrier frequency ω_1.

5.3.1   The up-link system

The CDMA single cell system consists of N mobile users transmitting to a BS receiver on the up-link. We consider a simplified mobile transmitter consisting of a BPSK modulator, formed by multiplying the data sequence for the ith user $b_i(t)$, by a carrier $\cos \omega_1 t$. Spreading occurs when the BPSK signal is multiplied by the code $c_i(t)$. This is equivalent to multiplying the data signal, $b_i(t)$, by $c_i(t)$ and this spread data signal modulates the carrier $\cos \omega_1 t$. Figure 5.2 shows the arrangement.

Let us consider a particular user, say the zeroth one. The spread BPSK signal $s_0(t)$ is applied to the radio channel shown in Figure 5.3. We have separated this channel into a part that allows for path loss and slow fading and is represented by the multiplicative factor a_0. The fast fading is represented by a number of impulse responses $h_j(t - \tau_j), j = 0, 1, \ldots, L$. The input of the receiver consists of: interference from the other users in the cell and is known as *intracellular interference*; the receiver noise $n(t)$; and the received signal for the zeroth user. The sum of these signals, $R_{up}(t)$, is demodulated by multiplying by a recovered carrier having the same frequency but different phase, relative to the transmitted carrier.

The resulting signal is applied to a RAKE receiver that may be considered to be composed of L matched filters, one for each significant path in the impulse response of the channel. We note that in general the number of matched filters and the number of channels will not be the same, but it is desirable if there are at least as many matched filters as there are significant paths in the channel. The RAKE receiver is a maximum ratio diversity system if it can obtain accurate estimates of the complex impulse responses $h_j(t - \tau_j)$. The RAKE receiver is described in Section 2.3.2.6.

A CDMA system has other attributes to combat the effects of fast fading on the signal $s_0(t)$. These include symbol interleaving, forward error correction (FEC) coding, space diversity reception, power control, and so forth. Using this battery of techniques we can effectively compensate for the effects of fast fading. The channel model is now reduced to the multiplicative factor a_0 which accounts for path loss and slow fading. The BS receiver may now be configured for our analysis as one having despreading followed by a matched filter, i.e. single stage RAKE, which is an integrator and dump circuit for each mobile. Our simplified model of the radio channel and the BS receiver is depicted in Figure 5.4.

Each user has a unique spreading code that is known to the BS. The spreading codes are

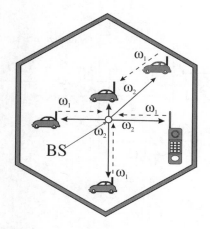

Figure 5.1: Single cell mobile radio communications in a hexagonal cell.

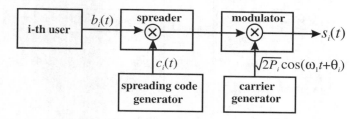

Figure 5.2: A mobile's CDMA transmitter diagram.

of length M chips, or an M-chip segment from the long psuedo noise (PN) sequence [1,2]. As the mobiles are in different locations within the cell, the transmission delay for each mobile is different. The signal transmitted from the ith user to its BS is

$$s_i(t) = \sqrt{2P_i}b_i(t)c_i(t)\cos(\omega_1 t + \theta_i), \tag{5.1}$$

where P_i is the transmitted power of the ith user, $b_i(t)$ is the data sequence of the ith user where each bit has an amplitude of ± 1 and a duration of T_b, $c_i(t)$ is the spreading code sequence of ith user and each of the M chips per code has a duration T_c, and θ_i is the random phase of the ith mobile carrier and is uniformly distributed in $[0, 2\pi)$. All the mobiles transmit their signals to the BS receiver over the same radio channel, and the received signal at the BS receiver is

$$
\begin{aligned}
R_{up}(t) &= \sum_{i=0}^{N-1} a_i s_i(t - \tau_i) + n(t) \\
&= \sum_{i=0}^{N-1} a_i \sqrt{2P_i} b_i(t - \tau_i) c_i(t - \tau_i) \cos\left[\omega_1(t - \tau_i) + \theta_i\right] \\
&\quad + n(t), \tag{5.2}
\end{aligned}
$$

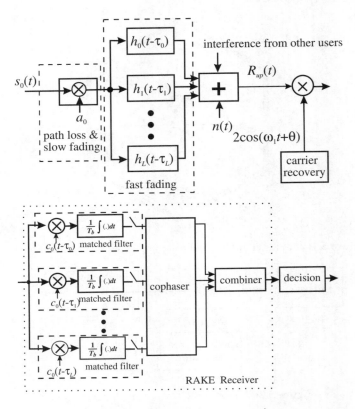

Figure 5.3: The up-link representation.

where a_i represents the path loss and slow fading of the ith user, τ_i is the random delay of the ith user signal at the receiver and is uniformly distributed in $[0, T_b)$, and $n(t)$ is the additive white Gaussian noise (AWGN) of the receiver noise. The signal at the output of the zeroth matched filter is given by

$$
\begin{aligned}
Z(T_b) &= \frac{1}{T_b} \int_{\tau_0}^{T_b+\tau_0} R_{up}(t) c_0(t - \tau_0) 2\cos(\omega_1 t + \theta)\,dt \\
&= \left\{ \frac{1}{T_b} \int_{\tau_0}^{T_b+\tau_0} \sum_{i=0}^{N-1} a_i \sqrt{2P_i}\, b_i(t - \tau_i) c_i(t - \tau_i) \cos[\omega_1(t - \tau_i) + \theta_i] \right. \\
&\quad \left. \times\, c_0(t - \tau_0) 2\cos(\omega_1 t + \theta)\,dt \right\} + \left\{ \frac{1}{T_b} \int_{T_0}^{T_b+\tau_0} n(t) c_0(t - \tau_0) \right. \\
&\quad \left. \times\, 2\cos(\omega_1 t + \theta)\,dt \right\},
\end{aligned}
\tag{5.3}
$$

where θ is the carrier phase angle in the receiver. Note that due to the propagation delay

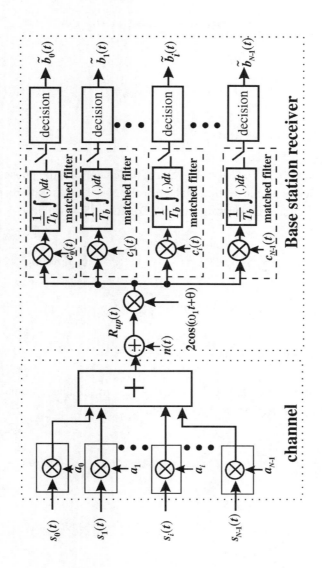

Figure 5.4: Simplified radio channel and BS receiver block diagram.

of the radio path associated with the zeroth user, the integration is done from $t = \tau_0$ to $t = \tau_0 + T_b$. Letting $t = t + \tau_0$ in Equation (5.3), we have

$$
Z(T_b) = \left\{ \frac{1}{T_b} \int_0^{T_b} \sum_{i=0}^{N-1} a_i \sqrt{2P_i} b_i(t + \tau_0 - \tau_i) c_i(t + \tau_0 - \tau_i) \right.
$$
$$
\times \cos(\omega_1 t + \hat{\theta}_i)
$$
$$
\left. \times c_0(t) 2 \cos(\omega_1 t + \hat{\theta}) dt \right\} + \left\{ \frac{1}{T_b} \int_0^{T_b} n(t + \tau_0) c_0(t) \right.
$$
$$
\left. \times 2 \cos(\omega_1 t + \hat{\theta}) dt \right\},
$$

(5.4)

where $\hat{\theta} = \omega_1 \tau_0 + \theta$ and $\hat{\theta}_i = \omega_1(\tau_0 - \tau_i) + \theta_i$ are the changes in the phase angle of θ and θ_i, respectively. We define a relative time delay for the $N-1$ users with respect to the zeroth user of $\tau_{i0} = \tau_i - \tau_0$, and on substituting into Equation (5.4) we have

$$
Z(T_b) = \left\{ \frac{1}{T_b} \int_0^{T_b} \sum_{i=0}^{N-1} a_i \sqrt{2P_i} b_i (t - \tau_{i0}) c_i (t - \tau_{i0}) \cos(\omega_1 t + \hat{\theta}_i) \right.
$$
$$
\left. \times c_0(t) 2 \cos(\omega_1 t + \hat{\theta}) dt \right\} + \left\{ \frac{1}{T_b} \int_0^{T_b} n(t + \tau_0) c_0(t) \right.
$$
$$
\left. \times 2 \cos(\omega_1 t + \hat{\theta}) dt \right\}.
$$

(5.5)

Owing to the stationary property of the AWGN, $n(t + \tau_0)$ in the above equation can be substituted by $n(t)$, and Equation (5.5) can be rewritten as

$$
Z(T_b) = \left\{ \frac{1}{T_b} \int_0^{T_b} \sum_{i=0}^{N-1} a_i \sqrt{2P_i} b_i (t - \tau_{i0}) c_i (t - \tau_{i0}) \cos(\omega_1 t + \hat{\theta}_i) \right.
$$
$$
\left. \times c_0(t) 2 \cos(\omega_1 t + \hat{\theta}) dt \right\}
$$
$$
+ \frac{1}{T_b} \int_0^{T_b} n(t) c_0(t) 2 \cos(\omega_1 t + \hat{\theta}) dt.
$$

(5.6)

Assuming the receiver is chip synchronised to the zeroth mobile, then for the zeroth mobile, $c_i(t - \tau_{i0})$ becomes $c_0(t)$ and from Equation (5.6) $c_i(t - \tau_{i0})$ for $i = 0$, multiplied by $c_0(t)$ yields unity, and therefore the wanted component of $Z(T_b)$ is

$$
Z_w(T_b) = \frac{1}{T_b} \int_0^{T_b} a_0 \sqrt{2P_0} b_0(t) \left[\cos \phi_0 + \cos(2\omega_1 t + \hat{\theta}_0 + \hat{\theta}) \right] dt
$$

$$= a_0\sqrt{2P_0}b_0\cos\phi_0 \tag{5.7}$$

since the average of a cosine wave over many periods is zero, b_0 is the value of $b(t)$ over a symbol period having values of $+1$ or -1, and $\phi_0 = \hat{\theta}_0 - \hat{\theta}$. Note that the term $Z_w(T_b)$ contains the original data sequence b_0 scaled by $a_0\sqrt{2P_0}\cos\phi_0$.

The intracellular interference at the output of the matched filter is

$$Z_{int}(T_b) = \sum_{i=1}^{N-1}\frac{1}{T_b}\int_0^{T_b} a_i\sqrt{2P_i}b_i(t-\tau_{i0})c_i(t-\tau_{i0})c_0(t)$$
$$\times\{\cos(\phi_i)+\cos[2\omega_1 t+\hat{\theta}+\hat{\theta}_i]\}\,dt \tag{5.8}$$

and since the term involving $\cos 2\omega_1 t$ is averaged to zero,

$$Z_{int}(T_b) = \sum_{i=1}^{N-1}\frac{1}{T_b}\int_0^{T_b} a_i\sqrt{2P_i}b_i(t-\tau_{i0})c_i(t-\tau_{i0})c_0(t)\cos\phi_i dt, \tag{5.9}$$

where

$$\phi_i = \hat{\theta}_i - \hat{\theta}. \tag{5.10}$$

We may express $Z_{int}(T_b)$ as

$$Z_{int}(T_b) = \frac{1}{T_b}\int_0^{T_b} n_{int}(t)c_0(t), \tag{5.11}$$

where

$$n_{int}(t) = \sum_{i=1}^{N-1} a_i\sqrt{2P_i}b_i(t-\tau_{i0})c_i(t-\tau_{i0})\cos\phi_i \tag{5.12}$$

is the equivalent baseband intracellular interference. The receiver noise term is, from Equation (5.6),

$$Z_n(T_b) = \frac{2}{T_b}\int_0^{T_b} n(t)c_0(t)\cos(\omega_1 t+\hat{\theta})dt. \tag{5.13}$$

Let us consider the intracellular interference shown in Equation (5.12) that comes from the other $N-1$ users. We are cognisant that $c_i(t-\tau_{i0})$ are the independent spreading codes for different users and that the relative time offset of the data transmitted from each mobile is a random variable, i.e. τ_{i0} is an independent random variable that is uniformly distributed over $[0, T_b)$. We further assume that $b_i(t)$ represents random independent binary data, and as a consequence the intracellular interference is a stationary random process. From the Central Limit Theorem, the summation of $N-1$ independent random process means that n_{int} can be approximated as a Gaussian random variable [3,4].

5.3.1.1   Perfect power control

Since all users are sharing the same radio frequency, a strong signal from mobiles close to the BS will mask weak signals from distant users. To reduce this so-called near–far problem, as well as to reduce the interference from other users, it is important to exercise a power control on the up- link of CDMA transmissions so that the received signal power levels from all users remain close to a target power, P_{tar}. Identically, the received power from each user at the BS is controlled to be the constant target power, P_{tar}, namely

$$a_i^2 P_i = P_{tar}, \qquad \text{for } i = 0, 1, \ldots, N-1. \tag{5.14}$$

With the aid of Equations (5.14), (5.7), (5.9) and (5.13) we may express Equation (5.3) as

$$Z(T_b) = \sqrt{2P_{tar}} b_0 \cos \phi_0 + Z_{int}(T_b) + Z_n(T_b), \tag{5.15}$$

where the first term is the desired signal, the second term is the interference from the $N-1$ users in the cell, and the last term is the AWGN component. The bit error probability at the output of the bit regeneration circuit depends upon the bit-energy-to-total-interference power spectral density (PSD) ratio or signal-to-total-interference power ratio (*SIR*). According to Equation (5.15), the average power of the wanted signal component is

$$S = P_{tar} b_0^2 = P_{tar}, \tag{5.16}$$

while the total noise power is the sum of the interference power coming from other users and the AWGN power of the receiver. The AWGN power at the output of the matched filter is given as

$$\eta = \text{var}[Z_n(T_b)] = \text{var}\left[\frac{2}{T_b} \int_0^{T_b} n(t) c_0(t) \cos(\omega_1 t + \hat{\theta}) dt\right], \tag{5.17}$$

where $n(t)$ is the AWGN component. Consequently,

$$
\begin{aligned}
\eta &= \frac{4}{T_b^2} E\left[\int_0^{T_b} n(t) c_0(t) \cos(\omega_1 t + \hat{\theta}) dt\right]^2 \\
&= \frac{4}{T_b^2} E\left[\int_0^{T_b} \int_0^{T_b} n(t) n(u) c_0(t) c_0(u) \cos(\omega_1 u + \hat{\theta}) \cos(\omega_1 t + \hat{\theta}) du dt\right] \\
&= \frac{4}{T_b^2} \int_0^{T_b} \int_0^{T_b} E[n(t) n(u)] E\left[c_0(t) c_0(u) \cos(\omega_1 u + \hat{\theta})\right. \\
&\quad \left. \times \cos(\omega_1 t + \hat{\theta})\right] dt \, du.
\end{aligned}
\tag{5.18}
$$

But for an AWGN having a double-sided PSD of $\frac{1}{2} N_0$,

$$E[n(t) n(u)] = \frac{1}{2} N_0 \delta(t - u), \tag{5.19}$$

where $\delta(t-u)$ is a delta function at $t = u$. So,

$$
\begin{aligned}
\eta &= \frac{4}{T_b^2} \int_0^{T_b} \int_0^{T_b} \frac{1}{2} N_0 \delta(t-u) \, \mathrm{E} \left\{ c_0(t) c_0(u) \cos(\omega_1 u + \hat{\theta}) \right. \\
&\qquad \left. \times \cos(\omega_1 t + \hat{\theta}) \right\} \mathrm{d}u \mathrm{d}t \\
&= \frac{2N_0}{T_b^2} \int_0^{T_b} c_0^2(u) \cos^2(\omega_1 u + \hat{\theta}) \mathrm{d}u \\
&= \frac{2N_0}{T_b^2} \frac{1}{2} \left[u + \frac{\sin(2\omega_1 u + 2\hat{\theta})}{2\omega_1} \right]_0^{T_b} \\
&= \frac{2N_0}{T_b^2} \frac{1}{2} T_b = N_0 R_b = \left(\frac{N_0}{T_c} \right) T_c R_b = \frac{N_0 W}{G_p},
\end{aligned}
\tag{5.20}
$$

where $R_b = 1/T_b$ is the bit rate of the message sequence $b_i(t)$, $W = 1/T_c$ is the chip rate and we assume it is also the bandwidth of the CDMA signal, and $W N_0$ is the noise power at the receiver input. Thus, after despreading, the noise power η is the input noise power decreased by the processing gain $G_p = T_b/T_c$. The intracellular interference power is

$$
\begin{aligned}
I_{int} &= \mathrm{var} \left[\frac{1}{T_b} \int_0^{T_b} n_{int}(t) c_0(t) \mathrm{d}t \right] \\
&= \frac{1}{T_b^2} \mathrm{E} \left[\int_0^{T_b} n_{int}(t) c_0(t) \mathrm{d}t \right]^2,
\end{aligned}
\tag{5.21}
$$

since $\mathrm{E}[n_{int}(t)c_0(t)] = 0$. Because the $n_{int}(t)$ in Equation (5.21) is a Gaussian random variable, its variance can be found as

$$
I_{int} = \frac{1}{T_b^2} \int_0^{T_b} \int_0^{T_b} \mathrm{E}[n_{int}(t)n_{int}(u)] \, \mathrm{E}[c_0(t)c_0(u)] \mathrm{d}u \mathrm{d}t.
\tag{5.22}
$$

Since $n_{int}(t)$ is Gaussian distributed having a power of $\mathrm{E}[n_{int}^2(t)]$ and a double-sided bandwidth of W, its double-sided PSD is

$$
\mathrm{E}[n_{int}(t)n_{int}(u)] = \frac{\mathrm{E}[n_{int}^2(t)]}{W} \delta(t-u).
\tag{5.23}
$$

Substituting the above equation into Equation (5.22):

$$
\begin{aligned}
I_{int} &= \frac{1}{T_b^2} \int_0^{T_b} \int_0^{T_b} \frac{\mathrm{E}[n_{int}^2(t)]}{W} \delta(t-u) \, \mathrm{E}[c_0(t)c_0(u)] \mathrm{d}u \mathrm{d}t \\
&= \frac{1}{T_b^2} \frac{\mathrm{E}[n_{int}^2(t)]}{W} \int_0^{T_b} \mathrm{d}u = \frac{\mathrm{E}[n_{int}^2(t)]}{G_p},
\end{aligned}
\tag{5.24}
$$

where the intracellular interference power is

$$
\mathrm{E}[n_{int}(t)]^2 = \mathrm{E}\left[\sum_{i=1}^{N-1} 2a_i^2 P_i b_i^2(t-\tau_{i0})c_i^2(t-\tau_{i0})\cos^2\phi_i\right] = \sum_{i=1}^{N-1} a_i^2 P_i \tag{5.25}
$$

because the expectation of $\cos^2\phi_i$ is 0.5.

By applying voice activity detection (VAD) and thereby discontinuous transmitting (DTX), the mobiles transmit only when speech signal is present. We introduce a voice activity variable v_i which is equal to 1 with probability of μ, and to 0 with probability of $1-\mu$, where μ is defined as the voice activity factor (VAF). By multiplying Equation (5.25) by v_i, and with the aid of Equations (5.14) and (5.16),

$$
\frac{I_{int}}{S} = \frac{1}{G_p}\sum_{i=1}^{N-1}\frac{v_i a_i^2 P_i}{S} = \frac{1}{G_p}\sum_{i=1}^{N-1} v_i , \tag{5.26}
$$

where S is the target power P_{tar} of Equation (5.16) for this case of perfect power control. Thus the intracellular interference-to-signal power ratio given by the summation term in Equation (5.26) is also reduced by a factor of G_p after the process of matched filtering.

The energy per bit E_b measured at the output of the matched filter is a random variable because of the variations in the path loss, slow fading and fast fading of the mobile channel. The interference PSD I_0 measured at the output of the matched filter is also a random variable because it depends on the interference being generated by mobiles roaming within the cell. We therefore need to take the expectation of the ratio of E_b to I_0, namely E_b/I_0, in determining the probability of symbol error. Now $E_b = ST_b$ and $I_0 = I/R_b = IT_b$, where I is the total interference power at the output of the matched filter. Consequently,

$$
\frac{E_b}{I_0} = \frac{S}{I} = SIR. \tag{5.27}
$$

Now the I is the sum of I_{int} and η, enabling us to express

$$
\frac{E_b}{I_0} = \frac{1}{\frac{I_{int}}{S} + \frac{\eta}{S}} . \tag{5.28}
$$

In Equation (5.26), the summation of v_i over $(N-1)$ users may be expressed, upon taking its expectation, as

$$
\mathrm{E}\left[\sum_{i=1}^{N-1} v_i\right] = \mu(N-1) \tag{5.29}
$$

so that

$$
\frac{E_b}{I_0} = \frac{1}{\frac{\mu(N-1)}{G_p} + \frac{\eta}{S}} \tag{5.30}
$$

From Equation (5.30), the bit error rate (BER) for the BPSK can be expressed as

$$p_b = \frac{1}{2}\operatorname{erfc}\left(\sqrt{\frac{E_b}{I_0}}\right), \tag{5.31}$$

where erfc(σ) is the complementary error function [5]. For a required BER, a required E_b/I_0, namely $(E_b/I_0)_{req}$ can be determined from Equation (5.31). Given $(E_b/I_0)_{req}$, the maximum number of active users, other than the zeroth user, that can be supported by the system is

$$m = \sum_{i=1}^{N-1} v_i = \left\lfloor \frac{G_p}{\left(\frac{E_b}{I_0}\right)_{req}} - \frac{G_p}{\frac{S}{\eta}} \right\rfloor, \tag{5.32}$$

where $\lfloor x \rfloor$ represents the largest integer that is smaller than x. Provided the number of active users does not exceed m, the required BER is secured. However, when the number of active users is larger than m, the BER will be greater than the required BER, and this situation is referred to as *system outage*. The outage probability of the single cell system is defined as

$$p_o = \Pr(BER > BER_{req}) = \Pr\left(\frac{E_b}{I_0} < \left(\frac{E_b}{I_0}\right)_{req}\right). \tag{5.33}$$

Since users in a cell are not active all the time, the number of active users is less than the number of potential users. Consequently, a cell can support more than m users, but the system will experience outage at those instances when the number of active users exceeds m. The outage probability is then the probability of the number of active users being greater than m, i.e.

$$p_o = \Pr\left(\sum_{i=1}^{N-1} v_i > m\right), \tag{5.34}$$

and because v_i is a random variable having a binomial distribution, the outage probability is

$$p_o = \Pr\left(\sum_{i=1}^{N-1} v_i > m\right) = \sum_{i=1}^{N-1} \binom{N-1}{i} \mu^i (1-\mu)^{N-1-i}. \tag{5.35}$$

5.3.1.2 Imperfect power control

In practice, the received signal power P_R from the ith mobile at its BS will differ from the target power level P_{tar} by δ_i dB. This error power δ_i is a random variable that is normally distributed with a standard deviation σ_e and is discussed in detail in Section 5.6 and in References [6]–[8]. There are several reasons for δ_i being non-zero, such as the inaccuracies in measuring the received power, S, at a BS, and the inability to adjust the mobile transmitted power sufficiently fast to force δ_i to zero. The relationship between P_R and P_{tar} for the ith mobile may be expressed as

$$P_R = a_i^2 \, P_i 10^{\frac{\delta_i}{10}} = P_{tar} 10^{\frac{\delta_i}{10}} .$$ (5.36)

According to Equation (5.36), the signal power at the output of the matched filter for the wanted zeroth mobile is

$$S' = P_{tar} 10^{\frac{\delta_0}{10}} ,$$ (5.37)

and the intracellular interference is

$$I_{int} = \frac{1}{G_p} \sum_{i=1}^{N-1} v_i P_{tar} \, 10^{\frac{\delta_i}{10}} .$$ (5.38)

Consequently, the intracellular interference-to-signal ratio at the output of the matched filter becomes

$$\frac{I'_{int}}{S'} = \frac{1}{G_p} \sum_{i=1}^{N-1} v_i 10^{\frac{\delta_i - \delta_0}{10}} ,$$ (5.39)

where δ_0 and δ_i are two mutually independent random variables of power control errors for the signal and the intracellular interferers, respectively. By setting $\varepsilon = \delta_i - \delta_0$, we have from Equations (5.39),

$$\begin{aligned}
\frac{I'_{int}}{S'} &= \frac{1}{G_p} \sum_{i=1}^{N-1} v_i 10^{\frac{\varepsilon}{10}} \\
&= \frac{I_{int}}{S} 10^{\frac{\varepsilon}{10}} ,
\end{aligned}$$ (5.40)

where ε is a random variable with zero mean and a normal distribution having a standard deviation of $\sigma_\varepsilon = \sqrt{2}\sigma_e$. Following the same procedure as in the perfect power control case, the signal-to-interference power ratio can be written as

$$\left(\frac{E_b}{I_0} \right)_{im} = \frac{1}{\frac{I'_{int}}{S'} + \frac{\eta}{S'}} = \frac{1}{\frac{I_{int}}{S} 10^{\frac{\varepsilon}{10}} + \frac{\eta}{S'}} .$$ (5.41)

Owing to imperfect power control, there is an error, ε, in $(E_b/I_0)_{im}$. Because ε is a normally distributed random variable, the BER varies accordingly. In order to evaluate the system performance, we introduce the outage probability that is defined as the probability of a system's BER being greater than 10^{-3}, i.e.

$$\begin{aligned}
P_o &= \Pr(BER > 10^{-3}) \\
&= \Pr\left\{ \left(\frac{E_b}{I_0} \right)_{im} = \frac{S'}{I'_{int} + \eta} < \gamma_{req} \right\}
\end{aligned}$$

$$= \text{Pr} \left(\frac{I'_{int} + \eta}{S'} > \frac{1}{\gamma_{req}} \right)$$

$$= \text{Pr} \left(\frac{1}{G_p} 10^{\frac{\varepsilon}{10}} \sum_{i=0}^{N-1} v_i > \frac{1}{\gamma_{req}} - \frac{\eta}{S'} \right)$$

$$= \text{Pr} \left\{ 10^{\frac{\varepsilon}{10}} \sum_{i=0}^{N-1} v_i > G_p \left(\frac{1}{\gamma_{req}} - \frac{\eta}{S'} \right) \right\}, \tag{5.42}$$

where γ_{req} is the required E_b/I_0 to ensure that the BER is less than 10^{-3}. If the number of active users inside the cell is k, i.e. $\sum_{i=0}^{N-1} v_i = k$, then Equation (5.42) can be rewritten as

$$p_o = \text{Pr} \left\{ \left(k10^{\frac{\varepsilon}{10}} > G_p \left(\frac{1}{\gamma_{req}} - \frac{\eta}{S'} \right) \right) \middle| \left(\sum_{i=0}^{N-1} v_i = k \right) \right\}$$

$$\times \text{Pr} \left(\sum_{i=0}^{N-1} v_i = k \right) = p_1 \, p_2. \tag{5.43}$$

The outage probability p_o is the product of two probabilities, p_1 and p_2. We will first consider the probability that there are k active intracellular users,

$$p_2 = \text{Pr} \left(\sum_{i=0}^{N-1} v_i = k \right). \tag{5.44}$$

The variable v_i is either 1 or 0 depending on whether the ith mobile user is active or not. The probability that a user is active is μ, and is called the voice activity factor (VAF). It will be recalled that for the case of tossing a coin the probability of k heads in $(N-1)$ tossings is

$$\binom{N-1}{k} p^k (1-p)^{N-1-k}, \tag{5.45}$$

where p here is the probability of a head being tossed. In our case we replace p by the VAF, μ, and observe that k can range from 0 to $N-1$. Hence,

$$p_2 = \sum_{k=0}^{N-1} \binom{N-1}{k} \mu^k (1-\mu)^{N-1-k}, \tag{5.46}$$

where

$$\binom{N-1}{k} = \frac{(N-1)(N-2)\ldots(N-k)}{k!}. \tag{5.47}$$

Turning our attention to probability p_1 in Equation (5.43), we note that as ε is a Gaussian random variable,

$$p_1 = Q \left[\frac{G_p \left(\frac{1}{\gamma_{req}} - \frac{\eta}{S'} \right) - kE\left[10^{\frac{\varepsilon}{10}} \right]}{\sqrt{k \, \text{var} \left[10^{\frac{\varepsilon}{10}} \right]}} \right], \tag{5.48}$$

where

$$Q(\theta) \equiv \frac{1}{\sqrt{2\pi}} \int_{\theta}^{\infty} e^{-\frac{\lambda^2}{2}} d\lambda \tag{5.49}$$

is the Q-function. This follows for a Gaussian random variable X with mean μ and variance σ^2:

$$\Pr[X > x] = Q\left(\frac{x - \mu}{\sigma}\right). \tag{5.50}$$

Because ε is a random variable with normal distribution, the mean of the term $10^{\frac{\varepsilon}{10}}$ in Equation (5.48) can be derived by following the same procedure as used in Section 3.3, i.e.

$$
\begin{aligned}
\mathrm{E}\left[10^{\frac{\varepsilon}{10}}\right] &= \int_{-\infty}^{\infty} \exp\left[\frac{\varepsilon \ln(10)}{10}\right] \frac{\exp\left(\frac{-\varepsilon^2}{4\sigma_e^2}\right)}{\sqrt{4\pi\sigma_e^2}} d\varepsilon \\
&= \exp\left(\sigma_e \frac{\ln(10)}{10}\right)^2 \int_{-\infty}^{\infty} \frac{\exp\left\{-\frac{1}{2}\left[\frac{\varepsilon}{\sqrt{2}\sigma_e} - \frac{\ln(10)\sqrt{2}\sigma_e}{10}\right]^2\right\}}{\sqrt{4\pi\sigma_e^2}} d\varepsilon \\
&= \exp\left(\sigma_e \frac{\ln(10)}{10}\right)^2 \{1 - Q[\infty]\} = \exp\left(\sigma_e \frac{\ln(10)}{10}\right)^2, \tag{5.51}
\end{aligned}
$$

where σ_e is the standard deviation of the power control error. The variance of the $10^{\frac{\varepsilon}{10}}$ is

$$
\begin{aligned}
\mathrm{var}\left[10^{\frac{\varepsilon}{10}}\right] &= \mathrm{E}\left[10^{\frac{\varepsilon}{10}} - \mathrm{E}\left[10^{\frac{\varepsilon}{10}}\right]\right]^2 \\
&= \mathrm{E}\left[10^{\frac{\varepsilon}{10}}\right]^2 - \left\{\mathrm{E}\left[10^{\frac{\varepsilon}{10}}\right]\right\}^2. \tag{5.52}
\end{aligned}
$$

From Equation (5.51) we have

$$\mathrm{var}\left[10^{\frac{\varepsilon}{10}}\right] = \mathrm{E}\left[10^{\frac{\varepsilon}{5}}\right] - \left[\exp\left(\frac{\sigma_e \ln(10)}{10}\right)^2\right]^2, \tag{5.53}$$

where

$$
\begin{aligned}
\mathrm{E}\left[10^{\frac{\varepsilon}{5}}\right] &= \int_{-\infty}^{\infty} \exp\left[\frac{\varepsilon \ln(10)}{5}\right] \frac{\exp\left(\frac{-\varepsilon^2}{4\sigma_e^2}\right)}{\sqrt{4\pi\sigma_e^2}} d\varepsilon \\
&= \exp\left(\frac{\sigma_e \ln(10)}{5}\right)^2 \left\{\int_{-\infty}^{\infty} \frac{\exp\left\{-\frac{1}{2}\left[\frac{\varepsilon}{\sqrt{2}\sigma_e} - \frac{\sqrt{2}\sigma_e \ln(10)}{5}\right]^2\right\}}{\sqrt{4\pi\sigma_e^2}} d\varepsilon\right\} \\
&= \exp\left(\frac{\sigma_e \ln(10)}{5}\right)^2 \{1 - Q[\infty]\} = \exp\left(\frac{\sigma_e \ln(10)}{5}\right)^2. \tag{5.54}
\end{aligned}
$$

From Equations (5.53) and (5.54), the variance of the $10^{\frac{\varepsilon}{10}}$ becomes

$$\text{var}\left[10^{\frac{\varepsilon}{10}}\right] = \exp\left(\frac{\sigma_e\ln(10)}{5}\right)^2 - \left[\exp\left(\frac{\sigma_e\ln(10)}{10}\right)^2\right]^2. \qquad (5.55)$$

The outage probability may now be expressed as

$$P_o = \sum_{k=0}^{N-1}\left\{\binom{N-1}{k}\mu^k(1-\mu)^{N-1-k}\right.$$
$$\left. \times Q\left[\frac{G_p\left(\frac{1}{\gamma_{req}}-\frac{\eta}{S'}\right)-kE\left[10^{\varepsilon/10}\right]}{\sqrt{k\text{var}\left[10^{\varepsilon/10}\right]}}\right]\right\}. \qquad (5.56)$$

5.3.1.3 Performance of the up-link

The performance of the up-link in a single cell CDMA system having a processing gain of 128 was evaluated over a channel having an inverse fourth power path loss law and slow fading whose standard deviation was 8 dB. A signal-to-AWGN ratio of 20 dB at the output of the matched filter was assumed and a BER outage threshold of 10^{-3} was used in the calculations. Figure 5.5 shows the outage probability from Equation (5.35) for perfect power control and VAFs of 3/8 and 1/2. For an outage probability of 2%, the single cell CDMA system can support 48 users and 38 users for a VAF of 3/8 and 1/2, respectively.

The outage probability of the imperfect power controlled system having different standard deviations of power control error in E_b/I_0 is show in Figures 5.6 and 5.7 for a VAF of 3/8 and 1/2, respectively. We observe that a standard deviation of the measured E_b/I_0 was found to be 1.7 dB in a particular set of measurements [7]. For an outage probability of 2% and a standard deviation of power control errors in E_b/I_0 of 2 dB, the single cell CDMA system can support 37 users and 28 users per cell for a VAF of 3/8 and 1/2, respectively. The capacity degradation due to imperfect power control is about 46%. This highlights the need for an accurate power control technique for the up-link in this type of CDMA system.

5.3.2 The down-link system

The CDMA down-link, namely the forward link, has a coherent BPSK communication system where the coherent demodulation is facilitated by a pilot signal. As shown in the system arrangement of Figure 5.8, the BS transmitter adds the CDMA signals from the $N-1$ traffic channels with a CDMA pilot, then transmits this combined signal to all the mobile users in its cell. A mobile can recover the portion of the signal intended for itself by coherently demodulating and despreading the signal with its own code. The signal transmitted from

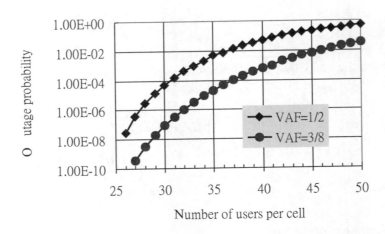

Figure 5.5: Outage probability of a single cell CDMA system in the presence of a perfectly power controlled up-link, with VAFs of 3/8 and 1/2.

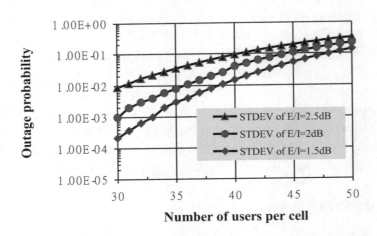

Figure 5.6: Outage probability of the single cell CDMA system in the presence of imperfect power controlled up-link, a VAF of 3/8, and different values of the standard deviation of power control errors in E_b/I_0.

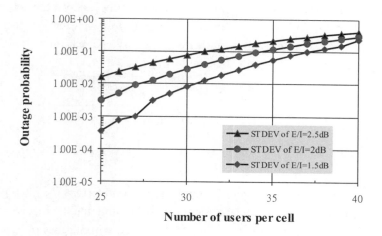

Figure 5.7: Outage probability of the single cell CDMA system in the presence of imperfect power controlled up-link, a VAF of 1/2, and different values of the standard deviation of power control errors in E_b/I_0.

the BS is given by

$$s_{dn}(t) = \sum_{i=0}^{N-1} \sqrt{2P_i} b_i(t - \tau_i) c_i(t - \tau_i) \cos(\omega_2 t + \theta)$$
$$+ \sqrt{2P_p} c_p(t - \tau_p) \cos(\omega_2 t + \theta), \quad (5.57)$$

where P_i and P_p are the transmitted power allocated for the ith mobile and the pilot signal, respectively, τ_i is the random time offset of the ith user, ω_2 is the down-link carrier frequency, $c_p(t)$ is the pilot code sequence, τ_p is the time offset of the pilot signal and θ is an arbitrary phase angle. Let us assume that the pilot signal is transmitted on the Nth channel, then Equation (5.57) can be simplified to

$$s_{dn}(t) = \sum_{i=0}^{N} \sqrt{2P_i} b_i(t - \tau_i) c_i(t - \tau_i) \cos(\omega_2 t + \theta), \quad (5.58)$$

where

$$P_p = P_N,$$
$$c_p(t) = c_N(t),$$
$$\tau_p = \tau_N,$$

and

$$b_N(t - \tau_N) = 1.$$

During the down-link transmission there is no relative time delay between each user's CDMA signal. For convenience we will set the signal delay on the down-link to zero. While the signal is transmitted by the zeroth BS to its service area, the signal received by one of its users, say, the zeroth mobile, has the form

$$
\begin{aligned}
R_{dn}(t) &= a_0 s_{dn}(t) + n(t) \\
&= a_0 \sqrt{2P_0} b_0(t - \tau_0) c_0(t - \tau_0) \cos(\omega_2 t + \theta) \\
&\quad + \sum_{i=1}^{N} a_0 \sqrt{2P_i} b_i(t - \tau_i) c_i(t - \tau_i) \cos(\omega_2 t + \theta) + n(t), \qquad (5.59)
\end{aligned}
$$

where the first term is the signal for zeroth mobile, the second term is the intracellular interference, and the last term is the AWGN component. Assuming that the receiver is correctly chip synchronised to the zeroth user, we can set τ_0 to zero without loss of generality. After demodulating and despreading, the signal at the output of the matched filter is, after following a similar procedure to that in Section 5.3.1,

$$
\begin{aligned}
Z(T_b) &= a_0 \sqrt{2P_0} b_0 + \frac{1}{T_b} \int_0^{T_b} n_{int}(t) c_0(t) dt \\
&\quad + \frac{1}{T_b} \int_0^{T_b} n(t) c_0(t) 2 \cos(\omega_2 t + \theta) dt , \qquad (5.60)
\end{aligned}
$$

where

$$
n_{int}(t) = a_0 \sum_{i=1}^{N} \sqrt{2P_i} b_i(t - \tau_i) c_i(t - \tau_i) \qquad (5.61)
$$

is the equivalent baseband intracellular interference. The first term in Equation (5.60) is the desired signal, the second term is the intracellular interference, while the last term is the AWGN component.

The performance of the down-link can be obtained by following the same procedure as used in the up-link. From Equation (5.60), the received signal power component for the zeroth mobile receiver is

$$
S = 2a_0^2 P_0 , \qquad (5.62)
$$

and the power in the received pilot is

$$
S_p = 2a_0^2 P_p. \qquad (5.63)
$$

If discontinuous transmission is applied to all the traffic channels, then the interference is the summation of $2a_0^2 P_i \nu_i$ for i ranging from 0 to $N - 1$. The pilot channel is usually transmitted at a higher power level than a traffic channel and also at a constant power level. The intracellular interference is therefore

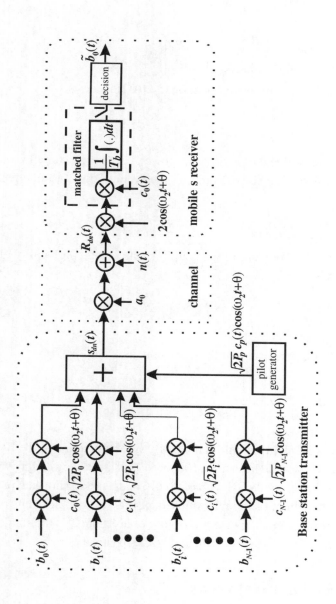

Figure 5.8: Block diagram of the single cell down-link system.

$$I_{int} = \frac{2a_0^2 \sum_{i=1}^{N-1} v_i P_i}{G_p} + \frac{2a_0^2 P_p}{G_p} \tag{5.64}$$

and

$$\frac{I_{int}}{S} = \frac{1}{G_p} \left[\sum_{i=1}^{N-1} v_i \frac{P_i}{P_0} + \frac{P_p}{P_0} \right]. \tag{5.65}$$

The AWGN power, η, is exactly the same as in Equation (5.20). Hence,

$$\frac{E_b}{I_0} = \frac{1}{\frac{I_{int}}{S} + \frac{\eta}{S}}. \tag{5.66}$$

If each traffic channel and the pilot signal have the same power, i.e. $P_i = P_p$ for all i, we obtain the average bit-energy-to-interference PSD ratio, or the average signal-to-interference power ratio, as

$$\frac{E_b}{I_0} = \frac{1}{\frac{1}{G_p} \sum_{i=1}^{N} v_i \frac{P_i}{P_0} + \frac{\eta}{S}} = \frac{1}{\frac{1}{G_p} \sum_{i=1}^{N} v_i + \frac{\eta}{S}}. \tag{5.67}$$

Adopting a similar approach to the one used for the up-lnk, the outage probability of the single cell down-link system is

$$p_o = \Pr \left(\sum_{i=1}^{N} v_i > m \right) = \sum_{i=1}^{N} \binom{N}{i} \mu^i (1-\mu)^{N-i}, \tag{5.68}$$

where the maximum number of active users for the down-link is

$$m = \sum_{i=1}^{N} v_i = \left\lfloor \frac{G_p}{\left(\frac{E_b}{I_0} \right)_{req}} - \frac{G_p}{\frac{S}{\eta}} \right\rfloor. \tag{5.69}$$

The performance of the down-link in a single cell CDMA system in terms of the BER is calculated using Equation (5.35). For an inverse fourth power loss law, a slow fading whose standard deviation is 8 dB, a signal-to-AWGN ratio of 20 dB, and a processing gain of 128, the outage probability as a function of the number of users per cell for two different values of VAF is displayed in Figure 5.9. For an outage probability of less than 2%, the single cell system can support 47 and 37 users for VAFs of 3/8 and 1/2, respectively.

5.4   CDMA Macrocellular Networks

In the previous section we addressed the performance of the single cell CDMA system. We now consider the performance of the multiple cellular arrangement shown in Figure 5.10. In addition to the intracellular interference, there is now interference from neighbouring cells. This interference is referred to as *intercellular interference*. The effects of intercellular

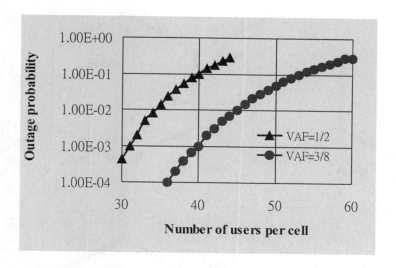

Figure 5.9: Outage probability of a single cell down-link system.

interference must be determined for both the up-link and the down-link communication systems.

5.4.1 The up-link system

The received signal at a BS includes the desired signal, intracellular interference, the AWGN at the receiver input, and intercellular interference. Figure 5.11 shows the up-link communication system where the arrangement for the mobile transmitter and BS receiver are exactly the same as those shown in Figures 5.2 and 5.4, respectively. The signal received at the zeroth BS is given by

$$R_{up}(t) = \sum_{i=0}^{N-1} a_i s_i(t-\tau_i) + \sum_{j=1}^{J-1} \sum_{i=0}^{N-1} a_{ij} s_{ij}(t-\tau_{ij}) + n(t)$$

$$= \sum_{i=0}^{N-1} a_i \sqrt{2P_i} b_i(t-\tau_i) c_i(t-\tau_i) \cos[\omega_1(t-\tau_i) + \theta_i] + n(t)$$

$$+ \sum_{j=1}^{J-1} \sum_{i=0}^{N-1} a_{ij} \sqrt{2P_{ij}} b_{ij}(t-\tau_{ij}) c_{ij}(t-\tau_{ij})$$

$$\times \cos[\omega_1(t-\tau_{ij}) + \theta_{ij}], \tag{5.70}$$

where the intercellular interference from the $J-1$ surrounding cells is

$$\sum_{j=1}^{J-1} \sum_{i=0}^{N-1} a_{ij} s_{ij}(t-\tau_{ij}), \tag{5.71}$$

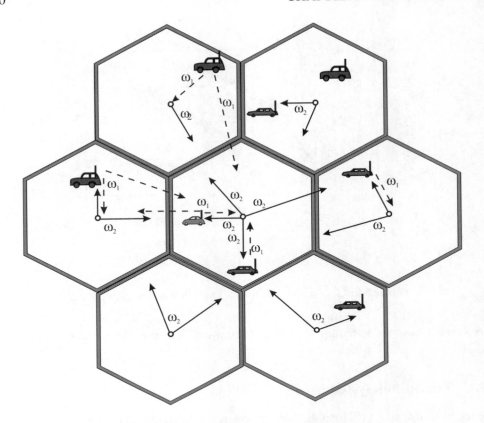

Figure 5.10: Hexagonal multicell arrangement.

and where a_{ij} represents the effects of path loss and slow fading, τ_{ij} is the random time delay of the ith mobile in the jth cell, and $s_{ij}(t)$ is the signal transmitted by the ith mobile in the jth cell. Assuming that the receiver is correctly chip synchronised to the zeroth user, we can set τ_0 to zero without loss of generality. After the received signal goes through the process of demodulation and despreading, the matched filter output is calculated following the methodology given in Section 5.2.1, as

$$Z(T_b) = a_0\sqrt{2P_0}b_0\cos\phi_0 + Z_{int}(T_b) + Z_{ext}(T_b) + Z_n(T_b), \qquad (5.72)$$

where ϕ_0 is the carrier phase difference. The first term is the desired signal, the second term is the intracellular interference component, the third term is the intercellular interference component, while the last term is the AWGN component. In Equation (5.72), $Z_{int}(T_b)$ and $Z_n(T_b)$ are given by Equations (5.9) and (5.13), respectively, while

$$Z_{ext}(t) = \frac{1}{T_b}\int_0^{T_b} n_{ext}(t)c_0(t)\mathrm{d}t, \qquad (5.73)$$

where $n_{ext}(t)$ is the equivalent baseband intercellular interference defined as

$$n_{ext}(t) = \sum_{j=1}^{J-1} \sum_{i=0}^{N-1} a_{ij}\sqrt{2P_{ij}}b_{ij}(t - \tau_{ij})c_{ij}(t - \tau_{ij})\cos\phi_{ij}, \qquad (5.74)$$

and ϕ_{ij} is the carrier phase difference between the interference signal from the ith mobile in the jth cell and the zeroth mobile in zeroth cell. Similar to the intracellular interference, $n_{ext}(t)$ is also a random variable with a Gaussian distribution.

5.4.1.1   Perfect power control in macrocellular networks

Power control, discussed in Section 5.2.1, is also applied in multicellular systems. For perfect power control, we can find the signal power S, the AWGN power η, and the intracellular interference-to-signal ratio I_{int}/S from Equations (5.16), (5.20), and (5.26), respectively, namely

$$\begin{aligned} S &= P_{tar}, \\ \eta &= N_0 R_b, \\ \frac{I_{int}}{S} &= \frac{1}{G_p}\sum_{i=1}^{N-1} v_i. \end{aligned} \qquad (5.75)$$

Similar to the approach in deriving the intracellular interference power, the intercellular interference power at the output of the matched filter, I_{ext}, can be shown to be (see Equations (5.24) and (5.25)),

$$\begin{aligned} I_{ext} &= E\left\{[Z_{ext}(T_b)]^2\right\} \\ &= E\left\{\left[\frac{1}{T_b}\int_0^{T_b} n_{ext}(t)c_0(t)dt\right]^2\right\} \\ &= \frac{1}{G_p}\sum_{j=1}^{J-1}\sum_{i=0}^{N-1} a_{ij}^2 P_{ij}. \end{aligned} \qquad (5.76)$$

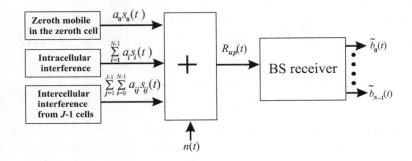

Figure 5.11: Block diagram of the multicellular CDMA up-link system.

By applying voice activity detection (VAD) and thereby discontinuous transmitting (DTX), the mobiles transmit only when speech is present. We introduce a voice activity variable v_{ij} to the intercellular interference power. Then the intercellular interference-to-signal ratio can be derived from Equation (5.76) as

$$\frac{I_{ext}}{S} = \frac{1}{G_p} \sum_{j=1}^{J-1} \sum_{i=0}^{N-1} v_{ij} \frac{a_{ij}^2 P_{ij}}{S}$$

$$= \frac{1}{G_p} \sum_{j=1}^{J-1} \frac{I_j}{S} , \tag{5.77}$$

where I_j/S is the interference-to-signal power ratio from the jth surrounding cell:

$$\frac{I_j}{S} = \sum_{i=0}^{N-1} v_{ij} \frac{a_{ij}^2 P_{ij}}{S} . \tag{5.78}$$

In our model, all mobiles, irrespective of which cell they are in, communicate on the up-link via carrier frequency ω_1. Hence mobiles cause interference at neighbouring base sites. Consider the centre cell in Figure 5.12. The mobiles in the surrounding cells, or first tier cells, cause the most interference at the zeroth cell site. The second tier cells are less impor-tant from an intercellular interference stand point. Let us consider one of these interfering cells, the jth cell say, where the cell site is a distance D_j from the zeroth cell site as shown in Figure 5.12. The interference term I_j in Equation (5.78) is the interference power from all the mobiles in the jth cell to the zeroth BS. We will calculate this interference power I_j and then sum the interference for all the $J-1$ significant interfering cells.

From Equation (5.78), I_j/S is found as the summation of N terms corresponding to N mobiles in the jth cell. We will replace this summation by an integration over the area of the jth cell, assuming that the mobiles are uniformly distributed. The active mobiles in the jth cell produce an interference power of $I(r_j, r)$ at the zeroth cell site. Under perfect power control, the mobiles in the jth cell have their power controlled by their own BS to be P_{tar}. In order to track the relative path loss and slow fading variations, the transmitted power P_{ij} from the ith mobile in the jth cell is made inversely proportional to a_{ij}^2, whence

$$P_{ij} = \frac{P_{tar}}{a_{ij}^2} = P_{tar} r^\alpha 10^{\frac{-\lambda_{ij}}{10}} = S r^\alpha 10^{\frac{-\lambda_{ij}}{10}} , \tag{5.79}$$

where α is the path loss exponent, and λ_{ij} is a normally distributed random variable with standard deviation σ and zero mean, while r is the distance from the interfering mobile to its own BS. The arrangement is shown in Figure 5.13. Consequently the interference-to-signal power ratio at the zeroth cell site due to the mobiles in area da who are communicating to

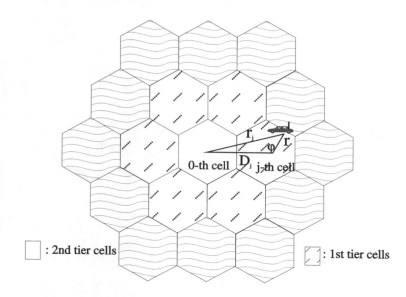

Figure 5.12: Intercellular interference geometry in hexagonal cells.

the jth cell site is [9]

$$\frac{I(r_j,r)}{S} = \frac{a_{0j}^2 P_{ij}}{S} = \frac{1}{S}P_{tar}\frac{r^\alpha}{r_j^\alpha}10^{\frac{(\lambda_0-\lambda_{ij})}{10}}$$

$$= \left(\frac{r}{r_j}\right)^\alpha 10^{\frac{\zeta}{10}}, \tag{5.80}$$

where r_j, the distance between the interfering mobile and zeroth BS, is

$$r_j = \sqrt{D_j^2 + r^2 - 2D_jr\cos(\varphi)}, \tag{5.81}$$

and ζ is the difference between λ_{ij} and λ_0, the independent random variables with zero mean and standard deviation σ. Hence ζ is also a random variable with zero mean and variance of $\sigma_\zeta^2 = 2\sigma^2$, while D_j is the distance between the zeroth BS and the jth co-channel BS. Note that a_{0j}^2 in Equation (5.80) is the path loss and slow fading between the interfering mobile in the jth cell and the zeroth BS.

Since the total intercellular interference-to-signal ratio is the sum of all the interference from the $J-1$ surrounding cells, then due to the Central Limit Theorem, the interfering power tends to be Gaussian distributed with a non-zero mean. In other words, the intercellular interfering power varies around a mean power. It is necessary to calculate the mean and variance of the intercellular interference power in order to calculate the outage probability. We commence by replacing Equation (5.78) by

$$\frac{I_j}{S} = \int_0^{2\pi}\int_0^R v_{ij}\frac{I(r_j,r)}{S}\phi\left(\zeta,\frac{r}{r_j}\right)\rho da, \tag{5.82}$$

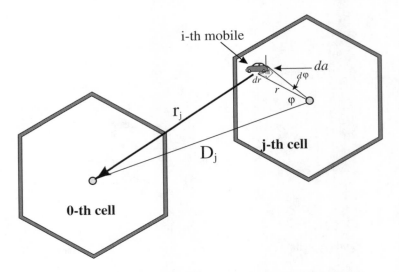

Figure 5.13: Intercellular interference in a multicell environment.

where ρ is the density of mobiles over the jth cell; $da = rdrd\varphi$ is the unit area in Figure 5.13; $\phi(\zeta, r/r_j)$ is the constraint function for the interfering users in the jth cell; and R is the cell radius. Because each mobile communicates with the cell site having the smallest path loss and slow fading attenuation, the constraint function can be defined as

$$\phi\left(\zeta, \frac{r}{r_j}\right) = \begin{cases} 1, & \text{if } \left(\frac{r}{r_j}\right)^{\alpha} 10^{\frac{\zeta}{10}} \leq 1, \\ 0, & \text{otherwise}. \end{cases} \tag{5.83}$$

By substituting Equation (5.80) into Equation (5.82), and upon taking the expectation, we obtain

$$\begin{aligned} E\left[\frac{I_j}{S}\right] &= \rho\mu \int_0^{2\pi} \int_0^R \left(\frac{r}{r_j}\right)^{\alpha} E\left[10^{\frac{\zeta}{10}} \phi\left(\zeta, \frac{r}{r_j}\right)\right] rdrd\varphi \\ &= \rho\mu \int_0^{2\pi} \int_0^R f\left(\frac{r_j}{r}, \sigma\right) \left(\frac{r}{r_j}\right)^{\alpha} rdrd\varphi, \end{aligned} \tag{5.84}$$

where $E[v_{ij}] = \mu$, $f(r_j/r, \sigma)$ is the expectation of the product of $10^{\zeta/10}$ and the constraint function $\phi(\zeta, r/r_j)$. The term $f(r_j/r, \sigma)$ can be derived noting that ζ is log-normally distributed, namely

$$\begin{aligned} f\left(\frac{r_j}{r}, \sigma\right) &= E\left[10^{\frac{\zeta}{10}} \phi\left(\zeta, \frac{r}{r_j}\right)\right] \\ &= \int_{-\infty}^{10\alpha \log\left(\frac{r_j}{r}\right)} \exp\left[\frac{\zeta\ln(10)}{10}\right] \frac{\exp\left(\frac{-\zeta^2}{4\sigma^2}\right)}{\sqrt{4\pi\sigma^2}} d\zeta. \end{aligned}$$

Note that the upper integral limit $10\alpha \log(r_j/r)$ is a consequence of $\phi\,(\zeta, r/r_j)$ being set to unity; see Equation (5.83). Hence,

$$
f\left(\frac{r_j}{r}, \sigma\right) = \exp\left(\sigma \frac{\ln(10)}{10}\right)^2 \int_{-\infty}^{10\alpha \log\left(\frac{r_j}{r}\right)} \frac{\exp\left\{-\frac{1}{2}\left[\frac{\zeta}{\sqrt{2}\sigma} - \frac{\ln(10)\sqrt{2}\sigma}{10}\right]^2\right\}}{\sqrt{4\pi\sigma^2}} d\zeta
$$

$$
= \exp\left(\sigma \frac{\ln(10)}{10}\right)^2
$$

$$
\times \left\{1 - Q\left[\frac{10\alpha \log\left(\frac{r_j}{r}\right)}{\sqrt{2}\sigma} - \sqrt{2}\sigma \frac{\ln(10)}{10}\right]\right\}. \tag{5.85}
$$

The variance of I_j/S from the jth cell is

$$
\mathrm{var}\left[\frac{I_j}{S}\right] = \mathrm{var}\left[\int_0^{2\pi}\int_0^R v_{ij}\frac{I(r_j, r)}{S}\phi\left(\zeta, \frac{r}{r_j}\right)\rho r\,dr\,d\varphi\right].
$$

From Equation (5.80),

$$
\mathrm{var}\left[\frac{I_j}{S}\right] = \int_0^{2\pi}\int_0^R \left\{E\left[v_{ij}\left(\frac{r}{r_j}\right)^\alpha 10^{\frac{\zeta}{10}}\phi\left(\zeta, \frac{r}{r_j}\right)\right]^2 \right.
$$

$$
\left. - \left\{E\left[v_{ij}\left(\frac{r}{r_j}\right)^\alpha 10^{\frac{\zeta}{10}}\phi\left(\zeta, \frac{r}{r_j}\right)\right]\right\}^2\right\}\rho r\,dr\,d\varphi
$$

$$
= \int_0^{2\pi}\int_0^R \left\{E\left[v_{ij}^2\right]\left(\frac{r}{r_j}\right)^{2\alpha} E\left[10^{\frac{\zeta}{10}}\phi\left(\zeta, \frac{r}{r_j}\right)\right]^2 \right.
$$

$$
\left. - \left\{E[v_{ij}]\left(\frac{r}{r_j}\right)^\alpha E\left[10^{\frac{\zeta}{10}}\phi\left(\zeta, \frac{r}{r_j}\right)\right]\right\}^2\right\}\rho r\,dr\,d\varphi,
$$

where v_{ij} and ζ are independent random variables. But $E\left[v_{ij}^2\right] = \mu$, the voice activity factor, so

$$
\mathrm{var}\left[\frac{I_j}{S}\right] = \int_0^{2\pi}\int_0^R \left\{\mu\left(\frac{r}{r_j}\right)^{2\alpha} E\left[10^{\frac{\zeta}{10}}\phi\left(\zeta, \frac{r}{r_j}\right)\right]^2 - \mu^2\left(\frac{r}{r_j}\right)^{2\alpha}\right.
$$

$$
\left. \times \left\{E\left[10^{\frac{\zeta}{10}}\phi\left(\zeta, \frac{r}{r_j}\right)\right]\right\}^2\right\}\rho\,dr\,d\varphi
$$

$$
= \int_0^{2\pi}\int_0^R \left(\frac{r}{r_j}\right)^{2\alpha}
$$

$$
\times \left[\mu g\left(\frac{r_j}{r}, \sigma\right) - \mu^2 f^2\left(\frac{r_j}{r}, \sigma\right)\right]\rho r\,dr\,d\varphi, \tag{5.86}
$$

where $f(r_j/r, \sigma)$ is given by Equation (5.85) and $g(r_j/r, \sigma)$ is

$$g\left(\frac{r_j}{r}, \sigma\right) = E\left[10^{\frac{\zeta}{10}}\phi\left(\zeta, \frac{r}{r_j}\right)\right]^2.$$
(5.87)

Using Equation (5.83) enables $g(r_j/r, \sigma)$ to be found as

$$
\begin{aligned}
g\left(\frac{r_j}{r}, \sigma\right) &= \int_{-\infty}^{10\alpha\log\left(\frac{r_j}{r}\right)} \exp\left[\frac{\zeta\ln(10)}{5}\right] \frac{\exp\left(\frac{-\zeta^2}{4\sigma^2}\right)}{\sqrt{4\pi\sigma^2}} d\zeta \\
&= \exp\left(\frac{\sigma\ln(10)}{5}\right)^2 \int_{-\infty}^{10\alpha\log\left(\frac{r_j}{r}\right)} \frac{\exp\left\{-\frac{1}{2}\left[\frac{\zeta}{\sqrt{2}\sigma} - \frac{\sqrt{2}\sigma\ln(10)}{5}\right]^2\right\}}{\sqrt{4\pi\sigma^2}} d\zeta \\
&= \exp\left(\frac{\sigma\ln(10)}{5}\right)^2 \\
&\quad \times \left\{1 - Q\left[\frac{10\alpha\log\left(\frac{r_j}{r}\right)}{\sqrt{2}\sigma} - \frac{\sqrt{2}\sigma\ln(10)}{5}\right]\right\}.
\end{aligned}
$$
(5.88)

By knowing the mean and variance of I_j/S we can now calculate the mean and variance of the total interference-power-to-signal-power ratio for all the surrounding cells. The expression I_{ext}/S in Equation (5.77) has a mean and variance of

$$E\left[\frac{I_{ext}}{S}\right] = \frac{1}{G_p}\sum_{j=1}^{J-1} E\left[\frac{I_j}{S}\right];$$
(5.89)

and

$$\text{var}\left[\frac{I_{ext}}{S}\right] = \left(\frac{1}{G_p}\right)^2 \sum_{j=1}^{J-1} \text{var}\left[\frac{I_j}{S}\right],$$
(5.90)

respectively. The bit energy-to-interference PSD ratio at the output of the matched filter (the input of the decision circuit) is

$$\frac{E_b}{I_0} = \frac{S}{I} = \frac{1}{\frac{I_{int}}{S} + \frac{I_{ext}}{S} + \frac{\eta}{S}}.$$
(5.91)

Note that η/S is a constant, and I_{int}/S is a function of the voice activity variable which is binomially distributed, while I_{ext}/S is a Gaussian random variable with a mean and variance of

$$E\left[\frac{I_{ext}}{S}\right] = \frac{1}{G_p}\sum_{j=1}^{J-1} \rho\mu \int_0^{2\pi} \int_0^R f\left(\frac{r_j}{r}, \sigma\right)\left(\frac{r}{r_j}\right)^\alpha r\, dr\, d\varphi$$
(5.92)

and

$$\text{var}\left[\frac{I_{ext}}{S}\right] = \frac{1}{G_p^2}\sum_{j=1}^{J-1}\rho\int_0^{2\pi}\int_0^R \left(\frac{r}{r_j}\right)^{2\alpha}$$

$$\times \left[\mu g\left(\frac{r_j}{r},\sigma\right) - \mu^2 f^2\left(\frac{r_j}{r},\sigma\right)\right] r\,dr\,d\varphi.\tag{5.93}$$

The system performance in terms of the outage probability that has a BER greater than 10^{-3} is

$$\begin{aligned}
P_o &= \Pr(\text{BER} > 10^{-3})\\
&= \Pr\left(\frac{S}{I} < \gamma_{req}\right) = \Pr\left(\frac{I}{S} > \frac{1}{\gamma_{req}}\right)\\
&= \Pr\left(\frac{I_{ext}}{S} > \frac{1}{\gamma_{req}} - \frac{I_{int}}{S} - \frac{\eta}{S}\right),\tag{5.94}
\end{aligned}$$

where γ_{req} is the E_b/N_0 required to have a BER $< 10^{-3}$. Substituting for I_{int}/S gives

$$P_o = \Pr\left[\frac{I_{ext}}{S} > \left(\frac{1}{\gamma_{req}} - \frac{\eta}{S} - \frac{1}{G_p}\sum_{i=1}^{N-1}v_i\right)\right],\tag{5.95}$$

and if the number of active users in the centre cell of Figure 5.13 is k, then we proceed as in Section 5.2.1, namely

$$\begin{aligned}
P_o &= \Pr\left[\frac{I_{ext}}{S} > \left(\frac{1}{\gamma_{req}} - \frac{\eta}{S} - \frac{k}{G_p}\right)\Bigg|\sum_{i=1}^{N-1}v_i = k\right]\\
&\quad \times \Pr\left[\sum_{i=0}^{N-1}v_i = k\right].\tag{5.96}
\end{aligned}$$

We note that

$$\Pr\left[\sum_{i=0}^{N-1}v_i = k\right]\tag{5.97}$$

has been determined before; see Equation (5.46). We now consider the first probability term in Equation (5.96). On applying Equation (5.50),

$$\begin{aligned}
&\Pr\left[\frac{I_{ext}}{S} > \left(\frac{1}{\gamma_{req}} - \frac{\eta}{S} - \frac{k}{G_p}\right)\right]\\
&= Q\left[\frac{\left(\frac{1}{\gamma_{req}} - \frac{\eta}{S} - \frac{k}{G_p}\right) - \text{E}\left[\frac{I_{ext}}{S}\right]}{\sqrt{\text{var}\left[\frac{I_{ext}}{S}\right]}}\right],\tag{5.98}
\end{aligned}$$

where $\text{E}[I_{ext}/S]$ and $\text{var}[I_{ext}/S]$ are given by Equations (5.89) and (5.90), respectively. Hence the outage probability is

$$= \sum_{k=0}^{N-1} \frac{(N-1)!}{(N-1-k)!k!} \mu^k (1-\mu)^{N-k-1}$$

$$\times Q \left[\frac{\frac{1}{\gamma_{req}} - \frac{\eta}{S} - \frac{k}{G_p} - \mathrm{E}\left[\frac{I_{ext}}{S}\right]}{\sqrt{\mathrm{var}\left(\frac{I_{ext}}{S}\right)}} \right]. \tag{5.99}$$

5.4.1.2 Imperfect power control in macrocellular networks

The received signal power S' from a mobile at its BS will differ from the target power level P_{tar} by δ_0 dB. This error power δ_0 is a random variable that is normally distributed with standard deviation σ_e. The received signal S' at the BS for the zeroth mobile, and the intracellular interference-to-signal ratio are given by Equations (5.37) and (5.40), respectively. Using Equations (5.37) and (5.80), the mobiles in area da in the jth cell produce an interfering power-to-received-signal-power ratio at the zeroth BS of

$$\frac{I'(r_j, r)}{S'} = \frac{a_{0j}^2 P_{ij}}{P_{tar} 10^{\delta_0/10}}$$

$$= \left(\frac{r}{r_j}\right)^\alpha 10^{\frac{\zeta}{10}} 10^{\frac{\delta_{ij}-\delta_0}{10}}$$

$$= \left(\frac{r}{r_j}\right)^\alpha 10^{\frac{\zeta}{10}} 10^{\frac{\varepsilon}{10}}, \tag{5.100}$$

where δ_{ij} is the power error for the ith mobile in the jth cell, and ε is a random variable having zero mean and normal distribution with a standard deviation of $\sigma_\varepsilon = \sqrt{2}\sigma_e$. From Equations (5.82) and (5.100), the total intercellular interference-to-signal ratio is

$$\frac{I'_{ext}}{S'} = \frac{1}{G_p} \sum_{j=1}^{J-1} \int_0^{2\pi} \int_0^R v_{ij} \left(\frac{r}{r_j}\right)^\alpha 10^{\frac{\zeta}{10}} \phi\left(\zeta, \frac{r}{r_j}\right) 10^{\frac{\varepsilon}{10}} \rho r \, dr \, d\varphi$$

$$= \frac{I_{ext}}{S} 10^{\frac{\varepsilon}{10}}. \tag{5.101}$$

From Equations (5.37), (5.40), and (5.101), the bit energy-to-interference PSD ratio at the input of the decision circuit for imperfect power control is

$$\left(\frac{E_b}{I_0}\right)_{im} = \frac{S'}{I'} = \frac{1}{\frac{I'_{int}}{S'} + \frac{I'_{ext}}{S'} + \frac{\eta}{S'}}$$

$$= \frac{1}{\frac{I_{int}}{S} 10^{\varepsilon/10} + \frac{I_{ext}}{S} 10^{\varepsilon/10} + \frac{\eta}{S'}}, \tag{5.102}$$

where ε is the error in E_b/I_0 due to imperfect power control. Following the same procedure as employed for the perfect power control case, the outage probability is found as

$$
\begin{aligned}
p_o &= \Pr(\mathrm{BER} > 10^{-3}) \\
&= \Pr\left(\frac{I'_{int}}{S'} + \frac{I'_{ext}}{S'} + \frac{\eta}{S'} > \frac{1}{\gamma_{req}}\right) \\
&= \Pr\left(\frac{I'_{int}}{S'} + \frac{I'_{ext}}{S'} > \frac{1}{\gamma_{req}} - \frac{\eta}{S'}\right).
\end{aligned}
\tag{5.103}
$$

In Equation (5.103), I'_{int}/S' and I'_{ext}/S' are two independent Gaussian distributed random variables, whose mean and variance may be expressed as

$$
\mathrm{E}\left[\frac{I'_{int}}{S'} + \frac{I'_{ext}}{S'}\right] = \mathrm{E}\left[\frac{I_{int}}{S} 10^{\frac{\varepsilon}{10}}\right] + \mathrm{E}\left[\frac{I_{ext}}{S} 10^{\frac{\varepsilon}{10}}\right]
$$

and

$$
\mathrm{var}\left[\frac{I'_{int}}{S'} + \frac{I'_{ext}}{S'}\right] = \mathrm{var}\left[\frac{I_{int}}{S} 10^{\frac{\varepsilon}{10}}\right] + \mathrm{var}\left[\frac{I_{ext}}{S} 10^{\frac{\varepsilon}{10}}\right].
\tag{5.104}
$$

Before we derive the outage probability, we have to calculate the mean and variance of the intracellular interference-to-signal ratio, and the intercellular interference-to-signal ratio. The mean and variance of the intracellular interference-to-signal power ratio are

$$
\mathrm{E}\left[\frac{I'_{int}}{S'}\right] = \sum_{i=1}^{N-1} v_i \mathrm{E}\left[10^{\frac{\varepsilon}{10}}\right] = \sum_{i=1}^{N-1} v_i \exp\left[\left(\frac{\ln(10)\sqrt{2}\sigma_e}{10}\right)^2\right]
$$

and

$$
\begin{aligned}
\mathrm{var}\left[\frac{I'_{int}}{S'}\right] &= \sum_{i=1}^{N-1} v_i \, \mathrm{var}\left[10^{\frac{\varepsilon}{10}}\right] \tag{5.105}\\
&= \sum_{i=1}^{N-1} v_i \left\{ \exp\left[\left(\frac{\sqrt{2}\sigma_e \ln(10)}{5}\right)^2\right]\right.\\
&\quad \left. - \exp\left[\left(\frac{\ln(10)\sqrt{2}\sigma_e}{10}\right)^4\right]\right\}. \tag{5.106}
\end{aligned}
$$

In Equation (5.101), the random variables ζ and ε are uncorrelated. Consequently the expectation of Equation (5.101) can be expressed as

$$
\mathrm{E}\left[\frac{I'_{ext}}{S'}\right] = \mathrm{E}\left[\frac{I_{ext}}{S}\right] \mathrm{E}\left[10^{\frac{\varepsilon}{10}}\right],
\tag{5.107}
$$

and from Equations (5.92) and (5.51) the expectation can be calculated. The variance of Equation (5.101) is

$$\text{var}\left[\frac{I'_{ext}}{S'}\right] = \text{E}\left[\left(\frac{I_{ext}}{S}\right)^2\right]\text{E}\left[\left(10^{\frac{\varepsilon}{10}}\right)^2\right] - \left\{\text{E}\left[\frac{I_{ext}}{S}\right]\text{E}\left[10^{\frac{\varepsilon}{10}}\right]\right\}^2, \qquad (5.108)$$

where $\text{E}[I_{ext}/S]$, $\text{E}\left[10^{\varepsilon/10}\right]$, and $\text{E}\left[10^{\varepsilon/5}\right]$ are given in Equations (5.92), (5.51) and (5.54), respectively, while

$$
\begin{aligned}
\text{E}\left[\left(\frac{I_{ext}}{S}\right)^2\right] &= \frac{1}{G_p^2}\sum_{j=1}^{J-1}\int_0^{2\pi}\int_0^R \text{E}\left[v_{ij}\left(\frac{r}{r_j}\right)^\alpha 10^{\frac{\zeta}{10}}\,\phi\left(\zeta,\frac{r}{r_j}\right)\right]^2 \rho r\,dr\,d\varphi \\
&= \frac{1}{G_p^2}\sum_{j=1}^{J-1}\int_0^{2\pi}\int_0^R \mu\text{E}\left[\left(\frac{r}{r_j}\right)^\alpha 10^{\frac{\zeta}{10}}\,\phi\left(\zeta,\frac{r}{r_j}\right)\right]^2 \rho r\,dr\,d\varphi \\
&= \frac{1}{G_p^2}\sum_{j=1}^{J-1}\int_0^{2\pi}\int_0^R \mu\left(\frac{r}{r_j}\right)^{2\alpha} g\left(\frac{r_j}{r},\sigma\right)\rho r\,dr\,d\varphi, \qquad (5.109)
\end{aligned}
$$

and $g(r_j/r,\sigma)$ is given in Equation (5.88). From Equations (5.107) and (5.108) the expectation and variance of I'_{ext}/S' can be calculated.

The outage probability of Equation (5.103) can be written following the same procedure as used for the perfect power control case:

$$
\begin{aligned}
P_o &= \sum_{k=0}^{N-1}\frac{(N-1)!}{(N-1-k)!k!}\mu^k(1-\mu)^{N-1-k} \\
&\quad \times Q\left(\frac{\frac{1}{\gamma_{req}} - \frac{\eta}{S'} - \text{E}\left[\frac{I'_{int}}{S'}\right] - \text{E}\left[\frac{I'_{ext}}{S'}\right]}{\sqrt{\text{var}\left[\frac{I'_{int}}{S'}\right] + \text{var}\left[\frac{I'_{ext}}{S'}\right]}}\right). \qquad (5.110)
\end{aligned}
$$

For a processing gain of 128 and a signal-to-AWGN ratio of 20 dB at the output of the matched filter, the performance of the up-link CDMA system is shown in Figure 5.14 for a VAF of 1/2, and in Figure 5.15 for a VAF of 3/8. For an outage probability of 2%, the perfect power controlled CDMA system can support 23 users and 30 users for a VAF of 1/2 and 3/8, respectively. The number of users per cell for different values of the standard deviation of power control errors in E_b/I_0, and the percentage decrease in users due to imperfect power control, are displayed in Table 5.1 for an outage of 2%.

From Table 5.1, the imperfect power control system having a standard deviation of 2 dB can only support 22 users and 28 users for VAFs of 1/2 and 3/8, respectively. The reduction in capacity caused by power control error is about 6.3% and 4.3% for VAFs of 3/8 and 1/2, respectively, for a standard deviation of 2 dB. For a 2.5 dB standard deviation of power control error, the percentage of capacity loss increases to 13%.

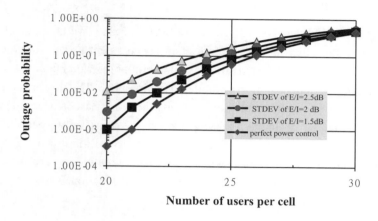

Figure 5.14: Performance of the multicellular up-link system with a VAF of 1/2.

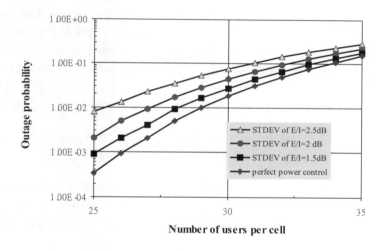

Figure 5.15: Performance of the multicellular up-link system with a VAF of 3/8.

Table 5.1: Number of users per cell for different values of the standard deviation of power control errors in E_b/I_0. The outage probability is 2%.

STD of power error	VAF = 3/8	% decrease	VAF = 1/2	% decrease
0 dB	30	0 %	23	0%
1.5 dB	29	3.3 %	22	4.3 %
2 dB	28	6.3 %	22	4.3 %
2.5 dB	26	13.3 %	20	13 %

5.4.2   The down-link system

The single cell down-link has been discussed in Section 5.2.2. We now consider the multicell down-link system that has the same system arrangement as the single cell system.
However, unlike the single cell system, there is intercellular interference from neighbouring
BSs. This interference depends on the mobile's location [10]. The nearer the mobile is to
its BS, the better the performance, and consequently the worst case is when the mobiles are
located at the cell boundaries. In the following analysis we consider two particular locations
at the hexagonal cell boundary to examine the performance of the down-link CDMA system with power control and without power control. The two locations are shown in Figure
5.16, where location A is at the corner of a hexagon, and location B is at the middle of a
hexagonal periphery.

5.4.2.1   Without power control

The CDMA down-link system is a coherent BPSK communication system where the coherent carrier is provided by sending a pilot signal. Figure 5.8 shows the system arrangement.
The BS sums up all the signals for all its users, together with a CDMA pilot signal, and
transmits the combined signal to the users in its cell. A mobile recovers the portion of the
signal intended for it by coherently demodulating and despreading the received CDMA signal with its own spreading code. The transmitted signal $s_{dn}(t)$ from its own BS (zeroth cell)
is given by Equation (5.57), while the signal transmitted from the jth neighbouring BS is
given by

$$s_{dn}^j(t) = \sum_{i=0}^{N} \sqrt{2P_{ij}} b_{ij}(t - \tau_{ij}) c_{ij}(t - \tau_{ij}) \cos(\omega_2 t + \theta_j), \qquad (5.111)$$

where P_{ij} is the transmitted power allocated for the ith channel and ω_2 is the down-link radio
frequency carrier. When the BS transmits the signal to all its mobiles within its coverage
area, the signal received by one of its users, say, the zeroth mobile in the zeroth cell, is

$$\begin{aligned}
R_{dn}(t) &= a_0 s_{dn}(t) + n(t) + \sum_{j=1}^{J-1} a_j s_{dn}^j(t) \\
&= a_0 \sqrt{2P_0} b_0(t - \tau_0) c_0(t - \tau_0) \cos(\omega_2 t + \theta) \\
&\quad + \sum_{i=1}^{N} a_0 \sqrt{2P_i} b_i(t - \tau_i) c_i(t - \tau_i) \cos(\omega_2 t + \theta) + n(t) \\
&\quad + \sum_{j=1}^{J-1} a_j \sum_{i=1}^{N} \sqrt{2P_{ij}} b_{ij}(t - \tau_{ij}) c_{ij}(t - \tau_{ij}) \cos(\omega_2 t + \theta_j), \qquad (5.112)
\end{aligned}$$

where τ_0, τ_i and τ_{ij} are the random time delay for the zeroth mobile in the zeroth cell, the ith
mobile in the zeroth cell, and the ith mobile in the jth cell, respectively. In Equation (5.112),

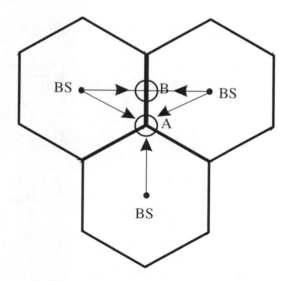

Figure 5.16: Multicellular down-link interference geometry.

the first term is the signal for zeroth mobile, the second term is the intracellular interference from the $N-1$ mobiles, while the third term and the last term are AWGN noise and intercellular interference, respectively. Note that a_0 and a_j represent the path loss and slow fading for the paths between the zeroth BS and the zeroth mobile in the zeroth cell, and for the jth BS to the zeroth mobile in the zeroth cell, respectively.

Assuming that the receiver is correctly chip synchronised to the zeroth user, we can set τ_0 to zero without loss of generality. After demodulating and despreading, and on applying the same procedure as used in Section 5.3.1, the signal at the output of the matched filter is

$$Z(T_b) = a_0\sqrt{2P_0}b_0 + \frac{1}{T_b}\int_0^{T_b} n_{int}(t)c_0(t)dt$$

$$\times \frac{1}{T_b}\int_0^{T_b} n(t)c_0(t)\cos(\omega_2 t + \theta_0)dt$$

$$+ \frac{1}{T_b}\int_0^{T_b} n_{ext}(t)c_0(t)dt, \tag{5.113}$$

where $n_{int}(t)$ is given by Equation (5.61) and $n_{ext}(t)$ is the equivalent baseband intercellular interference. The first term in Equation (5.113) is the wanted message signal, the second term is the intracellular interference due to the sum of the other $N-1$ users and the pilot signal, the third term is the AWGN component, and the last term is the intercellular interference. The intercellular interference, $n_{ext}(t)$, is defined as

$$n_{ext}(t) = \sum_{j=1}^{J-1}\sum_{i=0}^{N} a_j\sqrt{2P_{ij}}b_{ij}(t - \tau_{ij})c_{ij}(t - \tau_{ij})\cos\phi_j, \tag{5.114}$$

where ϕ_j is the phase difference between the neighbouring BS carrier and the zeroth BS carrier.

The performance of the down-link can be analysed by following the procedure used in the up-link analysis. The signal power and the AWGN power are given by

$$S = 2a_0^2 P_0 \tag{5.115}$$

and

$$\eta = N_0 R_b. \tag{5.116}$$

With reference to Equation (5.65), the intracellular interference-to-signal power ratio is, on the assumption that the pilot channel has the same power as the traffic channels,

$$\frac{I_{int}}{S} = \frac{1}{G_p} \sum_{i=1}^{N} v_i \frac{P_i}{P_0} = \frac{1}{G_p} \sum_{i=1}^{N} v_i . \tag{5.117}$$

Similarly, the intercellular interference-to-signal power ratio can be shown to be

$$\frac{I_{ext}}{S} = \frac{1}{2G_p} \sum_{j=1}^{J-1} \sum_{i=0}^{N} v_{ij} \frac{a_j^2 P_{ij}}{a_0^2 P_0}, \tag{5.118}$$

and with the aid of Equation (5.79) we can find $(a_j/a_0)^2$ to give

$$\frac{I_{ext}}{S} = \frac{1}{2G_p} \sum_{j=1}^{J-1} \sum_{i=0}^{N} v_{ij} \frac{P_{ij}}{P_0} \left(\frac{r_0}{r_j}\right)^\alpha 10^{\frac{\lambda_j - \lambda_0}{10}}. \tag{5.119}$$

Since there is no power control considered on the down-link, $p_{ij} = p_0$, and hence

$$\frac{I_{ext}}{S} = \frac{1}{2} \frac{1}{G_p} \sum_{j=1}^{J-1} \sum_{i=0}^{N} v_{ij} \left(\frac{r_0}{r_j}\right)^\alpha 10^{\frac{\zeta_j}{10}}, \tag{5.120}$$

where P_{ij} is the transmitted power from the jth BS for the ith mobile in the jth cell, v_{ij} is the mobile's voice activity variable, N is the number of users in the cell, R is the cell radius, r_j is the distance between the neighbouring jth BS and the zeroth mobile in the zeroth cell, and r_0 is the distance between the zeroth mobile and the zeroth BS. Note that there is a factor of 1/2 in I_{ext}/S, due to the carrier incoherence between the zeroth and jth BSs.

Because the system should be designed to give a required performance for any mobile within its cell area, the performance of a mobile located at the cell boundary is particularly critical in the system analysis. Therefore we consider two locations, A and B, at the boundary, as shown in Figure 5.16, to examine the performance in the down-link. By combining Equations (5.115), (5.116), (5.117), and (5.120), the output of the matched filter has a bit-energy- to-interference PSD of

$$\frac{E_b}{I_0} = \frac{S}{I} = \frac{1}{\frac{1}{G_p}\left[\sum_{i=1}^{N} v_i + \sum_{j=1}^{J-1} \sum_{i=0}^{N} \frac{1}{2} v_{ij} 10^{\frac{\zeta_j}{10}} \left(\frac{r_0}{r_j}\right)^\alpha\right] + \frac{\eta}{S}}, \tag{5.121}$$

where ζ_j is a normal random variable with standard deviation of $\sqrt{2}\sigma$.

Performance at location A Consider a mobile at location A, say, the zeroth mobile in the zeroth cell. As shown in Figure 5.17, it has a distance of r_j from the first tier of the jth group BSs, and has a distance of r_k from the second tier kth group BSs, and it also has a distance of r_l from the second tier lth group BSs. Assume that the power allocated to each mobile by the BS is equal to P_m, then Equation (5.121) can be rewritten as the interference-to-signal power ratio:

$$
\begin{aligned}
\frac{I}{S} &= \frac{I_{int}}{S} + \frac{I_{ext}}{S} + \frac{\eta}{S} \\
&= \frac{1}{G_p}\sum_{i=1}^{N} v_i + \frac{1}{G_p}\left\{ \sum_{j=1}^{2}\sum_{i=0}^{N}\frac{1}{2}v_{ij}\left(\frac{R}{r_j}\right)^{\alpha}10^{\frac{\zeta_j}{10}}\phi_j\left(\zeta_j,\frac{R}{r_j}\right) \right. \\
&\quad + \sum_{k=1}^{3}\sum_{i=0}^{N}\frac{1}{2}v_{ik}\left(\frac{R}{r_k}\right)^{\alpha}10^{\frac{\zeta_k}{10}}\phi_k\left(\zeta_k,\frac{R}{r_k}\right) \\
&\quad \left. + \sum_{l=1}^{6}\sum_{i=0}^{N}\frac{1}{2}v_{il}\left(\frac{R}{r_l}\right)^{\alpha}10^{\frac{\zeta_l}{10}}\phi_l\left(\zeta_l,\frac{R}{r_l}\right) \right\} + \frac{\eta}{S},
\end{aligned}
$$
(5.122)

where

$$
\phi_j\left(\zeta_j,\frac{R}{r_j}\right) = \begin{cases} 1, & \text{if } \left(\frac{R}{r_j}\right)^{\alpha}10^{\frac{\zeta_j}{10}} \leq 1, \\ 0, & \text{otherwise}, \end{cases}
$$

$$
\phi_k\left(\zeta_k,\frac{R}{r_k}\right) = \begin{cases} 1, & \text{if } \left(\frac{R}{r_k}\right)^{\alpha}10^{\frac{\zeta_k}{10}} \leq 1, \\ 0, & \text{otherwise}, \end{cases}
$$

$$
\phi_l\left(\zeta_l,\frac{R}{r_l}\right) = \begin{cases} 1, & \text{if } \left(\frac{R}{r_l}\right)^{\alpha}10^{\frac{\zeta_l}{10}} \leq 1, \\ 0, & \text{otherwise}, \end{cases}
$$
(5.123)

are constraint functions, and ζ_j, ζ_k and ζ_l are Gaussian random variables each having a standard deviation of $\sqrt{2}\sigma$. If $\phi_{(\cdot)}(\cdot)$ is not 1, then handover occurs. Because I_{ext}/S is a function of the random variables ζ_j, ζ_k and ζ_l, its mean can be derived as

$$
\begin{aligned}
E\left[\frac{I_{ext}}{S}\right] &= \frac{1}{G_p}\left[\sum_{j=1}^{2}\sum_{i=0}^{N}\frac{1}{2}E[v_{ij}]E\left[\phi_j\left(\zeta_j,\frac{R}{r_j}\right)10^{\zeta_j/10}\right]\left(\frac{R}{r_j}\right)^{\alpha} \right. \\
&\quad + \sum_{k=1}^{3}\sum_{i=0}^{N}\frac{1}{2}E[v_{ik}]E\left[\phi_k\left(\zeta_k,\frac{R}{r_k}\right)10^{\zeta_k/10}\right]\left(\frac{R}{r_k}\right)^{\alpha} \\
&\quad \left. + \sum_{l=1}^{6}\sum_{i=0}^{N}\frac{1}{2}E[v_{il}]E\left[\phi_l\left(\zeta_l,\frac{R}{r_l}\right)10^{\zeta_l/10}\right]\left(\frac{R}{r_l}\right)^{\alpha}\right].
\end{aligned}
$$
(5.124)

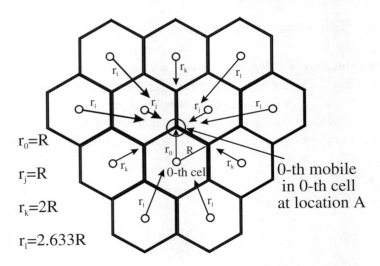

$r_0=R$

$r_j=R$

$r_k=2R$

$r_l=2.633R$

0-th mobile
in 0-th cell
at location A

Figure 5.17: Down-link intercellular interference geometry for mobiles at location A.

Now $E[v_{ij}] = \mu$, the voice activity factor (VAF), $\phi_j(\zeta_j, R/r_j) = \phi(\zeta_j, R/R)$ for $j = 1,2$, $\phi_k(\zeta_k, R/r_k) = \phi(\zeta_k, R/2R)$ for $k = 1,2,3$, $\phi_l(\zeta_l, R/r_l) = \phi(\zeta_l, R/2.633R)$ for $l = 1,2,3,4,5,6$, gives

$$
E\left[\frac{I_{ext}}{S}\right] = \frac{1}{G_p}\left\{ \left[\mu N \left(\frac{R}{R}\right)^\alpha E\left[\phi\left(\zeta_j, \frac{R}{R}\right) 10^{\frac{\zeta_j}{10}}\right] \right. \right.
$$
$$
+ \left[\frac{3}{2} \mu N \left(\frac{R}{2R}\right)^\alpha E\left[\phi\left(\zeta_k, \frac{R}{2R}\right) 10^{\frac{\zeta_k}{10}}\right] \right]
$$
$$
\left. \left. + \left[\frac{6}{2} \mu N \left(\frac{R}{2.633R}\right)^\alpha E\left[\phi\left(\zeta_l, \frac{R}{2.633R}\right) 10^{\frac{\zeta_l}{10}}\right] \right] \right\}. \quad (5.125)
$$

From Equation (5.85)

$$
E\left[\phi\left(\zeta_j, \frac{R}{r_j}\right) 10^{\frac{\zeta}{10}}\right] = \exp\left(\frac{\sigma \ln(10)}{10}\right)^2
$$
$$
\times \left\{ 1 - Q\left[\frac{10\,\alpha \log\left(\frac{r_j}{R}\right)}{\sqrt{2}\,\sigma} - \frac{\sqrt{2}\sigma \ln(10)}{10} \right] \right\} \quad (5.126)
$$

for $r_j = R$, $2R$, or $2.633R$. Hence $E[I_{ext}/S]$ can be evaluated.

The variance of I_{ext}/S is

$$
\text{var}\left[\frac{I_{ext}}{S}\right] = \frac{1}{4G_p^2}\left\{ 2N \left(\frac{R}{R}\right)^{2\alpha} \text{var}\left[v_{ij}\phi\left(\zeta_j, \frac{R}{R}\right) 10^{\frac{\zeta_j}{10}}\right] \right.
$$

$$+ 3N \left(\frac{R}{2R} \right)^{2\alpha} \mathrm{var} \left[v_{ik} \phi \left(\zeta_k , \frac{R}{2R} \right) 10^{\frac{\zeta_k}{10}} \right]$$

$$+ 6N \left(\frac{R}{2.633R} \right)^{2\alpha} \mathrm{var} \left[v_{il} \phi \left(\zeta_l , \frac{R}{2.633R} \right) 10^{\frac{\zeta_l}{10}} \right] \Big\} \quad (5.127)$$

and

$$\mathrm{var} \left[v_{ij} \phi \left(\zeta_j, \frac{R}{r_j} \right) 10^{\frac{\zeta_j}{10}} \right] = \mu \exp \left(\frac{\sigma \ln(10)}{5} \right)^2$$

$$\times \left\{ 1 - Q \left[\frac{10\alpha \log \left(\frac{r_j}{R} \right)}{\sqrt{2}\sigma} - \frac{\sqrt{2}\sigma \ln(10)}{5} \right] \right\}$$

$$- \mu^2 \mathrm{E} \left[\phi_j \left(\zeta_j, \frac{R}{r_j} \right) 10^{\frac{\zeta}{10}} \right]^2, \quad (5.128)$$

for $r_j = R, 2R$ or $2.633R$. Armed with these variances in Equation 5.128 we can compute var $[I_{ext}/S]$.

While the intracellular interference-to-signal ratio in Equation (5.122) is not affected by shadowing, the intercellular interference-to-signal ratio is affected. As a result, the overall E_b/I_0 varies about a mean value. Applying the same method that led to the derivation of Equation (5.99), the outage probability of the down-link becomes

$$P_o = \mathrm{Pr} \left(\frac{S}{I} < \gamma_{req} \right) = \mathrm{Pr} \left(\frac{I}{S} > \frac{1}{\gamma_{req}} \right)$$

$$= \mathrm{Pr} \left(\frac{I_{ext}}{S} > \frac{1}{\gamma_{req}} - \frac{\eta}{S} - \frac{I_{int}}{S} \left| \sum_{i=0}^{N-1} v_i = k \right. \right) \mathrm{Pr} \left(\sum_{i=0}^{N-1} v_i = k \right)$$

$$= \sum_{k=1}^{N} \binom{N}{k} \mu^k (1 - \mu)^{N-k}$$

$$\times Q \left(\frac{1/\gamma_{req} - \eta/S - k/G_p - \mathrm{E}[I_{ext}/S]}{\sqrt{\mathrm{var}[I_{ext}/S]}} \right). \quad (5.129)$$

Since the pilot signal is transmitted all the time, the minimal number of active channels is one. As a result, the summation in Equation (5.129) is from 1 to N.

For a CDMA system having a processing gain of 128, a signal-to-AWGN ratio of 20 dB at the matched filter output, a radio channel with an inverse fourth power path loss law, and an 8 dB standard deviation for the slow fading, the performance of a mobile at location A for a VAF of 1/2 and 3/8 is shown in Figure 5.18. While maintaining an outage probability of 2%, the down-link can support up to 23 users and 30 users for a VAF of 1/2 and 3/8, respectively.

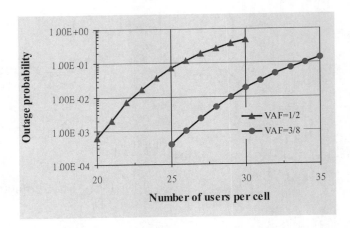

Figure 5.18: Down-link performance of mobiles at location A.

Performance at location B   Let us now consider the performance of the down-link system for a mobile located at position B in Figure 5.16. According to the intercellular interference geometry shown in Figure 5.19, the mobile is located at the midpoint on one of the hexagonal edges, where the distance from the jth interfering BS is r_j, the distance from the two kth interfering BSs is r_k, and the distance from the four lth interfering BSs is r_l. Following the procedure developed in the previous section, we calculate the signal-to-interference power ratio and the outage probability using Equations (5.121) and (5.129), respectively.

A mobile at location B is at a distance of $r_0 = \sqrt{3}/2\ R$ from its own BS, and $r_j = \sqrt{3}/2R$, $r_k = 1.5R$ and $r_l = 2.3R$. The outage probability as a function of the number of users is shown in Figure 5.20 for the down-link system having a G_p of 128, an inverse fourth power path loss law, an 8 dB standard deviation for the slow fading, and a 20 dB signal-to-AWGN ratio. For an outage probability of 2%, the down-link system can support up to 33 users and 43 users for a VAF of 1/2 and 3/8, respectively.

5.4.2.2 With power control

Power control in the down-link attempts to allocate the total transmitted power from the BS to its users according to their distances from the BS, while maintaining a E_b/I_0 for all users in its cell above a minimal level (typically ≥ 7 dB). Assume that the users are uniformly distributed over the cell coverage area with density ρ. The transmitted power for the ith mobile in a cell is controlled according to the following rule [10, 11]:

$$P_i = \begin{cases} P_m \left(\frac{R_0}{R}\right)^2, & \text{for } r_i \leq R_0, \\ P_m \left(\frac{r_i}{R}\right)^2, & \text{for } R \geq r_i > R_0, \end{cases} \tag{5.130}$$

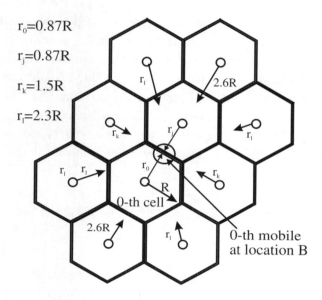

$r_0=0.87R$
$r_j=0.87R$
$r_k=1.5R$
$r_l=2.3R$

Figure 5.19: Down-link intercellular interference geometry for mobiles at location B.

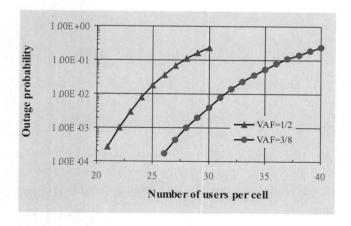

Figure 5.20: Down-link performance of a mobile at location B.

where P_m is the maximum designed power to be delivered to mobiles at the cell boundary and r_i is the distance between a mobile and its BS. We allocate to the near-in users (at a distance less than R_0) a minimum and constant transmitted power so that they can have the same E_b/I_0 as other users (at a distance larger than R_0) in the cell. Observe that for $R \geq r_i > R_0$ an inverse square power law is used, which we emphasise is not the propagation path loss law. If, say, we use an inverse fourth power law for the power control on the forward link, then R_0 would be close to R, and consequently the zone where the power control is exercised would be very small. Given the above power control condition, the average power transmitted by a BS can be found to be the summation of the power transmitted to all users within the circle of radius R_0, plus the power transmitted to all users in the doughnut ring from $r_i = R_0$ to R, namely

$$
\begin{aligned}
P_t &= 2\pi \int_0^{R_0} \frac{R_0^2}{R^2} P_m \rho r_i dr_i + 2\pi \int_{R_0}^R \frac{r_i^3}{R^2} P_m \rho dr_i \\
&= \frac{1}{2}\pi\rho P_m R^2 \left(1 + \frac{R_0^4}{R^4} \right),
\end{aligned}
\tag{5.131}
$$

where ρ, the user density, is

$$
\rho = \frac{N}{\pi R^2}
$$

and N is the total number of users in the cell. To determine R_0 we proceed as follows. The trasmitted power for the mobiles located in a ring area at a distance r from its BS is

$$
P(r) = P_m \left(\frac{r}{R} \right)^2,
\tag{5.132}
$$

and therefore the total power transmitted from the BS is given by

$$
P_t = N \int_0^R P(r)f(r)dr = \int_0^R P_m \left(\frac{r}{R} \right)^2 \frac{2r}{R^2} dr = \frac{1}{2}NP_m,
\tag{5.133}
$$

where $f(r)$ is the PDF of a mobile being in the ring of area $2\pi dr$ and has the form

$$
f(r) = \frac{2\pi r}{\pi R^2} = \frac{2r}{R^2}.
\tag{5.134}
$$

For mobiles near the BS, say at a distance $\leq R_0$, the interference from outside the cell is small compared with the intracellular interference within the cell due to the low path loss. By neglecting the intercellular interference, the signal-to-interference ratio (SIR) is

$$
\left(\frac{S}{I} \right)_{R_0} = \frac{P_m (R_0/R)^2}{NP_m/2} = \frac{(R_0/R)^2}{N/2}
\tag{5.135}
$$

where $NP_m/2$ is the intracellular interference. From Equation (5.20), the SIR at the cell boundary in position A in Figure 5.17 is the ratio of the signal power P_m to the interference power I. The latter has three components. The intracellular interference is given by Equation (5.133), the intracellular interference has three different interference terms depending on the location of the BS shown in Figure 5.17, and in addition there is the receiver noise power. The interference from one BS is, assuming an inverse fourth power path loss law,

$$\left(\frac{1}{2}\right)\left[\frac{1}{2}(N+1)P_m\right](r_j/R)^{-4},\tag{5.136}$$

where the first $(1/2)$ term is for carrier incoherence, $(N+1)$ is used rather than N since interfering BSs have N channels plus a pilot channel, and $r_j = R, 2R$ or $2.633R$. On neglecting the receiver noise power, the *SIR* at point A in Figure 5.17 is

$$\begin{aligned}
\left(\frac{S}{I}\right)_R &= \frac{P_m}{\frac{1}{2}P_m\left[N+\frac{1}{2}\{2(N+1)+3(N+1)(2)^{-4}+6(N+1)2.633^{-4}\}\right]}\\
&= \frac{1}{\frac{N}{2}+\frac{N+1}{2}+3\frac{N+1}{4}(2)^{-4}+3\frac{N+1}{2}(2.633)^{-4}}\\
&= \frac{1}{1.08N+0.58} \approx \frac{1}{1.08N}.
\end{aligned}\tag{5.137}$$

To ensure that mobiles near the BS have the same *SIR* as mobiles at the boundary,

$$\left(\frac{S}{I}\right)_{R_0} \geq \left(\frac{S}{I}\right)_R \quad \text{or} \quad \frac{(R_0/R)^2}{N/2} \geq \frac{1}{1.08N}.\tag{5.138}$$

After solving the above equation, we have $R_0 \geq 0.68R$, and we choose a value of $0.68R$ for R_0. By substituting $R_0 = 0.68R$ into Equation (5.131), we have

$$P_t = 0.6NP_m = N(0.6P_m).\tag{5.139}$$

Note that the power transmitted by a BS is reduced by a factor of 0.6 through the application of power control, and the average power allocated for each user should be reduced by the same factor as well. As a result, the total interference power should also be reduced by the same factor. The bit-energy-to-interference ratio in Equation (5.121) becomes, for a mobile in location A in Figure 5.17,

$$\frac{E_b}{I_0} = \frac{1}{\frac{1}{G_p}\left[\sum_{i=1}^{N}v_i\frac{P_i}{P_m}+\sum_{j=1}^{J-1}\sum_{i=0}^{N}\frac{1}{2}v_{ij}\frac{P_i}{P_m}10^{\frac{\zeta_j}{10}}\left(\frac{R}{r_j}\right)^{\alpha}\right]+\frac{\eta}{S}},\tag{5.140}$$

where for the intracellular interference of $\sum_{i=1}^{N}v_ip_i$ and signal power p_m for the mobile in position A, we obtain the first term in the denominator. The second term relates to the intercellular interference, where $\sum_{i=0}^{N}\frac{1}{2}v_{ij}P_i10^{\zeta_j/10}(R/r_j)^{\alpha}$ includes the VAF term v_{ij}, and we have previously set $\alpha = 4$ in our calculation of R_0. The term η/S represents the noise-to-signal ratio at the output of the matched filter. By employing forward power control, $P_t = N(0.6P_m)$, i.e. on average the users transmit only $0.6p_m$ instead of p_m. Consequently we set $P_i = 0.6P_m$ to get

$$\frac{E_b}{I_0} = \frac{1}{\frac{1}{G_p}\left[0.6\sum_{i=1}^{N} v_i + 0.6\sum_{j=1}^{J-1}\sum_{i=0}^{N} \frac{1}{2}v_{ij}10^{\frac{\zeta_j}{10}}\left(\frac{R}{r_j}\right)^{\alpha}\right] + \frac{\eta}{S}}. \tag{5.141}$$

For mobiles at locations A and B, the performance of the down-link system with power control can be found from Equations (5.141) and (5.129). For the down-link system having a processing gain of 128 and a signal-to-AWGN ratio of 20 dB at the matched filter output, the performance in terms of outage probability is shown in Figures 5.21 and 5.22. While maintaining an outage probability of 2%, for a mobile at location A, the down-link system can support 47 users and 54 users per cell for a VAF of 1/2 and 3/8, respectively. For a mobile at location B, the down-link system can support 45 users and 59 users for a VAF of 1/2 and 3/8, respectively. The capacity increase is more than 80% compared with the down-link system without power control, and therefore the power control should be applied on the down-link to increase the capacity of that link.

5.4.3   Down-link with orthogonal codes

In the previous section we considered down-link transmissions employing PN codes or any code having good autocorrelation and cross correlation properties, such as Gold codes. Since the spreading codes for users in the same cell are not orthogonal to each other, the intracellular interference is non-zero. In the IS-95 CDMA system, all users' CDMA channels are time aligned. Each users' coded symbol is replaced by a 64 chip Walsh code and then codes are orthogonal. The orthogonal property of the Walsh codes is only applicable when they are time synchronous. However, interference due to the reception of multipath signals from its own BS, and interference from neighbouring BS transmissions, have CDMA codes that are not orthogonal with the wanted signal, i.e. they are time asynchronous. When Walsh codes are not time synchronous, their autocorrelation and cross correlation properties are not as good as many other codes, e.g. Gold codes, and consequently the processing gain decreases. In order to improve the cross correlation property between the wanted signal and its interference, a short PN is added to the Walsh codes by means of exclusive-OR gates. The spreading code for the ith CDMA channel in the IS-95 down-link is a combination of Walsh code and the short PN code $C_{PN}(t)$,

$$c_i(t) = c_{w(i+1)}(t) \oplus c_{PN}(t), \tag{5.142}$$

where $\oplus$ represents an exclusive-OR operation, and $c_{w(i+1)}(t)$ is the $(i+1)$th Walsh code corresponding to the ith down-link channel. Note that the down-link pilot $c_{PN}(t)$ is trans-

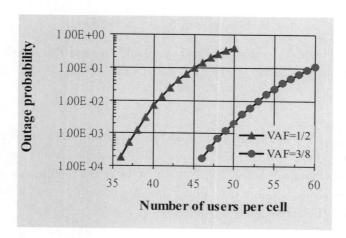

Figure 5.21: Performance of the power controlled down-link at location A.

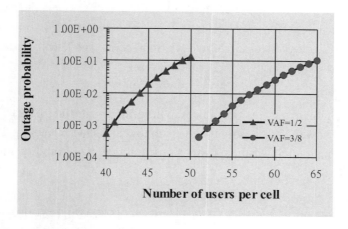

Figure 5.22: Performance of the power controlled down-link at location B.

mitted on the Nth channel and is in fact the code $C_{PN}(t)$ exclusively-ORed with the zeroth Walsh code $c_{w0}(t)$, which is a constant, i.e.

$$c_N(t) = c_{w0}(t) \oplus c_{PN}(t) = c_{PN}(t). \tag{5.143}$$

By arranging for a logical 1 to be represented by $+1$ and a logical 0 by -1, the exclusive-OR operation may be expressed as

$$c_i(t) = -c_{w(i+1)}(t)c_{PN}(t). \tag{5.144}$$

When the orthogonal spreading codes are used on the down-link, and the multipath interference is removed by the RAKE, FEC coding and symbol interleaving, the performance is improved, as we will now describe. Let us consider the zeroth mobile user who is communicating with the zeroth BS. The intracellular interference component at the output of the matched filter now becomes

$$
\begin{aligned}
Z_{int}(T_b) &= \frac{1}{T_b} \int_0^{T_b} n_{int}(t)c_0(t)dt \\
&= \frac{1}{T_b} \int_0^{T_b} n_{int}(t)[-c_{w1}(t)c_{PN}(t)]dt ,
\end{aligned}
\tag{5.145}
$$

where $c_{w1}(t)$ is the Walsh code for the zeroth channel. The equivalent baseband intracellular interference now becomes

$$
\begin{aligned}
n_{int}(t) &= a_0 \sum_{i=1}^{N} \sqrt{2P_i} b_i(t)c_i(t) \\
&= a_0 \sum_{i=1}^{N-1} \sqrt{2P_i} b_i(t) \left[-c_{w(i+1)}(t)c_{PN}(t)\right] \\
&\quad - a_0 \sqrt{2P_p} c_{w0}(t)c_{PN}(t)
\end{aligned}
\tag{5.146}
$$

where the second term in the equation relates to the pilot signal, since $b_N(t) = 1$ and $P_N = P_p$ is the transmitted power, and a_0 is due to the path loss and shadow fading. The pilot power P_p takes 20% of the total down-link power. Substituting $n_{int}(t)$ from Equation (5.146) into Equation (5.145) gives

$$
\begin{aligned}
Z_{int}(T_b) &= \frac{a_0}{T_b} \sum_{i=1}^{N-1} \sqrt{2P_i} \int_0^{T_b} b_i(t) \left[-c_{w(i+1)}(t)c_{PN}(t)\right] [-c_{w1}(t)c_{PN}(t)]dt \\
&\quad + \frac{a_0}{T_b} \sqrt{2P_p} \int_0^{T_b} [-c_{w0}(t)c_{PN}(t)][-c_{w1}(t)c_{PN}(t)]dt \\
&= \frac{a_0}{T_b} \sum_{i=1}^{N-1} \sqrt{2P_i} \int_0^{T_b} b_i(t) [c_{PN}(t)c_{PN}(t)] [c_{w1}(t)c_{w(i+1)}(t)] dt
\end{aligned}
$$

$$+ \frac{a_0}{T_b} \sqrt{2P_p} \int_0^{T_b} [c_{PN}(t)c_{PN}(t)][c_{w0}(t)c_{w1}(t)] dt$$

$$= \frac{a_0}{T_b} \sum_{i=1}^{N-1} \sqrt{2P_i} \int_0^{T_b} b_i(t) [c_{w1}(t)c_{w(i+1)}(t)] dt$$

$$+ \frac{a_0}{T_b} \sqrt{2P_p} \int_0^{T_b} [c_{w0}(t)c_{w1}(t)] dt. \tag{5.147}$$

In the above equation, $b_i(t)$ over the integral period 0 to T_b equals 1 or -1, and because the integrals of the products of the codes $c_{w1}(t)c_{w(i+1)}(t)$ and $c_{w0}(t)c_{w1}(t)$ are zero, $Z_{int}(T_b)$ is zero. In IS-95 each BS employs co-ordinated time via the global position salellite (GPS) system. Each BS transmits the same $c_{PN}(t)$, but with a phase shift. It is this phase shift that is used for BS identification. The intercellular interference experienced by a mobile in the zeroth cell from neighbouring BSs is finite. This is because the codes received from the interfering BSs are not orthogonal with the codes used in zeroth cell due to the phase offsets of the pilot signals in the adjacent BS and the presence of different propagation delays. Consequently, the intercellular interference and the AWGN components at the output of the matched filter remain the same as those derived in Section 5.2.2. By setting I_{int} in Equation (5.129) to zero, the down-link outage probability becomes

$$p_o = Q \left(\frac{1/\gamma_{req} - \eta/S - E[I_{ext}/S]}{\sqrt{\text{var}[I_{ext}/S]}} \right). \tag{5.148}$$

The VAF has no effect on the intracellular interference, but it is still an important factor in decreasing the intercellular interference. Using Equation (5.148), we calculate the performance of the orthogonal down-link system for a processing gain of 128, a signal-to-AWGN ratio of 20 dB at the output of the matched filter, a radio channel having an inverse fourth power path loss law and an 8 dB standard deviation for the slow fading. Let us consider mobiles located at positions A and B in Figure 5.16. Then in the absence of power control the down-link performances for mobiles at locations A and B are shown in Figures 5.23 and 5.24, respectively. A mobile at location A has a much higher outage probability than that experienced by a mobile at location B for the same number of users per cell. In order to provide the required outage probability for mobiles within a cell, the system capacity is limited by the outage probability at location A. For an outage probability of 2%, we found that the orthogonal down-link system can support 73 and 97 users per cell for a VAF of 1/2 and 3/8, respectively. Since this capacity is much higher than that on the up-link, the system capacity is limited by the up-link, and therefore we will refrain from calculating the performance for power control on the down-link in the presence of orthogonal codes.

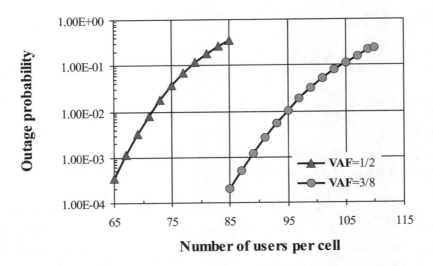

Figure 5.23: Down-link performance of mobiles at location A when the CDMA codes are orthogonal.

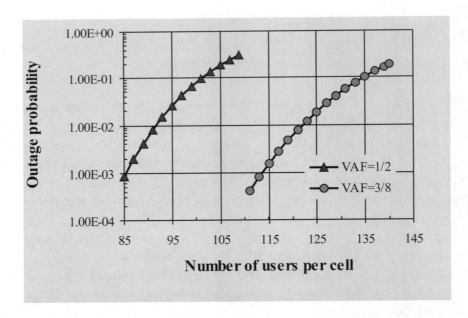

Figure 5.24: Down-link performance of mobiles at location B when the CDMA codes are orthogonal.

5.4.4 Effect of sectorisation

Sectorisation is used in CDMA to increase system capacity where each cell is divided into sectors and the same frequency spectrum is reused in all the sectors. Hence the capacity is multiplied accordingly. As shown in Figure 5.25, a cell is divided into N_s sectors by using sectorisation antennae at each cell site. Assuming that the sector antennae patterns are perfectly confined within the sector area, the interference on the up-link for the zeroth sector BS is from the mobiles in the shaded area and is $1/N_s$ of the whole area. Hence the interference is only $1/N_s$ of the unsectorised cell for the same number of users per cell. In other words, the number of users per cell can be increased N_s times to give the same level of interference. As a result, the capacity increases N_s times compared with the unsectorised cells.

On the down-link, for mobiles in any of the sectors in the central cell shown in Figure 5.26, there are six first tier interfering BSs which have the same number of interfering BSs as in the unsectorised cells. If the number of users per cell for the sectorised cell is the same as that of the unsectorised cells, namely N, then each sector of the BS handles N/N_s of the mobiles per cell. Consequently, the interference power from the six interfering sector BSs experienced by the mobiles in the central cell is only $1/N_s$ compared with a unsectorised cell arrangement. By assigning each sector with the same capacity as the unsectorised cell to give the same level of interference, the sectorised cell has a capacity equal to the capacity of the unsectorised cell multiplied by N_s.

5.4.4.1 Overlapping sectors

The radiation pattern of a sector antenna having an angle which is identical to the sector angle has been previously assumed. To ensure coverage and to provide soft handover zones between sectors we arrange for the sectorised antennae patterns slightly to overlap the adjacent ones. A three-sector cell arrangement having overlapping sectors is a single (idealised) lobe spanning an angle of $(2\pi/N_s) + \theta_0$, where θ_0 is the overlapping angle [12]. Since the system capacity is limited by the up-link, only the effect of imperfect sectorisation on the up-link performance is considered. If there is no overlapping, i.e. $\theta_0 = 0$, then $1/N_s$ of the total interference is received, and the capacity gain, due to sectorisation, is N_s times that of an unsectorised cell. In a sectorised cell with overlapping angle θ_0, only $(2\pi/N_s + \theta_0)/2\pi$ of the total interference from the surrounding is received. For this condition, the capacity gain due to sectorisation is $2\pi/(2\pi/N_s + \theta_0)$ times that of the unsectorised cell. We define the efficiency of sectorisation as the ratio of the capacity gain with the sector antennae pattern having an overlapping angle θ_0 to the non-overlapping antenna pattern of $2\pi/N_s$, namely

$$e_s \triangleq \frac{\frac{2\pi}{2\pi/N_s + \theta_0}}{\frac{2\pi}{2\pi/N_s}} = \frac{2\pi/N_s}{2\pi/N_s + \theta_0}; \tag{5.149}$$

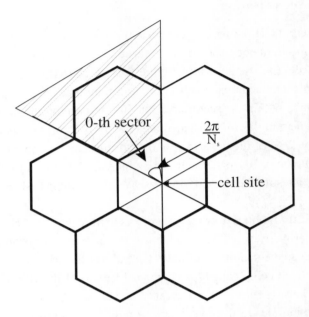

Figure 5.25: Hexagonal cell with N_s sectors per cell (shown here as $N_s = 6$). The up-link interference in the zeroth sector BS is from mobiles in the shaded area.

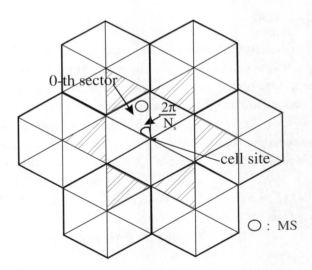

Figure 5.26: Hexagonal cell with N_s sectors per cell. The interference for MSs in the zeroth sector is from the shaded sector BSs.

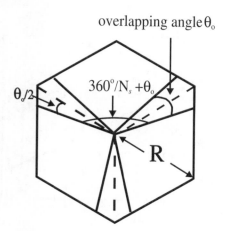

Figure 5.27: A model for imperfect sectorisation in a hexagonal cell, shown with a three sectors per cell arrangement.

the corresponding gain is $N_s e_s$.

For a three sectors per cell arrangement, the capacity gain due to sectorisation, and its corresponding sectorisation efficiency, for different values of overlapping angle θ_0 in degrees is shown in Figure 5.28 where $\theta_0 = 0$ denotes a nominal antenna pattern of 120°. The interference on the up-link increases as the overlapping angle θ_0 is increased, which causes the capacity gain and sectorisation efficiency to decrease proportionately. Increasing θ_0 from 0° to 5° causes the capacity gain to decrease from 3 to 2.88, while the corresponding sectorisation efficiency drops from 100% to 96%.

Observe that by using overlapping sectors soft handovers can be used where the transmitted power of MSs in the overlapping sectors is decreased due to both adjacent sector BSs handling the call. Soft handovers, often called softer handovers, are between sectors, and therefore decrease the total interference and thereby increase the capacity.

5.4.5 The capacity of the IS-95 CDMA in macrocells

In the previous subsections we investigated the performance of the up-link and down-link in terms of the number of users per cell. We will now evaluate the capacity of the CDMA system in terms of channels per cell per MHz and compare our findings with the capacity of the modern TDMA system described in Chapter 3.

For the down-link, when power control is not used, and a mobile is located at an apex of a cell (position A in Figure 5.17) the number of users per cell is the same as the performance of the up-link when this link has perfect power control. Consequently, when the up- link has imperfect power control, the performance of the down-link is superior. The down-link, or forward link, is enhanced when it has power control.

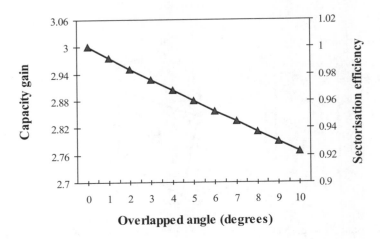

Figure 5.28: Capacity gain due to sectorisation and sectorisation efficiency as a function of the overlapping angle θ_0 for the three sectors per cell arrangement.

Therefore it is reasonable to say that the capacity of a CDMA system is limited by the up-link performance. With a message bit rate of 10 kb/s, a processing gain of 128, and three sectors per cell, the capacity of the CDMA system in terms of numbers of channels per cell per MHz is given as

$$k_{CDMA} = \frac{NN_s e_s}{W} = \frac{NN_s e_s}{R_b M} = \frac{NN_s e_s}{1.27}, \qquad (5.150)$$

where N is the number of users per sector, N_s is the number of sectors per cell, e_s is the sectorisation efficiency and W is the bandwidth in MHz. Note that we equate the number of users with the number of channels because of the soft capacity of CDMA. Consequently, N is the number of channels per sector that provides an acceptable outage probability.

For a standard deviation of power control errors in E_b/I_0 of 2 dB, the capacity of the unsectorised CDMA system is 22 and 17 channels per cell per MHz for VAFs of 3/8 and 1/2, respectively, while the capacity of the sectorised CDMA system having a sectorisation efficiency of 1, is 66 and 51 channels per cell per MHz for VAFs of 3/8 and 1/2, respectively. This capacity is much higher than that of a modern TDMA system which has 12 channels per cell per MHz for a three-cell cluster arrangement as described in Chapter 3.

5.4.6   The effect of channel coding on CDMA systems

In the previous sections, the IS-95 system without channel coding and bit interleaving has been addressed. The simplified block diagram of an uncoded system is shown in Figure 5.29. Each message bit is spread by an M-chip spreading code prior to RF modulation.

The received signal experiences an interference power of I, which includes a multiple access interference (MAI) power of I_{MAI}, and an AWGN power of η. The multiple access interference consists of intracellular and intercellular interference. The signal received at the receiver is demodulated and despread at the matched filter, and a recovered bit stream is obtained. Assuming that the received wanted signal power at the input to the receiver is P_R and that its chip rate is $R_c = 1/T_c$, than of an interference power $I = I_{MAI} + \eta$, the received E_b/I_0 at the output of the matched filter is

$$\gamma_b = \frac{E_b}{I_0} = \begin{cases} G_p \frac{P_R}{I} = G_p \frac{P_R}{I_{MAI}+\eta}, & \text{for non-coherent demodulation,} \\ \\ G_p \frac{2P_R}{I} = G_p \frac{2P_R}{I_{MAI}+\eta}, & \text{for coherent demodulation,} \end{cases} \qquad (5.151)$$

where $G_p = T_b/T_c = M$ is the processing gain. Because CDMA systems normally use BPSK and QPSK as their modulation schemes, the bit error probability for the E_b/I_0 shown in Equation (5.151) is

$$P_b = Q(\sqrt{\gamma_b}), \qquad (5.152)$$

where $Q(\cdot)$ is a Gaussian Q-function. This function becomes smaller as γ_b increases. Hence for a small bit error probability, a large γ_b is required for good communication quality. Channel coding, especially convolution coding, and bit interleaving are widely used to improve the communication quality in CDMA systems. A CDMA link having a convolutional codec is shown in Figure 5.30. The message bits are forward error corrected (FEC) coded and interleaved, followed by spreading and modulation to transform a bursty error channel into a random error channel when de-interleaving is performed at the receiver. In the FEC coder a $cc(n, k, K)$ convolutional encoder generates an n-bit code word for each k-bit input message and has a constraint length of K. The block diagram of the convolutional coder is shown in Figure 5.31. This channel coding scheme has a coding rate of $r = k/n$, where $k < n$. For the same spreading chip rate as we have in an uncoded system, the spreading factor is reduced to $M' = Mk/n$ because the bit rate has increased by n/k after channel coding.

At the receiver, the PSD of the total interference at the input of the channel decoder is I_0, while the received FEC coded bit energy becomes E_c instead of E_b. As a result, the coded bit-energy-to-interference PSD ratio is

$$\gamma_c = \frac{E_c}{I_0} = \frac{rR_b}{I_0} = r\gamma_b. \qquad (5.153)$$

By employing soft decision coding, the bit error probability after decoding becomes upper bound by [13, 14]

$$p_b \leq \frac{1}{k} \sum_{d=d_f}^{\infty} \beta_d Q\left(\sqrt{\gamma_c d}\right) = \frac{1}{k} \sum_{d=d_f}^{\infty} \beta_d Q\left(\sqrt{\gamma_b r d}\right), \qquad (5.154)$$

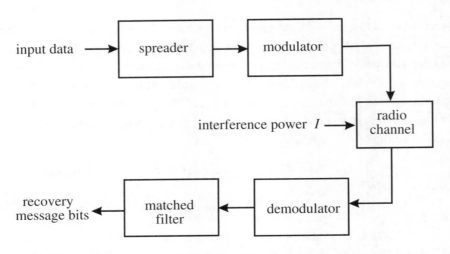

Figure 5.29: Simplified block diagram of a CDMA system without channel coding.

where d_f represents the minimal free distance, γ_b is E_b/I_0 given in Equation (5.151), k is the number of message bits at the input to the encoder that generates a code word of length n, and β_d is the coefficient of the transfer function $T(D,H)$ of the convolutional code. The bit error probability in the above equation mainly depends on the value of $Q(\sqrt{\gamma_b rd})$ which decreases with increasing $\sqrt{\gamma_b rd}$. Because the minimal value of d is d_f, the most significant term in the summation of P_b is $\beta_d Q(\sqrt{\gamma_b rd_f})$ for a high E_b/I_0. We may express, with the aid of Equation (5.151),

$$Q\left(\sqrt{\gamma_b rd_f}\right) = Q\left(\sqrt{\gamma_b G_c}\right) = Q\left(G_p G_c \frac{P_R}{I}\right), \tag{5.155}$$

where $G_c = rd_f$ is the asymptotic coding gain of the FEC code. As a consequence of FEC coding the Q-function becomes smaller due to the channel coding gain G_c, and this results in a lower P_b.

5.4.7 Summary

The performance of CDMA systems for the tessellated hexagonal cells in terms of number of users per cell was evaluated for both the up-link and down-link systems. We began by analysing a single cell CDMA system, and then progressed to a multicellular system. Both perfect power control and imperfect power control were addressed for the up-link, while only perfect power control was considered for the down-link. For a system having a processing gain of 128, a signal-to-AWGN ratio of 20 dB, and a 2% outage probability, the performance of the CDMA system was described as follows. The up-link system with perfect power control supported 23 and 30 users per cell for a VAF of 1/2 and 3/8, respectively.

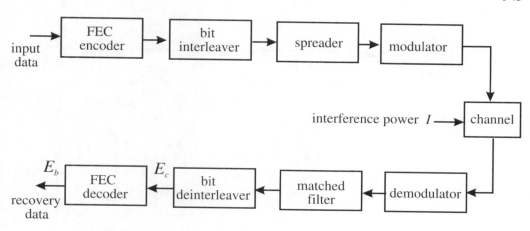

Figure 5.30: Simplified block diagram of a coded CDMA system.

The capacity reduction due to imperfect power control having a standard deviation of power control error in E_b/I_0 of 2.0 dB was 4.3% and 6.3% for VAFs of 1/2 and 3/8, respectively, while the capacity reduction was about 13% for a standard deviation of 2.5 dB, irrespective of the value of the VAFs. This highlights the importance of the good power control in CDMA systems.

For the down-link system, the performance was location dependent, and the performance of the two worst locations along a hexagonal cell boundary was considered. The down-link system at the corners of hexagonal cells supported 23 and 30 users for a VAF of 1/2 and 3/8, respectively, while the system at the edges of hexagonal cells supported 24 and 32 users per cell for a VAF of 1/2 and 3/8, respectively. The capacity of the down-link system was improved by at least 80% with power control. Without power control, the down-link capacity at the boundary, even at the hexagonal corners, was about 7% and 4.5% higher than that of the up-link system for a VAF of 3/8 and 1/2, respectively. Therefore, the capacity of CDMA systems was limited by the up-link performance.

Without sectorisation, the capacity of the cellular CDMA system was 22 and 17 channels per cell per MHz for VAFs of 3/8 and 1/2, respectively. This was about 1.5 times that of the unsectorised TDMA system described in Chapter 3 for a VAF of 3/8 and 1/2, respectively. By dividing each cell into N_s sectors, the capacity of the sectorised cellular system can be increased by N_s times compared with the unsectorised cellular system, assuming ideal antennae patterns. For three sectors per cell the capacity of the CDMA system was 66 and 51 users per cell per MHz for VAFs of 3/8 and 1/2, respectively. This capacity was 5.5 and 4.2 times that of a modern TDMA system described in Chapter 3 for a VAF of 3/8 and 1/2, respectively.

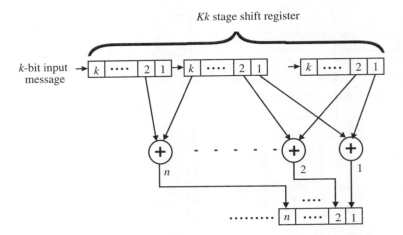

Figure 5.31: Block diagram of a $cc(n, k, K)$ convolutional encode.

5.5   IS-95 Street Microcellular Networks

High densities of teletraffic in city centres can be accommodated by a microcellular personal communications network (PCN) [15, 16]. Small base stations (BSs) located in the streets with their antennae below the urban skyline provide radio coverage that is dependent on the road topology and the cross-sectional area of the buildings. For cities with irregular street patterns, as exemplified by the city of London, the microcells also are irregularly shaped. Cities, such as Manhattan, that have grid patterns of streets and block-shaped buildings yield regularly shaped microcells. In this section we analyse the latter microcellular patterns because of their relative simplicity. However, our theory can be modified to accommodate less regular microcellular structures.

We consider two microcellular arrangements. Figure 5.32 shows tessellated cross-shaped city microcells formed by positioning the BSs at street intersections. Rectangular-shaped microcells that occupy a side of a city block are shown in Figure 5.33. CDMA systems are resistant to co-channel interference and as a consequence each BS uses the entire allotted spectrum. Nevertheless the capacity of CDMA systems is limited by co-channel interference from neighbouring microcells as well as from interference generated by other users in each microcell.

The city street microcells shown in Figures 5.32 and 5.33 are formed by controlling the radiated power levels to provide acceptable communications at the microcell boundaries. Although there will be penetration of electromagnetic energy into buildings, this does not affect the shape of the street microcell, assuming the buildings are of a similar type. While the microcell boundary is considered to be half a city block from its BS, the radiation from both the BS and mobile stations (MSs) will extend well beyond this microcellular boundary and thereby produce co-channel interference.

In this section we are interested in evaluating and comparing the capacity of IS-95 in the city street microcellular patterns shown in Figures 5.32 and 5.33. The CDMA system in our investigation will have BPSK modulation, and VAD. There will be power control on the up-link, but none on the down-link. The effect of imperfect power control on the system capacity will also be considered.

As the number of users that can be supported by the CDMA up-link and down-link may not be equal, we need to determine the number of users per microcell on both of these links and to use the smaller number as the representative of capacity. Imperfect power control is assumed on the up-link, while power control is not considered on the down-link. The capacity of the CDMA system in the city street microcells will be examined for both cross-shaped and rectangular-shaped microcells.

5.5.1 Up-link system and signal model

The block diagram of the CDMA system employing asynchronous BPSK modulation that is used in our deliberations is shown in Figure 5.34. We will use a similar procedure to that employed in Section 5.3.1 for the single cell, but here we have a multicellular arrangement. Although this does cause some duplication of previous text, we trust the reader will find it helpful. So, at the mobile transmitter, the data are processed by a spreading code, followed by BPSK modulation. The mobiles transmit to the BS under the influence of power control which attempts to maintain a received target power level at the BS that is the same for each mobile, irrespective of its distance from the BS and the slow fading on its up-link. The message data from the ith mobile communicating to its jth BS is $b_{ij}(t)$. Each bit has a duration $T_b = 1/R_b$, where R_b is the bit rate. The transmitted signal from the ith mobile to the jth BS is

$$s_{ij}(t) = \sqrt{2P_{ij}} b_{ij}(t) c_{ij}(t) \cos(\omega_1 t + \theta_{ij}), \tag{5.156}$$

where P_{ij} is the transmitted power, θ_{ij} is the carrier phase of the ith mobile in the jth microcell, ω_1 is the up-link carrier frequency, and $c_{ij}(t)$ is a spreading code sequence consisting of M chips having a chip duration $T_c = T_b/M$. The transmitted signal passes over the radio channel to the receiver at the microcell site.

Let us consider the communication from the zeroth mobile in the presence of the $N-1$ other mobiles. Suppose there are J BSs in the PCN. We consider the BS of index $j = 0$ when our desired communication is between the zeroth MS and the zeroth BS. The received signal at the zeroth BS is

$$a_{00} s_{00}(t), \tag{5.157}$$

where a_{00} includes both the effects of path loss and slow fading defined in Equation (3.54). Note that the variable t in Equation (5.157) is different from the t in Equation (5.156) as

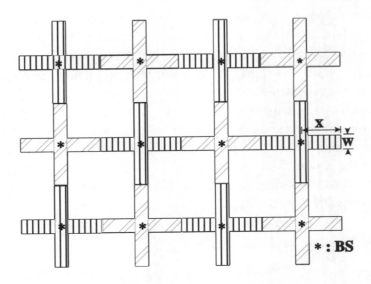

Figure 5.32: Cross-shaped city street microcells in a two-cell per cluster arrangement.

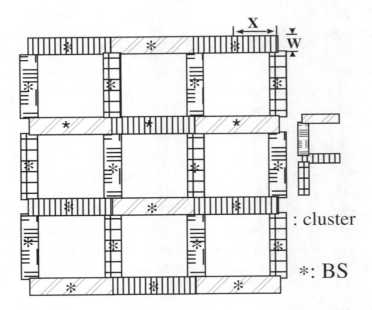

Figure 5.33: Rectangular-shaped city street microcells in a four cell per cluster arrangement.

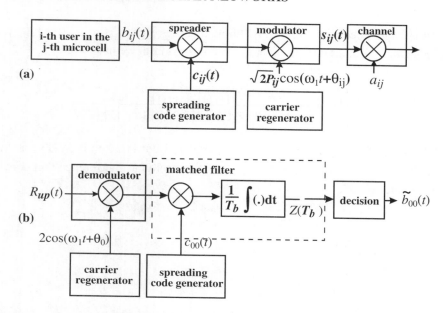

Figure 5.34: (a) Block diagram of the MS transmitter, and (b) BS receiver for the zeroth mobile.

it now includes propagation delay. We make this change to t for mathematical simplicity. This normalisation of t applies for all received signals at the BS. The other $N-1$ users in the zeroth microcell present an intracellular interference at the BS receiver of

$$\sum_{i=1}^{N-1} a_{i0} s_{i0}(t - \tau_{i0}) ,\qquad(5.158)$$

where τ_{i0} is the propagation delay of the received signal from the ith mobile in the zeroth microcell relative to the zeroth mobile in the zeroth microcell. The users in the $J-1$ neighbouring microcells also produce an intercellular interference for the zeroth mobile at the zeroth BS receiver of

$$\sum_{j=1}^{J-1} \sum_{i=0}^{N-1} a_{ij} s_{ij}(t - \tau_{ij}) ,\qquad(5.159)$$

where τ_{ij} is the propagation delay of the received signal from the ith mobile in the jth cell relative to the zeroth in the zeroth microcell. In addition to the intracellular and intercellular interference, the channel presents an additive white Gaussian noise (AWGN), $n(t)$, of double-sided power spectral density $N_0/2$. The signal received at the zeroth cell site is

$$R_{up}(t) = a_{00} s_{00}(t) + \sum_{i=1}^{N-1} a_{i0} s_{i0}(t - \tau_{i0}) + \sum_{j=1}^{J-1} \sum_{i=0}^{N-1} a_{ij} s_{ij}(t - \tau_{ij}) + n(t) .\qquad(5.160)$$

The BS receiver shown in Figure 5.34 consists of a carrier regenerator, a demodulator, a

matched filter, and a decision circuit. Assuming that the receiver is correctly chip synchronised to the zeroth user, after demodulating and despreading, we have

$$
\begin{aligned}
Z(T_b) &= \frac{1}{T_b} \int_0^{T_b} R_{up}(t) c_{00}(t) 2 \cos(\omega_1 t + \theta) dt \\
&= Z_w(T_b) + Z_n(T_b) + Z_{int}(T_b) + Z_{ext}(T_b),
\end{aligned}
\tag{5.161}
$$

where $Z_w(T_b)$, $Z_n(T_b)$, $Z_{int}(T_b)$, and $Z_{ext}(T_b)$ are the desired signal, the AWGN component,
the intracellular interference component, and the intercellular interference component, respectively, where

$$
Z_w(T_b) = \sqrt{2 P_{00}} a_{00} b_{00} \cos \phi_{00},
\tag{5.162}
$$

$$
Z_n(T_b) = \frac{1}{T_b} \int_0^{T_b} n(t) c_{00}(t) 2 \cos(\omega_1 t + \theta) dt,
\tag{5.163}
$$

$$
Z_{int}(T_b) = \frac{1}{T_b} \int_0^{T_b} n_{int}(t) c_{00}(t) dt,
\tag{5.164}
$$

and

$$
Z_{ext}(T_b) = \frac{1}{T_b} \int_0^{T_b} n_{ext}(t) c_{00}(t) dt.
\tag{5.165}
$$

In Equations (5.164) and (5.165), $n_{int}(t)$ and $n_{ext}(t)$ are the intracellular and intercellular
equivalent baseband interference components, respectively, defined as

$$
n_{int}(t) = \sum_{i=1}^{N-1} a_{i0} \sqrt{2 P_{i0}} b_{i0}(t - \tau_{i0}) c_{i0}(t - \tau_{i0}) \cos \phi_{i0}
\tag{5.166}
$$

and

$$
n_{ext}(t) = \sum_{j=1}^{J-1} \sum_{i=0}^{N-1} a_{ij} \sqrt{2 P_{ij}} b_{i0}(t - \tau_{ij}) c_{ij}(t - \tau_{ij}) \cos \phi_{ij},
\tag{5.167}
$$

where $\phi_{i0} = \theta_{i0} - \theta$ and $\phi_{ij} = \theta_{ij} - \theta$.

5.5.2 Performance of the up-link

The bit error probability at the output of the decision circuit depends on the bit-energy-to-
total-interference ratio in the signal, $Z(T_b)$. The power of the desired signal is

$$
S = E[Z_w^2(T_b)] = a_{00}^2 P_{00}
\tag{5.168}
$$

as $b_{00}^2 = (\pm 1)^2 = 1$ and $E[\cos^2 \phi_{00}] = 1/2$. The AWGN component at the output of the
matched filter from Equation (5.20) is

$$
\eta = \text{var}[Z_n(T_b)] = N_0 R_b .
\tag{5.169}
$$

In deriving the interference power, we are cognisant that the received carrier phase differs from the transmitted carrier phase, that the relative time delay of the data transmitted from each mobile is a random variable, i.e. that τ_{i0} and τ_{ij} are independent random variables uniformly distributed over $[0, T_b)$, and that ϕ_{i0} and ϕ_{ij} are independent random variables uniformly distributed over $[0, 2\pi)$. We further assume that $b_{ij}(t)$ is random independent binary data, $c_{ij}(t)$ is a random sequence, and as a consequence $n_{int}(t)$ and $n_{ext}(t)$ are two Gaussian random processes.

Similar to the approach adopted in Section 5.2.1, the intracellular interference power for,

$$I_{int} = \frac{1}{G_p} \sum_{i=1}^{N-1} v_{i0} a_{i0}^2 P_{i0} . \tag{5.170}$$

5.5.2.1 Perfect power control

With perfect power control, the received power from each mobile at the cell site remains at the constant target power level

$$P_{tar} = a_{ij}^2 P_{ij} . \tag{5.171}$$

Replacing $a_{00}^2 P_{00}^2$ in Equation (5.168) with P_{tar} gives the signal power

$$S = P_{tar} . \tag{5.172}$$

From Equations (5.170), (5.171), and (5.172), the intracellular interference-to-signal-power ratio is

$$\frac{I_{int}}{S} = \frac{1}{SG_p} \sum_{i=1}^{N-1} v_{i0} P_{tar} = \frac{1}{G_p} \sum_{i=1}^{N-1} v_{i0} . \tag{5.173}$$

The intracellular interference-to-signal ratio in Equation (5.173) is a function of the number of users per microcell irrespective of the shape of the microcell. The intercellular interference-to-signal ratio is given by Equation (5.77) where I_j/S is the interference-to-signal-power ratio from the jth adjacent cell. Assuming the users are uniformly distributed within their microcells, I_j/S can be calculated by integration instead of summation. Unlike the intracellular interference, the intercellular interference depends on the microcellular arrangement. The intercellular interference from the cross-shaped and the rectangular-shaped microcells is now calculated.

Cross-shaped microcells   In the cross-shaped microcells, the microcell sites are located at the cross-sections of streets, as shown in Figure 5.35. Owing to the sudden decrease in signal levels at street corners and the steeper path loss slopes associated with the OOS paths, the main intercellular interference comes from the LOS paths. For street blocks of side $2X$ and street width of W (see Figure 5.35), the number of users per area is

$$\rho = \frac{N}{4WX - W^2} \approx \frac{N}{4WX}, \tag{5.174}$$

for $W << X$. Since $W << X$, all the mobile users in area da in Figure 5.35 are at a distance r_j from the zeroth cell site. These users in area da are also at a distance r from the jth cell site. Mobiles in the jth microcell when active produce an interfering power of $I(r_j, r)$ at the zeroth microcell site. Under perfect power control, the mobiles in the jth cell have their power controlled by their own BS to be P_{tar}. In order to track the relative path loss and slow fading variations, the transmitted power P_{ij} from the ith mobile in the jth microcell is made inversely proportion to a_{ij}^2, whence

$$P_{ij} = \frac{P_{tar}}{a_{ij}^2} \tag{5.175}$$

and a_{ij} is the path loss and shadow fading between the zeroth BS and the ith mobile in the jth microcell. This power P_{ij} is the interference due to the ith mobile at the zeroth BS. Consequently, the interference-to-signal-power ratio at the zeroth microcell site due to mobiles in area da who are communicating to the jth microcell site is

$$\frac{I(r_j, r)}{S} = \left(\frac{a_{0j}}{a_{ij}}\right)^2. \tag{5.176}$$

where a_{0j}^2 is the path loss and shadow fading between the interfering mobile in the jth microcell and the zeroth BS. From Equation (3.54), $(a_{0j}/a_{ij})^2$ in Equation (5.176) has the form

$$\left(\frac{a_{0j}}{a_{ij}}\right)^2 = \begin{cases} \left(\frac{r}{r_j}\right)^2 10^{\frac{\lambda_j - \lambda}{10}} = \left(\frac{r}{r_j}\right)^2 10^{\frac{\zeta_j}{10}}, & \text{for } r \text{ and } r_j \leq X_b, \\ \left(\frac{r}{r_j}\right)^4 10^{\frac{\lambda_j - \lambda}{10}} = \left(\frac{r}{r_j}\right)^4 10^{\frac{\zeta_j}{10}}, & \text{for } r \text{ and } r_j > X_b, \\ \frac{X_b^2 r^2}{r_j^4} 10^{\frac{\lambda_j - \lambda}{10}} = \frac{X_b^2 r^2}{r_j^4} 10^{\frac{\zeta_j}{10}}, & \text{for } r \leq X_b \text{ and } r_j > X_b, \end{cases} \tag{5.177}$$

where λ_j and λ are random variables with zero mean and standard deviation σ, and $\zeta_j = \lambda_j - \lambda$ is also a random variable with zero mean and variance of $\sigma_\zeta^2 = 2\sigma^2$ (λ_j and λ are independent). In Equation (5.177),

$$r_j = D_j - r, \tag{5.178}$$

where D_j is the frequency reuse distance between the zeroth and jth microcellular BSs. The total intercellular interference-to-signal-power ratio at the zeroth microcell site due to all the mobiles in the jth adjacent microcell is [3],

$$\frac{I_j}{S} = \int\int_{\text{cell area}} v_{ij} \frac{I(r_j, r)}{S} \phi_l\left(\zeta, \frac{r}{r_j}\right) \rho \, d a, \tag{5.179}$$

where the subscript in $\phi_l(\zeta, r/r_j)$ is the constraint function of the interference from mobiles in the jth microcell. Owing to the two-slope path loss law, the constraint function is defined

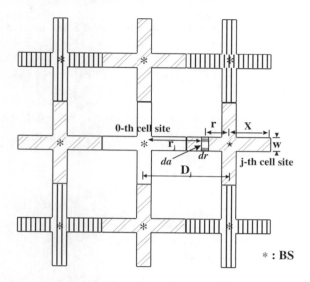

Figure 5.35: Up-link intercellular interference geometry for the cross-shaped microcells. Intercellular interfering microcells are shaded.

as

$$\phi_1\left(\zeta_j,\frac{r}{r_j}\right) = \begin{cases} 1, & \text{if } \left(\frac{r}{r_j}\right)^2 10^{\frac{\zeta_j}{10}} \le 1, \\ 0, & \text{otherwise}, \end{cases} \tag{5.180}$$

for r and $r_j \le X_b$,

$$\phi_2\left(\zeta_j,\frac{r}{r_j}\right) = \begin{cases} 1, & \text{if } \left(\frac{r}{r_j}\right)^4 10^{\frac{\zeta_j}{10}} \le 1, \\ 0, & \text{otherwise}, \end{cases} \tag{5.181}$$

for r and $r_j > X_b$, and

$$\phi_3\left(\zeta_j,\frac{r}{r_j}\right) = \begin{cases} 1, & \text{if } \frac{X_b^2 r^2}{r_j^4} 10^{\frac{\zeta_j}{10}} \le 1, \\ 0, & \text{otherwise}, \end{cases} \tag{5.182}$$

for $r \le X_b$ and $r_j > X_b$. From Equations (5.176)–(5.182), the expectation of I_j/S can be written as

$$E\left[\frac{I_j}{S}\right] = \int\int_{\text{cell area}} E\left[v_{ij}\right] E\left[\left(\frac{a_{0j}}{a_{ij}}\right)^2 \phi_l\left(\zeta_j,\frac{r}{r_j}\right)\right] \rho \, da. \tag{5.183}$$

From Figure 5.35, da has a rectangular area of Wdr. As a result the double integral in Equation (5.183) can be simplified to a single integral, namely

$$E\left[\frac{I_j}{S}\right] = \rho\mu W \int_{-X}^{X} E\left[\left(\frac{a_{0j}}{a_{ij}}\right)^2 \phi_l\left(\zeta_j,\frac{r}{r_j}\right)\right] dr. \tag{5.184}$$

Now ρ, given by Equation 5.174, is inserted into Equation 5.184 to give

$$E\left[\frac{I_j}{S}\right] = \frac{\mu N}{4X} \int_{-X}^{X} E\left[\left(\frac{a_{0j}}{a_{ij}}\right)^2 \phi_l\left(\zeta_j, \frac{r}{r_j}\right)\right] dr. \tag{5.185}$$

When $X_b \leq X$, the integration in Equation (5.185) can be divided into three sections, namely $-X$ to $-X_b$, $-X_b$ to X_b, and X_b to X. From Figure 5.36, in the first and third sections where both r and r_j are greater than X_b,

$$\left(\frac{a_{0j}}{a_{ij}}\right)^2 \phi_l\left(\zeta_j, \frac{r}{r_j}\right) = \left(\frac{r}{r_j}\right)^4 10^{\frac{\zeta_j}{10}} \phi_2\left(\zeta_j, \frac{r}{r_j}\right). \tag{5.186}$$

In the second section where $r \leq X_b$ and $r_j > X_b$, then

$$\left(\frac{a_{0j}}{a_{ij}}\right)^2 \phi_l\left(\zeta_j, \frac{r}{r_j}\right) = \frac{X_b^2 r^2}{r_j^4} 10^{\frac{\zeta_j}{10}} \phi_3\left(\zeta_j, \frac{r}{r_j}\right). \tag{5.187}$$

Substituting Equations (5.186) and (5.187) into Equation (5.185), yields

$$E\left[\frac{I_j}{S}\right] = \frac{N\mu}{4X} \int_{-X}^{-X_b} \left(\frac{r}{r_j}\right)^4 E\left[10^{\frac{\zeta_j}{10}} \phi_2\left(\zeta_j, \frac{r}{r_j}\right)\right] dr$$

$$+ \frac{N\mu}{4X} \int_{-X_b}^{X_b} \frac{X_b^2 r^2}{r_j^4} E\left[10^{\frac{\zeta_j}{10}} \phi_3\left(\zeta_j, \frac{r}{r_j}\right)\right] dr$$

$$+ \frac{N\mu}{4X} \int_{X_b}^{X} \left(\frac{r}{r_j}\right)^4 E\left[10^{\frac{\zeta_j}{10}} \phi_2\left(\zeta_j, \frac{r}{r_j}\right)\right] dr. \tag{5.188}$$

For $2X > X_b > X$ as shown in Figure 5.37, the integration in Equation (5.185) can be divided into two sections, namely $-X$ to $X_b - 2X$ and $X_b - 2X$ to X. In the first section, where both r and r_j are less than X_b, the interference is according to an inverse square path loss, namely

$$\left(\frac{a_{0j}}{a_{ij}}\right)^2 \phi_l\left(\zeta_j, \frac{r}{r_j}\right) = \left(\frac{r}{r_j}\right)^2 10^{\frac{\zeta_j}{10}} \phi_1\left(\zeta_j, \frac{r}{r_j}\right). \tag{5.189}$$

When $r \leq X_b$ and $r_j > X_b$, the situation is exactly the same as Equation (5.187). Substituting Equations (5.187) and (5.189) into (5.185), yields

$$E\left[\frac{I_j}{S}\right] = \frac{N\mu}{4X} \int_{-X}^{X_b-2X} \left(\frac{r}{r_j}\right)^2 E\left[10^{\frac{\zeta_j}{10}} \phi_1\left(\zeta_j, \frac{r}{r_j}\right)\right] dr$$

$$+ \frac{N\mu}{4X} \int_{X_b-2X}^{X} \frac{X_b^2 r^2}{r_j^4} E\left[10^{\frac{\zeta_j}{10}} \phi_3\left(\zeta_j, \frac{r}{r_j}\right)\right] dr. \tag{5.190}$$

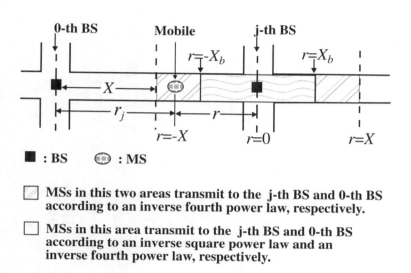

Figure 5.36: Path loss exponent for the interfering MSs in the jth microcells when $X_b \leq X$.

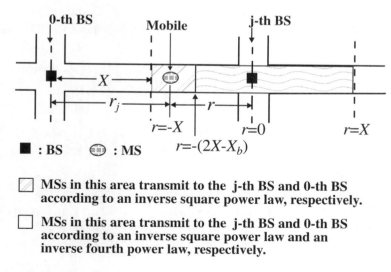

Figure 5.37: Path loss exponent for the interfering MS in the jth cell when $X_b > X$.

The term $E\left[10^{\frac{\zeta_j}{10}}\phi_l\left(\zeta_j,\frac{r}{r_j}\right)\right]$ in Equations (5.188) and (5.190) can be derived following the same procedure used in deriving Equation (5.85), namely

$$E\left[10^{\frac{\zeta_j}{10}}\phi_1\left(\zeta_j,\frac{r}{r_j}\right)\right] = f_1\left(\frac{r_j}{r},\sigma\right)$$

$$= \exp\left(\sigma\frac{\ln(10)}{10}\right)^2\left\{1-Q\left[\frac{20\log\left(\frac{r_j}{r}\right)}{\sqrt{2}\sigma}-\sqrt{2}\sigma\frac{\ln(10)}{10}\right]\right\}, \qquad (5.191)$$

$$E\left[10^{\frac{\zeta_j}{10}}\phi_2(\zeta_j,r/r_j)\right] = f_2\left(\frac{r_j}{r},\sigma\right)$$

$$= \exp\left(\sigma\frac{\ln(10)}{10}\right)^2\left\{1-Q\left[\frac{40\log\left(\frac{r_j}{r}\right)}{\sqrt{2}\sigma}-\sqrt{2}\sigma\frac{\ln(10)}{10}\right]\right\}, \qquad (5.192)$$

$$E\left[10^{\frac{\zeta_j}{10}}\phi_3(\zeta_j,r/r_j)\right] = f_3\left(\frac{r_j}{r},\sigma\right)$$

$$= \exp\left(\sigma\frac{\ln(10)}{10}\right)^2\left\{1-Q\left[\frac{10\log\left(\frac{r_j^4}{X_b^2 r^2}\right)}{\sqrt{2}\sigma}-\sqrt{2}\sigma\frac{\ln(10)}{10}\right]\right\}, \qquad (5.193)$$

for $l = 1, 2$, and 3, namely r and $r_j \le X_b$; r and $r_j > X_b$; and $r \le X_b$ and $r_j > X_b$, respectively. Substituting Equations (5.191), (5.192) and (5.193) into Equations (5.188) and (5.190), we have

$$E\left[\frac{I_j}{S}\right] = \frac{N\mu}{4X}\left\{\int_{-X}^{-X_b}\left(\frac{r}{r_j}\right)^4 f_2\left(\frac{r_j}{r},\sigma\right)dr\right.$$

$$\left. + \int_{-X_b}^{X_b}\frac{X_b^2 r^2}{r_j^4} f_3\left(\frac{r_j}{r},\sigma\right)dr + \int_{X_b}^{X}\left(\frac{r}{r_j}\right)^4 f_2\left(\frac{r_j}{r},\sigma\right)dr\right\} \qquad (5.194)$$

and

$$E\left[\frac{I_j}{S}\right] = \frac{N\mu}{4X}\left\{\int_{-X}^{X_b-2X}\left(\frac{r}{r_j}\right)^2 f_1\left(\frac{r_j}{r},\sigma\right)dr + \int_{X_b-2X}^{X}\frac{X_b^2 r^2}{r_j^4} f_3\left(\frac{r_j}{r},\sigma\right)dr\right\} \qquad (5.195)$$

for $X_b \le X$ and $X_b > X$, respectively.

The variance of I_j/S given by Equation (5.179) is

$$\text{var}\left[\frac{I_j}{S}\right] = \frac{N}{4X}\int_{-X}^{X}\text{var}\left[v_{ij}\left(\frac{a_{0j}}{a_{ij}}\right)^2 \phi_l\left(\zeta_j,\frac{r}{r_j}\right)\right]dr. \qquad (5.196)$$

Following the similar procedure used in deriving Equation (5.85), we have

$$
\text{var}\left[\frac{I_j}{S}\right] = \frac{N}{4X} \int_{-X}^{X} \left\{ E\left[v_{ij} \left(\frac{a_{0j}}{a_{ij}}\right)^2 \phi_l \left(\zeta_j, \frac{r}{r_j}\right) \right]^2 \right.
$$

$$
\left. - \left\{ E\left[v_{ij} \left(\frac{a_{0j}}{a_{ij}}\right)^2 \phi_l \left(\zeta_j, \frac{r}{r_j}\right) \right] \right\}^2 \right\} dr, \tag{5.197}
$$

where $E\left[10^{\frac{\zeta_j}{10}} \phi_l(\zeta_j, r/r_j)\right]$ is given in Equations (5.191), (5.192) and (5.193) for $l = 1$,

2 and 3, respectively. The term $E\left[10^{\frac{\zeta_j}{10}} \phi_l(\zeta_j, r/r_j)\right]^2$ can be derived by using the similar

procedure in deriving Equation (5.87), namely

$$
E\left[10^{\frac{\zeta_j}{10}} \phi_l(\zeta_j, r/r_j)\right]^2
$$

$$
= g_1\left(\frac{r_j}{r}, \sigma\right)
$$

$$
= \exp\left(\sigma \frac{\ln(10)}{5}\right)^2 \left\{ 1 - Q\left[\frac{20\log\left(\frac{r_j}{r}\right)}{\sqrt{2}\sigma} - \sqrt{2}\sigma\frac{\ln(10)}{5}\right] \right\}, \tag{5.198}
$$

$$
E\left[10^{\frac{\zeta_j}{10}} \phi_2(\zeta_j, r/r_j)\right]^2
$$

$$
= g_2\left(\frac{r_j}{r}, \sigma\right) = \exp\left(\sigma\frac{\ln(10)}{5}\right)^2 \left\{ 1 - Q\left[\frac{40\log(r_j/r)}{\sqrt{2}\sigma} - \sqrt{2}\sigma\frac{\ln(10)}{5}\right] \right\},
$$

and

$$
E\left[10^{\frac{\zeta_j}{10}} \phi_3(\zeta_j, r/r_j)\right]^2
$$

$$
= g_3\left(\frac{r_j}{r}, \sigma\right) = \exp\left(\sigma\frac{\ln(10)}{5}\right)^2
$$

$$
\times \left\{ 1 - Q\left[\frac{10\log\left(\frac{r_j^4}{X_b^2 r^2}\right)}{\sqrt{2}\sigma} - \sqrt{2}\sigma\frac{\ln(10)}{5}\right] \right\}, \tag{5.199}
$$

for $l = 1, 2$ and 3, respectively. From Equations (5.197), (5.198), and (5.199), the variance of I_j/S becomes

$$
\text{var}\left[\frac{I_j}{S}\right] = \frac{N\mu}{4X} \left\{ \int_{-X}^{-X_b} \left(\frac{r}{r_j}\right)^8 \left[\mu g_2\left(\frac{r_j}{r}, \sigma\right) - \mu^2 f_2^2\left(\frac{r_j}{r}, \sigma\right)\right] dr
$$

$$+ \int_{-X_b}^{X_b} \frac{X_b^4 r^4}{r_j^8} \left[\mu g_3 \left(\frac{r_j}{r}, \sigma \right) - \mu^2 f_3^2 \left(\frac{r_j}{r}, \sigma \right) \right] dr$$

$$\left. + \int_{X_b}^{X} \left(\frac{r}{r_j} \right)^8 \left[\mu g_2 \left(\frac{r_j}{r}, \sigma \right) - \mu^2 f_2^2 \left(\frac{r_j}{r}, \sigma \right) \right] dr \right\} \tag{5.200}$$

for $X_b \leq X$, and

$$\text{var} \left[\frac{I_j}{S} \right] = \frac{N\mu}{4X} \left\{ \int_{-X}^{X_b - 2X} \left(\frac{r_j}{r} \right)^8 \left[\mu g_1 \left(\frac{r_j}{r}, \sigma \right) - \mu^2 f_1^2 \left(\frac{r_j}{r}, \sigma \right) \right] dr \right.$$

$$\left. + \int_{X_b - 2X}^{X} \frac{X_b^4 r^4}{r_j^8} \left[\mu g_3 \left(\frac{r_j}{r}, \sigma \right) - \mu^2 f_3^2 \left(\frac{r_j}{r}, \sigma \right) \right] dr \right\} \tag{5.201}$$

for $2X > X_b > X$. From Equations (5.89), (5.90), (5.190) and (5.197), the mean and variance of the total intercellular interference-to-signal ratio are then expressed as

$$\text{E} \left[\frac{I_{ext}}{S} \right] = \frac{1}{G_p} \sum_{j=1}^{J-1} \text{E} \left[\frac{I_j}{S} \right]$$

and

$$\text{var} \left[\frac{I_{ext}}{S} \right] = \left(\frac{1}{G_p} \right)^2 \sum_{j=1}^{J-1} \text{var} \left[\frac{I_j}{S} \right]. \tag{5.202}$$

Rectangular-shaped microcells   Similarly, the intercellular interference from mobiles in rectangular-shaped microcells can be calculated according to the interfering geometry shown in Figure 5.38. There are six first tier interfering microcells surrounding the zeroth microcell, where the four interfering cells marked A are in the OOS streets and the other two cells marked B are in the LOS streets. As the interference created by the mobiles in the four OOS street microcells suffers a diffraction loss at the turning corner, the most significant interference to the zeroth BS comes from mobiles in the two LOS street microcells.

From Figure 5.38, the number of the significant interfering microcells in the rectangular-shaped microcells is two, while it is four in the cross-shaped microcells shown in Figure 5.35. Only half of the mobiles in a cross-shaped microcell can produce significant intercellular interference due to the blocking of the buildings, while all the mobiles in a rectangular-shaped microcell can contribute to the intercellular interference. As a result, the intercellular interference in the rectangular-shaped microcells is the same as that of rectangular-shaped microcells. Consequently, we do not explicitly calculate the $\text{E} [I_{ext} / S]$.

Outage probability   Since both cross-shaped and rectangular-shaped microcells have the same intracellular and intercellular interference-to-signal-power ratios, the probability of

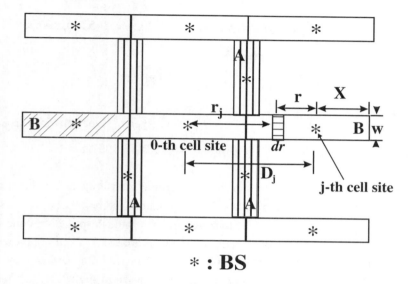

Figure 5.38: Up-link intercell interference geometry for rectangular-shaped microcells.

outage on the up-link of the city street microcells in terms of number of users per microcell is evaluated as follows. We first calculate the ratio of the bit energy E_b to the power spectral density (PSD) I_0, where I_0 is the summation of the intracellular interference PSD, the intercellular interference PSD, and the zeroth BS's AWGN PSD. We employ Equation (5.91) to get E_b/I_0, and thence the outage probability using Equation (5.99).

5.5.2.2 Imperfect power control

In Section 5.3.2 we calculated the outage probability p_o for macrocellular networks. The essential difference for the microcellular situation compared with the macrocellular one is the intercellular interference. Consequently, the outage probability for the microcellular network is given by Equation (5.110) but with (I'_{ext}/S') being different, namely

$$\frac{I'_{ext}}{S'} = \frac{I_{ext}}{S} 10^{\frac{\varepsilon}{10}} , \tag{5.203}$$

where the term I_{ext}/S for the microcellular networks is different from that in macrocellular networks. From Equation (5.107), the expectation of I'_{ext}/S' is

$$E\left[\frac{I'_{ext}}{S'}\right] = E\left[\frac{I_{ext}}{S}\right] E\left[10^{\frac{\varepsilon}{10}}\right], \tag{5.204}$$

where $E[I_{ext}/S]$ and $E\left[10^{\frac{\varepsilon}{10}}\right]$ are given in Equation (5.202) and (5.51), respectively. From Equation (5.108), the variance of I'_{ext}/S', is

$$\text{var}\left[\frac{I'_{ext}}{S'}\right] = E\left[\left(\frac{I_{ext}}{S}\right)^2\right]E\left[\left(10^{\frac{\varepsilon}{10}}\right)^2\right] - \left\{E\left[\frac{I_{ext}}{S}\right]E\left[10^{\frac{\varepsilon}{10}}\right]\right\}^2, \tag{5.205}$$

where $E\left[10^{\frac{\varepsilon}{10}}\right]$ is given in Equation (5.51), $E[I_{ext}/S]$ is given in Equation (5.202), $E\left[(I_{ext}/S)^2\right]$ can be calculated from Equations (5.109) and (5.198)–(5.199) and $E\left[10^{\frac{\varepsilon}{5}}\right]$ is given in Equation (5.54).

The outage probability is calculated for the intracellular interference plus the intercellular interference from mobiles in the four first tier co-channel microcells, where a signal-to-AWGN ratio of 20 dB is assumed at the matched filter output. A spreading code sequence of 128 chips per code and a microcell with $X = 200$ m are used in our analytical results. The variations of p_o versus the number of users per microcell for perfect power control and for different levels of imperfect power control are shown in Figures 5.39, 5.40, 5.41, 5.42, 5.43, and 5.44. Figures 5.39, 5.40, and 5.41 refer to a path loss break-distance of $X_b = 3/4X$, while Figures 5.42, 5.43, and 5.44 refer to a path loss break-distance of $X_b = 5/4X$. These values of X_b were selected because experimental results have reported that $150 < X_b < 250$ [17,18], and in our calculation we set $X = 200$ m. The system capacity can be estimated by the number of users per microcell for an outage probability below 2%. If the mobiles are perfectly power controlled, then for $X_b = 3/4X$, the CDMA system can support 35 and 45 users for VAFs of 1/2 and 3/8, respectively, and for $X_b = 5/4X$, the CDMA system can support 34 and 44 users for VAFs of 1/2 and for 3/8, respectively. The probability of outage increases as the standard deviation of the imperfect power control error increases, and the capacity of the CDMA system decreases accordingly. For an outage probability of less than 2%, the capacity and percentage of capacity degradation for different levels of standard deviation of the errors in E_b/I_0 are summarised in Tables 5.2 and 5.3 corresponding to $X_b = 3/4X$ and $X_b = 5/4X$, respectively.

For an outage probability of 2% and perfect power control, the CDMA system in the hexagonal cells can support 30 and 23 users per cell for a VAF of 3/8 and 1/2, respectively, as shown in Section 5.3. For the same conditions, the CDMA system in city street microcells can support 44 and 34 users for a VAF of 3/8 and 1/2, respectively, which is 60% higher than that in hexagonal cellular schemes. The probability of outage increases as the standard deviation of the imperfect power control in E_b/I_0 increases, and the capacity of the CDMA system decreases accordingly.

From Tables 5.2 and Table 5.3, we observe that the reduction in capacity for the best case of $X_b = 3/4X$ in the presence of imperfect power control is about 8.8%, 11.1%, and 13.3% for ε having a standard deviation of 1.5 dB, 1.7 dB, and 2.0 dB, respectively, while the corresponding reductions in capacity for $X_b = 5/4X$ is 8.8%, 11.8%, and 14.7%, respectively.

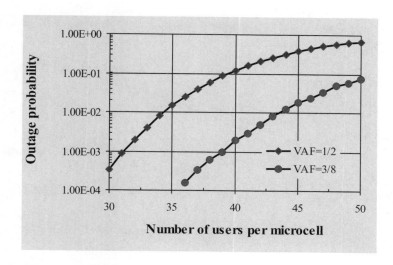

Figure 5.39: Outage probability versus the number of users per microcell as a function of VAF for perfect power control on the up-link and $X_b = 3/4X$.

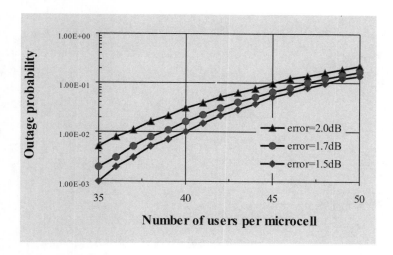

Figure 5.40: Outage probability versus the number of users per microcell as a function of the standard deviation of the power control errors in E_b/I_0 on the up-link for $X_b = 3/4X$ m and VAF= 3/8.

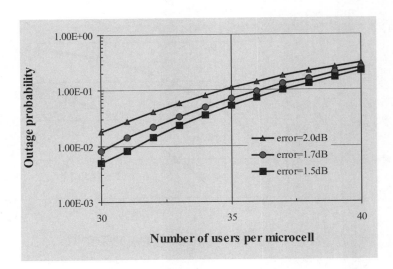

Figure 5.41: Outage probability versus the number of users per microcell as a function of the standard deviation of the power control errors in E_b/I_0 on the up-link for $X_b = 3/4X$ m and VAF= 1/2.

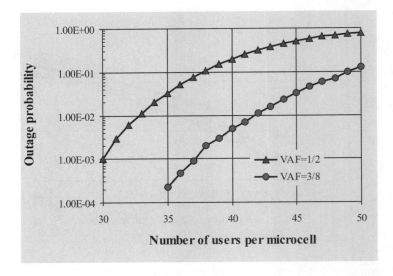

Figure 5.42: Outage probability versus the number of users per microcell as a function of VAF for perfect power control on the up-link and $X_b = 5/4X$ m.

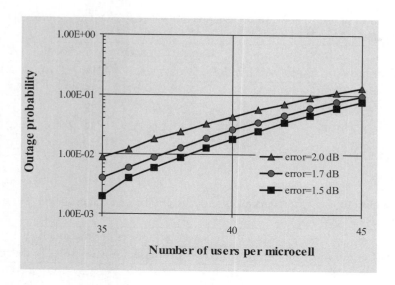

Figure 5.43: Outage probability versus the number of users per microcell as a function of the standard deviation of the power control errors in E_b/I_0 on the up-link for $X_b = 5/4X$ m and VAF= 3/8.

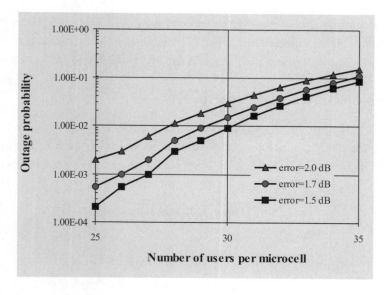

Figure 5.44: Outage probability versus the number of users per microcell as a function of the standard deviation of the power control errors in E_b/I_0 on the up-link for $X_b = 5/4X$ m and VAF= 1/2.

Table 5.2: Number of users per microcell as a function of the standard deviation (STD) of errors in E_b/I_0 for $X_b = 3/4X$.

STD of errors in E_b/I_0	Capacity in users per microcell			
	VAF = 3/8	% decrease	VAF = 1/2	% decrease
0 dB	45	0 %	35	0%
1.5 dB	41	8.8 %	32	8.6 %
1.7 dB	40	11.1 %	31	11.4 %
2.0 dB	39	13.3 %	30	14.3 %

Table 5.3: Number of users per microcell as a function of the standard deviation (STD) of errors in E_b/I_0 for $X_b = 5/4X$.

STD of errors in E_b/I_0	Capacity in users per microcell			
	VAF = 3/8	% decrease	VAF = 1/2	% decrease
0 dB	44	0 %	34	0 %
1.5 dB	40	9.1 %	31	8.8 %
1.7 dB	39	11.4 %	30	11.8 %
2.0 dB	37	14.7 %	29	14.7 %

Therefore, for capacity evaluation, we recommend the use of Table 5.3. For a standard deviation of ε of 1.7 dB, reported in Reference [7], the number of users that can be supported by the CDMA is 39 and 30 users per microcell for VAFs of 3/8 and 1/2, respectively.

5.5.3   Down-link system and signal model

The down-link system in the city street microcellular networks uses the same arrangement as that used in the macrocellular networks which is shown in Figure 5.8. The signal model was described in Sections 5.2.2 and 5.3.2, where the transmitted signal from the zeroth and jth BSs are

$$s_{dn}(t) = \sum_{i=0}^{N} \sqrt{2P_i} b_i(t - \tau_i) c_i(t - \tau_i) \cos(\omega_2 t + \theta) \qquad (5.206)$$

and

$$s_{dn}^{j}(t) = \sqrt{2P_{ij}} b_{ij}(t - \tau_{ij}) c_{ij}(t - \tau_{ij}) \cos(\omega_2 t + \theta_j), \qquad (5.207)$$

respectively. Considering the zeroth mobile in the zeroth microcell, and on allowing for propagation delays, the signal received from the zeroth BS is

$$R_{dn}(t) = a_0 s_{dn}(t) + \sum_{j=1}^{J-1} a_j s_{dn}^{j}(t), \qquad (5.208)$$

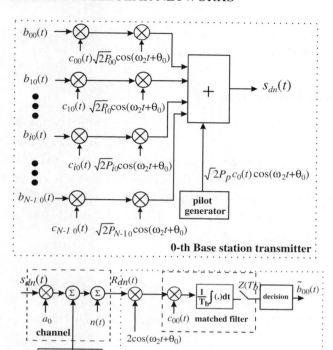

Figure 5.45: CDMA down-link system diagram.

where a_j is the path loss and the slow fading from the jth BS to the zeroth mobile. From Equation (5.113), the received signal, after being demodulated and despread, becomes

$$
\begin{aligned}
Z(T_b) &= \frac{1}{T_b} \int_0^{T_b} R_{dn}(t)c_{00}(t)2\cos(\omega_2 t + \theta_0)\mathrm{d}t \\
&= Z_w(T_b) + Z_n(T_b) + Z_{int}(T_b) + Z_{ext}(T_b), \quad (5.209)
\end{aligned}
$$

where $Z_w(T_b)$, $Z_n(T_b)$, $Z_{int}(T_b)$, and $Z_{ext}(T_b)$ are the desired signal, the AWGN, the intracellular interference, and the intercellular interference components, respectively, and

$$
\begin{aligned}
Z_w(T_b) &= \sqrt{2P_0}a_{00}b_{00}, \\
Z_n(T_b) &= \frac{1}{T_b} \int_0^{T_b} n(t)c_{00}(t)2\cos(\omega_2 t + \theta_0)\mathrm{d}t, \\
Z_{int}(T_b) &= \frac{1}{T_b} \int_0^{T_b} n_{int}(t)c_{00}(t)\mathrm{d}t, \\
Z_{ext}(T_b) &= \frac{1}{T_b} \int_0^{T_b} n_{ext}(t)c_{00}(t)\mathrm{d}t, \quad (5.210)
\end{aligned}
$$

and $n_{int}(t)$ and $n_{ext}(t)$ are the intracellular and intercellular equivalent baseband interference components defined in Equations (5.61) and (5.114), respectively.

5.5.4   Performance of the down-link

We assume that all the BSs transmit the same power without any power control. We will evaluate the performance of the down-link by using the Gaussian approximation method, i.e. that the interference being intracellular or intercellular tends to be Gaussian due to the Central Limit Theorem. We will calculate the interference power from which the bit energy-to-interference ratio can be found. Consider the zeroth mobile located at a distance r_o from the zeroth BS and at a distance of r_j from the jth BS as shown in Figure 5.46. From Section 5.2.2, the intracellular interference-to-signal ratio is

$$\frac{I_{int}}{S} = \frac{1}{G_p} \sum_{i=1}^{N} v_{i0}, \tag{5.211}$$

while from Section 5.4.2, the intercellular interference-to-signal ratio is

$$\frac{I_{ext}}{S} = \frac{1}{G_p} \sum_{j=1}^{J-1} \sum_{i=0}^{N} \frac{1}{2} v_{ij} \left(\frac{a_j}{a_0}\right)^2 \phi_l\left(\varsigma_j, \frac{r_0}{r_j}\right), \tag{5.212}$$

where $(a_j/a_0)^2$ is

$$\left(\frac{a_j}{a_0}\right)^2 = \begin{cases} \left(\frac{r_0}{r_j}\right)^2, & \text{for } r_0 \text{ and } r_j \le X_b, \\ \left(\frac{r_0}{r_j}\right)^4, & \text{for } r_0 \text{ and } r_j > X_b, \\ \left(\frac{X_b^2 r_0^2}{r_j^4}\right)^2, & \text{for } r_0 \le X_b \text{ and } r_j > X_b, \end{cases} \tag{5.213}$$

and

$$\phi_1\left(\varsigma_j, \frac{r_0}{r_j}\right) = \begin{cases} 1, & \text{if } \left(\frac{r_0}{r_j}\right)^2 10^{\frac{\varsigma_j}{10}} \le 1, \\ 0, & \text{otherwise,} \end{cases} \tag{5.214}$$

$$\phi_2\left(\varsigma_j, \frac{r_0}{r_j}\right) = \begin{cases} 1, & \text{if } \left(\frac{r_0}{r_j}\right)^4 10^{\frac{\varsigma_j}{10}} \le 1, \\ 0, & \text{otherwise,} \end{cases} \tag{5.215}$$

$$\phi_3\left(\varsigma_j, \frac{r_0}{r_j}\right) = \begin{cases} 1, & \text{if } \left(\frac{X_b^2 r_0^2}{r_j^4}\right)^{10^{\frac{\varsigma_j}{10}}} \le 1, \\ 0, & \text{otherwise,} \end{cases} \tag{5.216}$$

are the constraint function for r_0 and $r_j \le X_b$, r_0 and $r_j > X_b$, $r_0 \le X_b$, and $r_j > X_b$, respectively. By taking Equation (5.177) and $\phi_l(\zeta_j, r_j/r_0)$ into Equation 5.212, I_{ext}/S becomes a function of r_0, r_j, and the random variable ζ, and its mean and variance are given by

$$
E\left[\frac{I_{ext}}{S}\right] = \frac{1}{G_p} \sum_{j=1}^{J-1} \sum_{i=0}^{N} \frac{1}{2} E\left[v_{ij}\left(\frac{a_j}{a_0}\right)^2 \phi_l\left(\zeta_j, \frac{r_0}{r_j}\right)\right]
\tag{5.217}
$$

and

$$
\text{var}\left[\frac{I_{ext}}{S}\right] = \frac{1}{4G_p^2} \sum_{j=1}^{J-1} \sum_{i=0}^{N} \text{var}\left[v_{ij}\left(\frac{a_j}{a_0}\right)^2 \phi_l\left(\zeta_j, \frac{r_0}{r_j}\right)\right],
\tag{5.218}
$$

respectively.

Since the number of interfering microcells, $J - 1$, and their associated co-channel distances, r_j, are dependent on the microcellular arrangement, the intercellular interference will be different for the cross-shaped and rectangular-shaped microcells. In the cross-shaped street microcell, due to the structure of the streets, transmissions from some of the BSs are blocked by the buildings. As shown in the Figure 5.46, the most potential inference to an MS at a distance, r_0, from the zeroth microcell site originates from the BSs located in the two shaded microcells. In the rectangular-shaped microcells, there are four BSs interfering with an MS located at a distance, r_0, from the zeroth microcell site as shown in Figure 5.47. Consequently, Equation (5.212) applies, where $J - 1$ equals two and four for the cross-shaped and the rectangular-shaped microcells, respectively.

From Equation (5.129), the outage probability of the down-link is

$$
\begin{aligned}
p_o &= \Pr\left(\frac{S}{I} < \gamma_{req}\right) = \Pr\left(\frac{I}{S} > \frac{1}{\gamma_{req}}\right) \\
&= \sum_{k=1}^{N} \binom{N}{k} \mu^k (1 - \mu)^{N-k} Q\left(\frac{1/\gamma_{req} - \eta/S - k/G_p - E[I_{ext}/S]}{\sqrt{\text{var}[I_{ext}/S]}}\right).
\end{aligned}
\tag{5.219}
$$

Since the pilot signal is transmitted all the time, the minimal number of active channels is one. As a result, the summation in Equation (5.219) is from 1 to N.

For the street microcells with $X = 200$ m, the probability of outage for the down-link, p_o, is calculated by using Equation (5.219) for the cross-shaped and the rectangular-shaped microcells. The CDMA system has a processing gain of 128, a signal-to-AWGN ratio of 20 dB at the matched filter output, and a slow fading having a standard deviation of 4 dB for LOS paths. By considering the first and second tier interfering BSs, the variations in p_o versus the number of users per cell uniformly distributed throughout the cell, including the boundaries, are shown in Figures 5.48 and 5.49, where Figure 5.48 relates to the cross-shaped microcells, while Figure 5.49 refers to the rectangular-shaped microcells. Observe

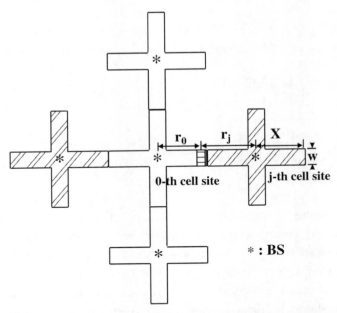

Figure 5.46: Down-link intercellular interference geometry for the cross-shaped microcells. Interfering BSs are shaded.

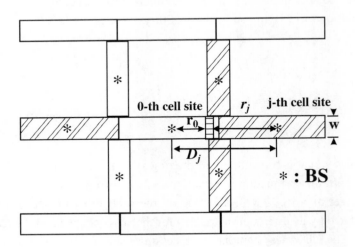

Figure 5.47: Down-link intercellular interference geometry for the rectangular-shaped microcells. Interfering BSs are shaded.

that the performance curves in Figures 5.48 and 5.49 are only relevant when there is at least one mobile at a boundary. If this is not the case, then more users can be supported. We found that the difference in performance due to path loss break distance is quite small. However, the performance for $X_b = 150$ m is slightly better than that for $X_b = 250$ m. Consider the worst case, namely $X_b = 250$ m, for an outage probability $P_o \leq 2\%$ at the microcellular boundary (where the interference from the neighbouring BS is strongest), the cross-shaped microcell can support 30 and 39 users for a VAF of $1/2$ and $3/8$, respectively. Since there are two more interfering BSs in the rectangular-shaped microcells compared with the cross-shaped microcell, the rectangular-shaped microcell can only support 32 and 25 users per microcell for a VAF of $3/8$ and $1/2$, respectively.

The main intercellular interference on the down-link comes from the nearest neighbour BSs, where the number of nearest interfering BSs depends on the cellular arrangement. For mobiles located at the corners of a hexagonal cell, there are two BSs at an equal distance from the mobiles, while there is one interfering BS that is closer than the other interfering BSs for mobiles located at the edges of a hexagonal cell. In city street microcells, there are either one or three nearest interfering BSs for the cross-shaped and the rectangular-shaped microcells, respectively.

The cross-shaped microcells can support more users than the rectangular-shaped microcells because they have two more significant interfering BSs. The capacity of the cross-shaped microcells is limited by the up-link for power control errors where the standard deviation exceeds 1 dB, while the capacity of the rectangular-shaped microcells is limited by the down-link transmissions.

Orthogonal spreading codes are used in the down-link of the IS-95 system as discussed in Section 5.3.3. We found that the down-link capacity of the IS-95 system in macrocellular networks is approximately three times that of a system using non-orthogonal codes. Similarly, by employing orthogonal codes in the down-link, the capacity of the CDMA system in the rectangular-shaped microcells is no longer down-link limited. Therefore, the capacity of the IS-95 system in city street microcells becomes up-link limited.

5.5.5 Capacity of IS-95 in street microcells

For a message bit rate of 10 kHz and a spreading factor of 128, the capacity of the CDMA system in terms of numbers of channels per cell per MHz is evaluated using Equation (5.120). For a standard deviation of ε of 1.7 dB, the CDMA system in the cross-shaped microcells can support 39 and 30 users per microcell for VAFs of $3/8$ and $1/2$, which is a capacity of 31 and 24 channels per microcell per MHz for VAFs of $3/8$ and $1/2$, respectively. The corresponding values for the rectangular-shaped microcells are 32 and 25 users per microcell for VAFs of $3/8$ and $1/2$, respectively, which is 25 and 20 channels per microcell per MHz for VAFs of $3/8$ and $1/2$, respectively. Compared with the capacity of CDMA systems

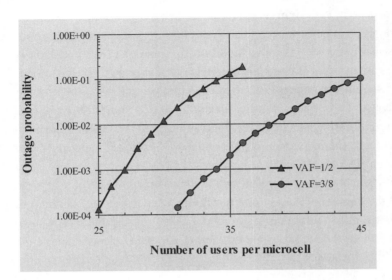

Figure 5.48: Outage probability of the down-link as a function of the number of users per microcell for cross-shaped microcells and for two values of X_b and VAF.

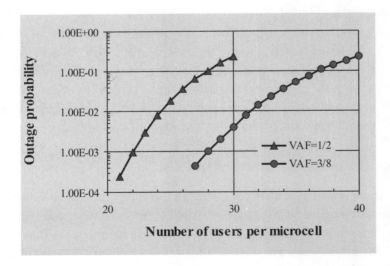

Figure 5.49: Outage probability of the down-link as a function of the number of users per microcell for rectangular-shaped microcells and for two values of X_b and VAF.

in hexagonal omni cells, CDMA systems for city street microcells have capacities 15% and 40% higher than CDMA systems in the hexagonal cells for rectangular-shaped and cross-shaped microcells, respectively. Since the coverage area of the rectangular-shaped microcells is only half of the cross-shaped microcells, the rectangular-shaped microcells have a capacity in terms of channels per MHz per area that is 1.65 and 1.66 times that of the cross-shaped microcells for a VAF of 3/8 and 1/2, respectively.

Since the IS-95 system uses orthogonal codes on its down-link, its capacity is basically up-link limited. The capacity of the IS-95 system in both the cross-shaped and rectangular-shaped microcellular networks has the same capacity, which is 31 and 24 channels per microcell per MHz for a VAF of 3/8 and 1/2, respectively.

5.5.6 Summary

For CDMA, the most significant interference on the up-link originated from users inside the microcell, as approximately half of the interference generated from outside the microcell was prevented from entering due to shielding from surrounding buildings. We observed that in hexagonal cellular structures all the mobiles were able to cause interference, and consequently a CDMA system for the city street microcells had a capacity that was 25% and 35% higher than CDMA systems in the hexagonal cells for rectangular-shaped and cross-shaped microcells, respectively. The intracellular interference on the down-link was the same as on the up-link, irrespective of cell shape, while dominant intercellular interference was from the nearest neighbour cells. In hexagonal cells, there were two and one dominant interfering BSs for mobiles located at the corners and edges of a hexagonal cells, respectively. In city street microcells, there were one and three dominant interfering BSs for the cross-shaped and the rectangular-shaped microcells, respectively. Unlike on the up-link, there was only a small decrease in the intercellular interference on the down-link in the cross-shaped microcells, while for the rectangular-shaped microcells there was an increase in the intercellular interference.

For perfect power control, the CDMA system supported 34 and 44 users for a VAF of 1/2 and for 3/8, respectively. There were errors in the E_b/I_0 on the up-link power control errors in the received E_b/I_0 had a standard deviation of 1.5dB, 1.7 dB, and 2.0 dB, the corresponding decrease in capacity was approximately 9% , 12%, and 15%, respectively. For E_b/I_0 having a standard deviation of 1.7 dB, the CDMA system using cross-shaped microcells supported 39 users and 30 per microcell for VAFs of 3/8 and 1/2, corresponding to a capacity of 30 and 24 channels per cell per MHz for VAFs of 3/8 and 1/2, respectively. With rectangular-shaped microcells whose capacity, on the other hand, is limited by the down-link, 32 and 25 users per microcell were supported, which was 25 and 20 channels per cell per MHz for VAFs of 3/8 and 1/2, respectively.

5.6 Power Control in CDMA Systems

CDMA technology has the potential to provide a significant improvement in the capacity of cellular mobile radio systems compared with FDMA and TDMA systems [19]. However, the improvement is dependent upon the effectiveness of the power control system [6], especially on the up-link. In the absence of power control, a BS would receive a much stronger signal from a mobile that is geographically close to it than from a mobile that is farther away. The consequence is a dramatic decrease in system capacity. This is the so-called 'near–far' problem. A power control scheme is needed not only to eliminate the near–far problem, but also to equalise and minimise the power transmitted by each mobile.

Power control on the up-link, or reverse link, in CDMA attempts to adjust the transmitted power of each mobile such that nominal received power from each mobile station (MS) at the base station (BS) is the same. To achieve this goal, a power control scheme is required to monitor and track the power transmitted by each mobile. Since the path loss and slow fading can be assumed to be identical on both up-link and down-link, the open-loop power control is used to estimate the path loss and slow fading from the CDMA pilot transmitted on the down-link, while the effect of fast fading is mitigated by a closed-loop power control scheme [20].

In a fixed-step-size, closed-loop power control system, the BS measures the received power from an MS and compares it with a reference target power. A command bit is then generated and sent to the MS. If the command bit is a logical 1 it signifies that the received power is less than the target power, while a logical 0 bit signifies the power is greater than the target power. The MS adjusts its transmitted power according to the command bit with a fixed increment or decrement of power. The closed-loop power control scheme used in IS-95 is a fixed-step-size power control with the transmitted power changing by +1 dB in response to each transmitted BS command bit [20].

In this section we consider an adaptive-step-size closed-loop power control scheme in which a mobile automatically adjusts its power step size according to the pattern of its received power control bits, without an increase in the control bit rate. Our results are compared to a fixed-step power control scheme. Multi-tap RAKE receivers are employed. Since the efficiency of the power control relies on how accurately we can measure the received power, power measurement plays an important role in the power control mechanism. We first describe our method of estimating the received power, and then describe our power control algorithm.

5.6.1 Channel model

The desired CDMA signal from an MS and the interference power received at the BS are key parameters in a power control system. Another major factor is the dispersive mobile radio channel. For a wideband fading channel, the baseband impulse response of the channel having L resolvable taps can be expressed as [6]

$$h_i(t) = \text{Re} \left\{ \sum_{l=0}^{L-1} a_i \beta_l (t - \tau_l) \exp[j\theta_l] \right\}, \tag{5.220}$$

where a_i is a factor that allows for slow fading and path loss over the link between the ith MS and the BS, β_l is the magnitude of the lth path of the fast fading channel impulse response at time τ_l and is Rayleigh or Rician distributed for macrocells and microcells, respectively, $\beta(t - \tau_l)$ is the amplitude gain of an impulse at $t = \tau_l$, and θ_l is the phase shift of the lth path relative to the zeroth path and is uniformly distributed. The resolvable path interval, τ_l, is equal to an integer number of chip duration T_c and $0 < \tau_l < T_m$, where T_m is the delay spread of the channel, and $T_m = LT_c$. Note that

$$a_i^2 = \frac{1}{r^\alpha} 10^{\frac{\lambda}{10}}, \tag{5.221}$$

where r is the distance from the transmitter to the receiver, α is the path loss exponent, and λ is the shadowing variable having a normal distribution with zero mean and a standard deviation of σ.

5.6.2 Estimation of the received signal power

Consider the up-link transmission in a BPSK CDMA system. The transmitted baseband signal from the ith MS to the jth BS is

$$x_{ij}(t) = \sqrt{P_{ij}} b_{ij} c_{ij}(t), \tag{5.222}$$

where P_{ij} is the transmitted power, b_{ij} is the message bit, and $c_{ij}(t)$ is the spreading code sequence. As this signal is transmitted to its BS over a wideband radio channel whose impulse response is $h_{ij}(t)$, the received signal at the jth BS is

$$
\begin{aligned}
r_{ij}(t) &= x_{ij}(t) * h_{ij}(t) \\
&= \sqrt{P_{ij}} b_{ij} a_{ij} \sum_{l=0}^{L-1} \beta_l c_{ij}(t - \tau_{ij} - \tau_l)[\cos\theta_l + j\sin\theta_l], \tag{5.223}
\end{aligned}
$$

where the symbol $*$ represents convolution and τ_{ij} is the transmission delay.

Let the desired communication be between the zeroth MS and the zeroth BS. The received wanted signal component from the zeroth MS is from Equation (5.223):

$$
\begin{aligned}
r_{00}(t) &= \sqrt{P_{00}} b_{00} a_{00} \beta_0 c_{00}(t - \tau_{00} - \tau_0)[\cos\theta_0 + j\sin\theta_0] \\
&\quad + \sqrt{P_{00}} b_{00} a_{00} \sum_{l=1}^{L-1} \beta_l c_{00}(t - \tau_{00} - \tau_l)[\cos\theta_l + j\sin\theta_l] \\
&= s_I(t) + j s_Q(t) + n_{I,m}(t) + j n_{Q,m}(t), \tag{5.224}
\end{aligned}
$$

where $s_I(t)$ and the $s_Q(t)$ are the inphase and quadrature components of the wanted signal, respectively, and $n_{I,m}(t)$ and $n_{Q,m}(t)$ are the inphase and quadrature components of the
multipath interference from itself, respectively. The intracellular interference, which is the
interference due to the signals received from the $N-1$ MSs in the zeroth cell, is

$$
\begin{aligned}
n_{int}(t) &= \sum_{i=1}^{N-1} r_{i0}(t) = \sum_{i=1}^{N-1} x_{i0}(t) * h_{i0}(t) \\
&= \sum_{i=1}^{N-1} \sqrt{P_{i0}} b_{i0} a_{i0} \sum_{l=0}^{L-1} \beta_l c_{i0}(t-\tau_{i0}-\tau_l)[\cos\theta_l + j\sin\theta_l] \\
&= n_{I,int}(t) + j n_{Q,int}(t),
\end{aligned}
\tag{5.225}
$$

where $n_{I,int}(t)$ and $n_{Q,int}(t)$ are the inphase and quadrature components of the intracellular
interference, respectively. The intercellular interference, which is the interference due to
signals received from the MSs in the surrounding cells, is

$$
\begin{aligned}
n_{ext}(t) &= \sum_{j=1}^{J}\sum_{i=0}^{N-1} r_{ij}(t) = \sum_{j=1}^{J}\sum_{i=0}^{N-1} x_{ij}(t) * h_{ij}(t) \\
&= \sum_{j=1}^{J}\sum_{i=0}^{N-1} \sqrt{P_{ij}} b_{ij} a_{ij} \sum_{l=0}^{L-1} \beta_l c_{ij}(t-\tau_{ij}-\tau_l)[\cos\theta_l + j\sin\theta_l] \\
&= n_{I,ext}(t) + j n_{Q,ext}(t),
\end{aligned}
\tag{5.226}
$$

where J is the number of significant intercellular interfering cells, and $n_{I,ext}(t)$ and $n_{Q,ext}(t)$
are the inphase and quadrature components of the intercellular interference. The additive
white Gaussian noise (AWGN) in the receiver is

$$
n(t) = n_I(t) + j n_Q(t),
\tag{5.227}
$$

where $n_I(t)$ and $n_Q(t)$ are the inphase and quadrature components of the AWGN, respectively. From Equations (5.224), (5.225), (5.226), and (5.227), the received baseband signal
at the zeroth BS becomes

$$
\begin{aligned}
R(t) &= r_{00}(t) + n_{int}(t) + n_{ext}(t) + n(t) \\
&= s_I(t) + j s_Q(t) + n_{I,m}(t) + j n_{Q,m}(t) + n_{I,int}(t) + j n_{Q,int}(t) \\
&\quad + n_{I,ext}(t) + j n_{Q,ext}(t) + n_I(t) + j n_Q(t) \\
&= R_I(t) + j R_Q(t),
\end{aligned}
\tag{5.228}
$$

where

$$
\begin{aligned}
R_I(t) &= s_I(t) + n_{I,m}(t) + n_{I,int}(t) + n_{I,ext}(t) + n_I(t), \\
R_Q(t) &= s_Q(t) + n_{Q,m}(t) + n_{Q,int}(t) + n_{Q,ext}(t) + n_Q(t).
\end{aligned}
\tag{5.229}
$$

Assuming that all the data from each mobile are spread with a different independent random code, then the total interference noise, which is the summation of all the interferences, tends to be Gaussian distributed due to the Central Limit Theorem. Representing the total interference noise by the inphase interference component $n_{I,total}(t)$ and the quadrature phase interference component $n_{Q,total}(t)$, yields

$$\begin{aligned} R_I(t) &= s_I(t) + n_{I,total}(t), \\ R_Q(t) &= s_Q(t) + n_{Q,total}(t). \end{aligned} \tag{5.230}$$

Since the received BPSK signal is separated into inphase and quadrature phase components, we use quadrature phase demodulation for our power estimation to eliminate the effect of phase shifts due to channel fading. Consider a single-tap CDMA RAKE receiver which is locked to the strongest path, say the zeroth path. Then the average wanted received signal power at the zeroth BS from the zeroth mobile is

$$P_R = E[s_I^2(t)] + E[s_Q^2(t)]$$

and as $c_{00}(t - \tau_{00} - \tau_0)$ squared is unity,

$$\begin{aligned} P_R &= E[a_{00}^2\beta_0^2 P_{00}\cos^2\theta_0] + E[a_{00}^2\beta_0^2 P_{00}\sin^2\theta_0] \\ &= a_{00}^2\beta_0^2 P_{00}, \end{aligned} \tag{5.231}$$

where P_{00} is the power transmitted by this zeroth mobile. The task of separating the wanted signal from the interference and AWGN components and measuring the signal power is not easy. Our scheme for estimating the received signal power is shown in Figure 5.50. The inphase and quadrature components of the received signal, namely I and Q, corrupted by interference noise, become $R_I(t)$ and $R_Q(t)$, respectively. They pass through a digital matched filter with a sampler to yield $Z_I(T_b)$ and $Z_Q(T_b)$. The moduli of $Z_I(T_b)$ and $Z_Q(T_b)$ are then averaged, squared, and summed to give an estimate of the average signal power. Specifically,

$$\begin{aligned} Z_I(T_b) &= \frac{1}{T_b}\int_0^{T_b} R_I(t)c_{00}(t)dt, \\ Z_Q(T_b) &= \frac{1}{T_b}\int_0^{T_b} R_Q(t)c_{00}(t)dt, \end{aligned} \tag{5.232}$$

where T_b is the bit duration. Substituting $R_I(t)$ of Equation (5.230) into $Z_I(T_b)$ in Equation (5.232), yields

$$\begin{aligned} Z_I(T_b) &= \frac{1}{T_b}\int_0^{T_b}\{s_I(t) + n_{I,total}(t)\}c_{00}(t)dt \\ &= a_{00}\beta_0\sqrt{P_{00}}b_{00}\cos\theta_0 + \frac{1}{T_b}\int_0^{T_b} n_{I,total}(t)c_{00}(t)dt \\ &= x + y, \end{aligned} \tag{5.233}$$

where the meanings of x and y are obvious assuming that $c_{00}(t - \tau_{00} - \tau_0)$ and $c_{00}(t)$ are time aligned. The first term in Equation (5.233) is the wanted inphase signal component having a magnitude $a_{00}\beta_0\sqrt{P_{00}}\cos\theta_0$ and a bipolar data bit of b_{00}. The second term is due to inphase interference noise, and as $c_{00}(t)$ is uncorrelated with the interference noise, this component is a random Gaussian process. Similarly we have

$$
\begin{aligned}
Z_Q(T_b) &= \frac{1}{T_b}\int_0^{T_b}\{s_Q(t) + n_{Q,total}(t)\}c_{00}(t)dt \\
&= a_{00}\beta_0\sqrt{P_{00}}b_{00}\sin\theta_0 + \frac{1}{T_b}\int_0^{T_b}n_{Q,total}(t)c_{00}(t)dt \quad (5.234)
\end{aligned}
$$

for the quadrature phase channel. The inphase component at the output of the digital matched filter in Equation (5.233) is x, representing the wanted signal, plus y, representing the interference component. For satisfactory performance, the magnitude of the wanted signal should be larger than the magnitude of the interference noise, i.e.

$$
|x| \geq |y|. \quad (5.235)
$$

From Equation (5.235), the absolute value of Equation (5.233) can be found to be

$$
|Z_I(T_b)| = |x + y| = \begin{cases} x + y, & \text{for } x \geq 0, \\ -(x + y), & \text{for } x < 0. \end{cases} \quad (5.236)
$$

Because y is a random variable with Gaussian distribution, the mean of y tends to be zero. Therefore Equation (5.236) becomes

$$
E[|Z_I(T_b)|] = E[|x|] = E[|a_{00}\beta_0\sqrt{P_{00}}\cos\theta_0|]. \quad (5.237)
$$

Similarly, the quadrature component at the output of the digital matched filter can be shown to be

$$
E[|Z_Q(T_b)|] = E[|a_{00}\beta_0\sqrt{P_{00}}\sin\theta_0|]. \quad (5.238)
$$

From Equations (5.231), (5.237), and (5.238), the received signal power prior to bit detection is

$$
P_R = a_{00}^2\beta_0^2 P_{00} = \{E[|Z_I(T_b)|]\}^2 + \{E[|Z_Q(T_b)|]\}^2. \quad (5.239)
$$

The expectations in Equation (5.239) are computed over a window of WT_b width, namely

$$
P_R = \left[\frac{1}{W}\sum_{k=1}^{W}|Z_I(T_b)|\right]^2 + \left[\frac{1}{W}\sum_{k=1}^{W}|Z_Q(T_b)|\right]^2 + \xi, \quad (5.240)
$$

where ξ is the estimation error due to finite window duration.

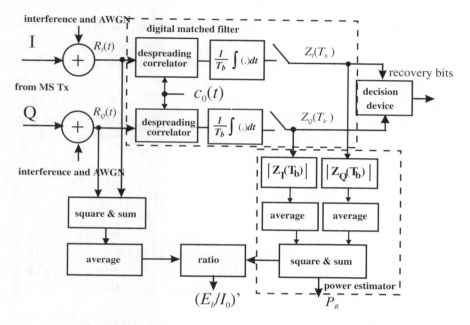

Figure 5.50: Block diagram of the power and E_b/I_0 estimation scheme.

5.6.3 Estimation of E_b/I_0

The bit-energy-to-interference power spectral density (PSD) ratio, E_b/I_0, is an important power control system parameter that has a profound effect on system capacity. As the wanted signal and the interference have the same band-width at the output of the digital matched filter, E_b/I_0 is equal to the signal-to-interference power ratio at the output of the digital matched filter. The received signal power at the output of the digital matched filter is given by Equation (5.239).

Since the interference noise is Gaussian distributed, the variance of the interference can be found from the sum of the variances of $|Z_I(T_b)|$ and $|Z_Q(T_b)|$, namely

$$
\begin{aligned}
I &= \text{var}\,[|Z_I(T_b)|] + \text{var}\,[|Z_Q(T_b)|] \\
&= \text{E}\,[|Z_I(T_b)| - \text{E}\,[|Z_I(T_b)|]]^2 + \text{E}\,[|Z_Q(T_b)| - \text{E}\,[|Z_Q(T_b)|]]^2 \\
&= \text{E}\,[Z_I^2(T_b)] - \text{E}\,[Z_I(T_b)|]^2 + \text{E}\,[Z_Q^2(T_b)] - \text{E}\,[|Z_Q(T_b)|]^2 .
\end{aligned}
\tag{5.241}
$$

From Equations (5.239) and (5.241), we have

$$
\frac{E_b}{I_0} = \frac{P_R}{I} = \frac{P_R}{\text{var}\,[|Z_I(T_b)|] + \text{var}\,[|Z_Q(T_b)|]} .
\tag{5.242}
$$

Since P_R and I in the above equation are the mean-square value and variance of the signal at the output of the matched filter, we therefore refer to it as the mean-square-to-variance ratio (MSVR) method. To calculate $\mathrm{var}[|Z_I(T_b)|]$ and $\mathrm{var}[|Z_Q(T_b)|]$ in the denominator of Equation (5.242), as shown in Equation (5.241), in addition to an error ξ in performing the mean square of $|Z_I(T_b)|$ and $|Z_Q(T_b)|$, there will also be another error in performing the mean of $Z_I^2(T_b)$ and $Z_Q^2(T_b)$.

In order to improve the accuracy of the estimation of E_b/I_0, we proposed a method to estimate the interference power, which will be derived as follows. As the interference signal passes through the digital matched filter, the total interference power is decreased by the processing gain, G_p, i.e.

$$I = \frac{1}{G_p}\left\{\mathrm{E}[n_{I,total}(t)]^2 + \mathrm{E}[n_{Q,total}(t)]^2\right\}. \tag{5.243}$$

Because the total received signal is the sum of all the signals from all the mobiles, both inside and outside the cell, the total received signal tends to be Gaussian distributed due to the Central Limit Theorem. Because the wanted signal, $s(t)$, is mutually independent of the interference component, the total average received power before detection is given by

$$\begin{aligned} \mathrm{E}[R^2(t)] &= \mathrm{E}[s_I^2(t)] + \mathrm{E}[s_Q^2(t)] + \mathrm{E}[n_{I,total}(t)]^2 + \mathrm{E}[n_{Q,total}(t)]^2 \\ &= P_R + \mathrm{E}[n_{I,total}(t)]^2 + \mathrm{E}[n_{Q,total}(t)]^2. \end{aligned} \tag{5.244}$$

From Equations (5.231), (5.243) and (5.244), the value of E_b/I_0 at the output of the digital matched filter can be represented by

$$\begin{aligned} \frac{E_b}{I_0} &= \frac{P_R}{\frac{1}{G_p}\left\{\mathrm{E}[n_{I,total}(t)]^2 + \mathrm{E}[n_{Q,total}(t)]^2\right\}} \\ &= \frac{F}{1 - \frac{F}{G_p}}, \end{aligned} \tag{5.245}$$

where

$$F = \frac{P_R}{\mathrm{E}[R^2(t)]/G_p}. \tag{5.246}$$

From Equation 5.246, we can estimate the E_b/I_0 from F. As an example, for a processing gain G_p of 128 or 64, an E_b/I_0 of 7 dB is obtained by F=6.82 or 7.0dB, respectively. By taking the ratio of the measured average received power at the output of the digital matched filter to the average total received power at received front-end we obtain the value of F.

5.6.4   Estimation of E_b/I_0 for RAKE receivers

The system performance is determined by E_b/I_0 at the output of the RAKE receiver, where the E_b/I_0 is dependent on the combining schemes. For a maximal ratio combining scheme the E_b/I_0 at the output of the RAKE receiver equals the summation of the E_b/I_0 from the L taps, and is equal to [21]

$$\left(\frac{E_b}{I_0}\right)_c = \sum_{l=0}^{L-1}\left(\frac{E_b}{I_0}\right)_l = \sum_{l=0}^{L-1}\frac{G_p P_{R,l}}{I_l}, \tag{5.247}$$

where $P_{R,l}$ is the received signal power of the lth tap and I_l is the interference power associated with the lth tap. Since the interference power in the L taps is approximately the same, I_l can be estimated as

$$I_l = E[R^2(t)] - E[P_{R,l}] = E[R^2(t)] - \frac{1}{L}\sum_{l=0}^{L-1} P_{R,l}. \tag{5.248}$$

Substituting Equation (5.248) into (5.247), yields

$$\left(\frac{E_b}{I_0}\right)_c = \frac{G_p \sum_{l=0}^{L-1} P_{R,l}}{E[R^2(t)] - 1/L\sum_{l=0}^{L-1} P_{R,l}} = \frac{\frac{G_p \sum_{l=0}^{L-1} P_{R,l}}{E[R^2(t)]}}{1 - \frac{\sum_{l=0}^{L-1} P_{R,l}}{E[R^2(t)]L}}. \tag{5.249}$$

Let the power sum of the L taps be the received signal power of an L-tap RAKE receiver, i.e.

$$P_R = \sum_{l=0}^{L-1} P_{R,l}, \tag{5.250}$$

then Equation (5.249) can be rewritten as

$$\left(\frac{E_b}{I_0}\right)_c = \frac{\frac{G_p P_R}{E[R^2(t)]}}{1 - \frac{P_R}{E[R^2(t)]L}} = \frac{F}{1 - \frac{F}{G_p L}}. \tag{5.251}$$

For an equal gain combining scheme, the E_b/I_0 at the output of the RAKE receiver is 1 dB less than that for a maximal ratio combining scheme [21, 22].

5.6.5 Power control scheme

The power control system diagram is shown in Figure 5.51. In the BS receiver there is a RAKE receiver, power estimator circuits based on the power estimation theory described in the previous section, a comparison circuit, and monitoring of the received bit energy-to-power spectral density ratio. In the mobile receiver, there is step-size logic that is used to adjust the transmitted power P_T of the MS in an attempt to ensure that the received baseband power at the BS, P_R, is close to the target power, P_{target}, at all times. P_{target} is determined from Equations (5.246) and (5.251) according to the required E_b/I_0 at the output of the RAKE receiver and the interference level.

The power P_R is compared with a target power P_{target}. A logical 0 command bit is generated if $P_R \geq P_{target}$, otherwise a logical 1 is formed. The command bit is embedded in the BS traffic stream for transmission on the down-link. When the MS receives the CDMA signal, it removes the command bit from the data stream and inserts it into a shift register which stores the current and six previous power command bits. For a fixed-step-size control scheme, only the present control bit, R0, is used, while for the adaptive-step-size scheme the logic pattern of the seven bits stored in the shift register are employed to adjust the MS transmitted power. A pattern of 0,1,0,1,0,1,0 or 1,0,1,0,1,0,1 means the received power P_R at the BS is approximately the same as the required target power, while consecutive groups of 1s or 0s means the transmitted power from an MS is much larger or smaller than the target power, respectively. Hence, we can predict the power control error according to the pattern of the previous control bits [23]. The performance of the power control scheme is also monitored by the received bit-to-interference PSD in the BS, where the statistics of the measured E_b/I_0 can be examined.

5.6.5.1   Adaptive-step-size control algorithm

In order to track the target power and to minimise the power control error at the BS, the step size Δ_i applied to the integrator whose output adjusts the transmitted power P_T should vary according to the power error. If a large power error occurs, the step size should increase proportionally. If a smaller power error occurs, then a small step should be assigned. The variable step size Δ_i is based on the previous control bits stored in the mobile receiver.

There are three factors that cause the received power at the BS to vary. These are the path loss, slow fading, and fast fading. As long as the power control bit rate is fast enough, the dominant factor becomes the fast fading. A simplified representation of the received signal level versus distance profile is shown in Figure 5.52. The fading rate is proportional to the vehicle speed, while the depth of each fade is a random variable. We divide the fast fading shown by the dotted line into two zones A and B shown in Figure 5.52, where zone A relates to the situation where the variation in the fading is relatively slow compared with that in zone B. In order to track the fast fading, the power should be adjusted according to the inverse of the fast fading characteristics as shown by the dashed line in Figure 5.52. This is achieved by formulating a variable g whose value is determined by the logical values in the shift register, namely R0, R1, R2, R3, R4, R5, and R6. The step size is

$$\Delta_i = \Phi + g\Lambda, \tag{5.252}$$

where $\Delta_i = \Phi$ when g is zero, and Λ is a system parameter. For the slow variation in zone A, a fixed step size of Φ is sufficient to track the power variation. For the rapid power variation in zone B, the step size Δ_i should be larger. This is achieved by increasing the value of g to compensate for the abrupt signal variations around the deep fades. The value of g for

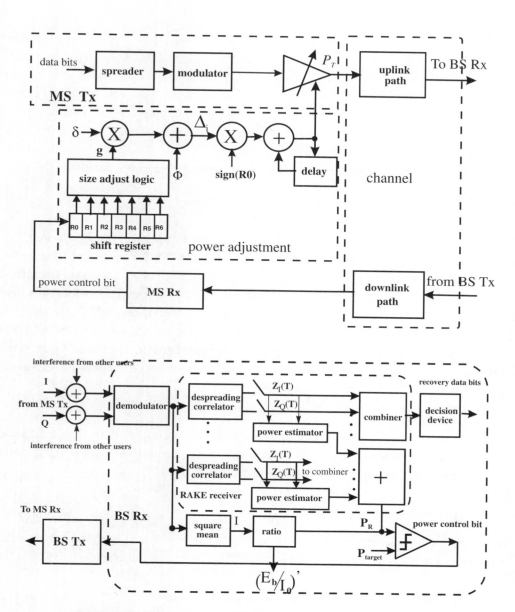

Figure 5.51: Block diagram of the closed-loop power control system.

received power in dB

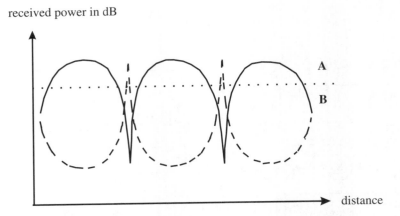

Figure 5.52: Characteristics of fast fading.

different control bit patterns is shown in Table 5.4. Since the depth of the fade is a random variable, the Δ_i for falling edges of deep fades is limited to $\Phi + 3\Lambda$ to prevent creating too much interference to other users. On the rising edge of the deep fades, the power is adjusted rapidly to decrease the interference. Therefore, we begin with a step size of $\Phi + 3\Lambda$, and then the step size is decreased gradually thereafter. Note that Δ_i depends on the status of present and previous power control bits. The sign(R0) in Figure 5.51 represents whether the current transmitted power is larger or smaller than the target power. The transmitting power of the mobile is the integration of the previous transmitting power level plus the current Δ_i, viz.: -

$$P_T = P_T^{-1} + \text{sign(R0)}\Delta_i, \qquad (5.253)$$

where P_T^{-1} is the previous transmitted power level. We do not show in Figure 5.51 the initial P_T generation at start-up. Once operating, power control errors $P_R - P_{target}$ form power control bits that adjust Δ_i to form P_T to drive P_R towards P_{target}.

5.6.5.2   Fixed-step-size algorithm

The adaptive closed-loop power control system shown in Figure 5.51 becomes a fixed-step-size system when g is set to zero in Equation (5.252), i.e. the step size is always Φ.

5.6.5.3   Combined open-loop and closed-loop power control

As the mobile moves, it experiences slow changes in received signal level due to path loss and log-normal shadow fading. The mobile pilot receiver is able to estimate the path loss and slow fading and to adjust the mobile's transmitted power. A simple arrangement is to average the signal power (average to remove the effect of fast fading) from the pilot

Table 5.4: Adaptive step size of the power adjustment. In any row x may be a logical 1 or 0. All other combinations not shown in the table have $g = 0$.

R0	R1	R2	R3	R4	R5	R6	g
1	1	1	0	x	x	x	1
1	1	1	1	0	x	x	3
1	1	1	1	1	0	x	2
1	1	1	1	1	1	0	1
1	1	1	1	1	1	1	0
0	1	1	1	1	1	1	3
0	1	1	1	1	1	0	2
0	0	1	1	1	1	1	3
0	0	0	1	1	1	1	2
0	0	0	1	x	x	x	1
0	0	0	0	1	x	x	1
0	0	0	0	0	1	x	2
0	0	0	0	0	0	1	0

correlator and subtract it from the previous average value [24]. The difference power signal is then added to the signal from the output closed-loop power adjustment sub-system, and the combined signal is used to modify the transmitted power of the mobile.

5.6.6 Simulations and results

This closed-loop power control system has been simulated for mobiles experiencing path loss profiles measured in city areas and fast fading when travelling at speeds of 12, 24, 48 and 96 km/h. We considered two closed-loop power control algorithms, the fixed-step-size (FS) algorithm and the adaptive-step-size (AS) algorithm. These control schemes could operate in a closed-loop incorporated with an open-loop mode (/C/O) and in a closed-loop mode (/C). The received power was measured every 1.25 ms. The fixed-step-size control scheme used a fixed step size of 1 dB for both the Rayleigh and Rician fading channels. The adaptive-step-size control scheme was simulated with a Φ of 1 dB, and a Λ which depends on the architecture of the RAKE receiver and types of fading. In fading channels where each component fades according to Rayleigh statistics, a Λ of 0.9 dB, 0.5 dB and 0.3 dB was used for a one-tap, two-tap, and three-tap RAKE receiver, respectively. In Rician fading channels having a K factor of 6 dB, a Λ of 0.7 dB, 0.4 dB, and 0.2 dB was used for a one-tap, two-tap, and three-tap RAKE receiver, respectively. In Rician fading channels having a K factor of 12 dB, a Λ of 0.1 was used. Since the transmitted power was updated every

1.25 ms, the received power at the BS was averaged over the same period. The required E_b/I_0 was set at 7 dB in the simulation and the received control bit at the MS was assumed to be error free. Owing to the limited number of samples in the averaging process, the estimation of the received power had errors. We first simulated a system where the mobile transmitted a constant power over a perfect channel in the presence of other users. The received power error due to power estimation at the BS receiver was then identified. Next, fixed-step-size and adaptive-step-size control schemes were introduced for a channel having path loss, slow fading, and fast fading. The effect of control bit errors on the down-link were also investigated.

5.6.6.1   Power and E_b/I_0 estimation error

The mobile only adjusts its transmitted power every 1.25 ms. Mobile traffic was therefore conveyed at a constant transmitted power in the 1.25 ms interval. The BS receiver measured the average received power during these periods. Since this averaging period was insufficient to cancel the interference noise completely, there were errors. We simulated the situation where the mobile transmitted a constant power through a perfect channel in the presence of interference. Both the proposed method and the mean-square-to-variance ratio (MSVR) method for estimating E_b/I_0 were simulated. With a one-tap RAKE receiver, the standard deviation of the errors in E_b/I_0 due to the limited period of averaging is shown in Figure 5.53 for different averaging periods measured in integer number of control bit periods. The corresponding mean value of E_b/I_0 is shown in Figure 5.54. From Figure 5.53, the standard deviation of the errors in E_b/I_0 decreased as the averaging period increased, where the value of the standard deviation of our proposed method was about half that of the MSVR method. The set-up value of E_b/I_0 in our simulation was 7.0 dB. The mean value of E_b/I_0 at the BS for the proposed method was about 7.06 dB irrespective of averaging period, while the mean value of E_b/I_0 for the MSVR method varied from 7.87 dB to 7.31 dB as the averaging period increased. For an averaging period of 2.5 ms, the error between the mean value of the estimated E_b/I_0 and the set-up value was 0.06 dB and 0.58 dB for the proposed method and the MSVR method, respectively. Since the proposed method was more accurate than the MSVR method, we used the proposed method in the following simulation.

From Figure 5.53 we observe that the longer the averaging period, the smaller the errors in E_b/I_0. However, a longer averaging period can wipe out the power variation due to fast fading. In order to maintain the power variation due to fading, while also providing a better estimated E_b/I_0, we use an averaging period of two control bits duration (2.5 ms) in the following simulation. The standard deviation of the errors in E_b/I_0 for an averaging period of 1.25 ms and 2.5 ms was found to be 1.28 dB and 0.77 dB, respectively. Consequently, for an average period of 2.5 ms, the lower bound of the power control error in E_b/I_0 was limited by this standard deviation of 0.77 dB.

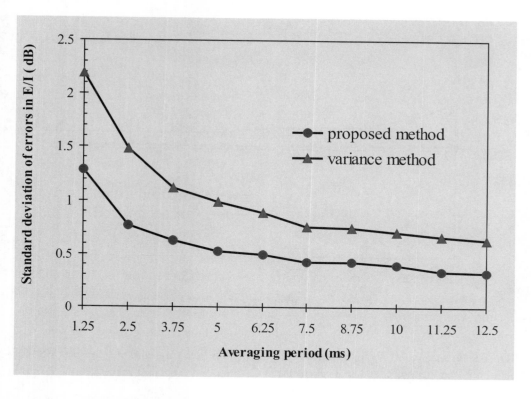

Figure 5.53: Standard deviation of the errors in E_b/I_0 for different averaging periods in the presence of a constant MS transmitter power and an ideal channel.

5.6.6.2   Performance of the power control scheme

An averaging period of 2.5 ms was used for estimating the signal power P_R and the interference power I in our closed-loop power control system. A channel with only fast fading and a channel with path loss, slow fading, and fast fading were simulated to examine the performance of the closed-loop control system with and without an open-loop control system. Both Rayleigh and Rician fading channels for mobiles having a speed of 12, 24, 48 and 60 km/h were simulated. A RAKE receiver with one, two, or three taps was simulated to investigate the effect of the RAKE receiver in the closed-loop power control system. Both the fixed-step-size and the adaptive-step-size closed-loop control algorithms were simulated for comparison purpose.

Performance in a Rayleigh fading channel   We first simulated the FS/C/O and AS/C/O schemes. In the AS/C/O scheme, the value of Λ in the simulation was 0.9, 0.5, and 0.3 for a one-tap, two-tap, and three-tap RAKE receiver, respectively. For a one-tap RAKE receiver,

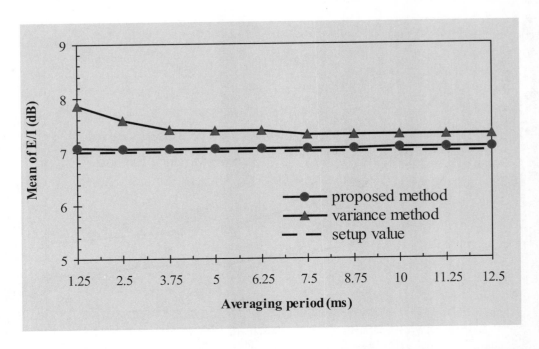

Figure 5.54: Mean of E_b/I_0 for different averaging periods in the presence of a constant MS transmitter power and an ideal channel.

the histograms of the received E_b/I_0 are shown in Figures 5.55 and 5.56 for a mobile speed of 48 and 96 km/h, respectively. Owing to power control errors, the received E_b/I_0 shown in Figures 5.55 and 5.56 tends to be log-normally distributed. For a mobile speed of 48 km/h, the standard deviation of errors in E_b/I_0 was 3.92 dB and 3.0 dB for the fixed-step and adaptive systems, respectively. For a mobile speed of 96 km/h, the standard deviation of the errors in E_b/I_0 was 4.28 dB and 4.08 dB for the fixed-step-size and adaptive-step-size algorithms, respectively. Next the FS/C and AS/C schemes were simulated. The simulation results in terms of E_b/I_0 are shown in Figures 5.57 and 5.58 for a mobile speed of 48 km/h and 96 km/h, respectively. For a mobile speed of 48 km/h, the standard deviation of the errors in E_b/I_0 was 3.98 dB and 3.32 dB for the FS/C and AS/C power control schemes, respectively. For a mobile speed of 96 km/h, the average E_b/I_0 was 7.2 dB and the standard deviation of the errors in E_b/I_0 was 4.44 dB and 4.22 dB for the FS/C and AS/C power control schemes, respectively.

Similarly we simulated our power control scheme with the two-tap and three-tap RAKE receivers, and with a mobile speed of 12, 24, 48, and 96 km/h. For a mobile speed of 48 and 96 km/h, the simulation results for the two-tap and three-tap RAKE receivers are shown in Figures 5.59–5.66. The standard deviations of the error in E_b/I_0 as a function of mobile

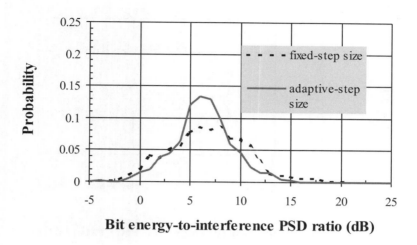

Figure 5.55: Histogram of the received E_b/I_0 for the FS/C/O and AS/C/O schemes having a one-tap RAKE at the BS with a mobile speed of 48 km/h.

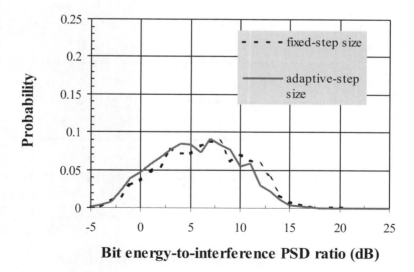

Figure 5.56: Histogram of the received E_b/I_0 for the FS/C/O and AS/C/O schemes having a one-tap RAKE at the BS with a mobile speed of 96 km/h.

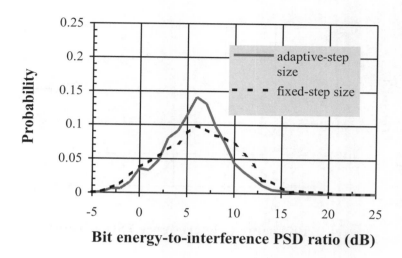

Figure 5.57: Histogram of the received E_b/I_0 for the FS/C and AS/C schemes having a one-tap RAKE at the BS with a mobile speed of 48 km/h.

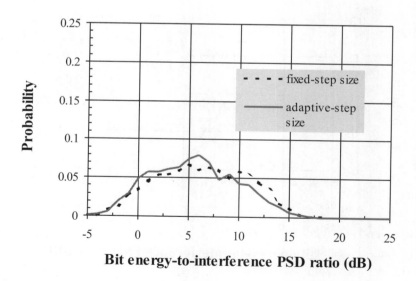

Figure 5.58: Histogram of the received E_b/I_0 for the FS/C and AS/C schemes having a one-tap RAKE at the BS with a mobile speed of 96 km/h.

speed are shown in Figures 5.67 and 5.68 for the FS/C/O and AS/C/O schemes and the FS/C and AS/C schemes, respectively. From Figures 5.67 and 5.68 we observe that the standard deviation of the power control error in E_b/I_0 increased as the mobile speed increased. For a three-tap RAKE receiver, the standard deviation of the power control error in E_b/I_0 was 1 dB for a mobile speed of 12 km/h, while it was approximately 2 dB for a mobile speed of more than 48 km/h.

For mobile speeds of 24 and 48 km/h in Figure 5.67 the AS/C/O scheme provided a decrease in the standard deviation of 0.9, 0.6, and 0.1 dB for the one-tap, two-tap, and three-tap RAKE receivers, respectively, compared with the FS/C/O scheme. For the FS/C and AS/C schemes, the decrease in the standard deviation was 0.8, 0.4, and 0.1 dB for the one-tap, two-tap, and three-tap RAKE receivers, respectively. There was little difference between using FS/C/O and FS/C for the fixed-step schemes. This to be expected because if the fast fading can be approximately tracked, then there is little difficulty in coping with much slower variations in path loss and slow fading. We note in Figure 5.68 and 5.69 that the standard deviation in E_b/I_0 of the AS/C scheme was 0.1, 0.2, and 0 dB, i.e. higher than that of the AS/C/O scheme for the one-tap, two-tap, and three-tap RAKE receivers, respectively. There was a decrease in the standard deviation of errors in E_b/I_0 for the adaptive scheme compared with the fixed-step scheme for the one-tap, two-tap, and three-tap RAKE receivers, of 0.2, 0.2, and 0 dB, respectively, when the mobile speed was 96 km/h. We observe that the decrease in the standard deviation due to the adaptive algorithm became smaller as the mobile speed increased from 48 km/h to 96 km/h.

In the fixed-step scheme, the decrease in the standard deviation of E_b/I_0 due to both the two-tap and three-tap RAKE receivers was larger than 1.1 dB. In the adaptive scheme, the decrease in the standard deviation of E_b/I_0 due to both the two-tap and three-tap RAKE receivers was larger than 0.9 dB. This highlights the importance of using a RAKE receiver having more than one tap in a CDMA system. With a RAKE receiver, the dynamic range of the received power decreased as the number taps increased, but the power variation became even more random as the number of taps increased. Since there was very little correlation in the received power as the number of taps was increased, the gain in using the adaptive-step-size algorithm compared with the fixed-size one became smaller.

Performance in a Rician fading channel   The power control scheme was also simulated in a Rician fading channel having a K factor of 6 dB and 12 dB. The fixed-step-size scheme still had a fixed power adjusting step size of 1 dB. In simulating the adaptive-step-size scheme in a Rician fading channel having a K factor of 6 dB, Λ was 0.7, 0.4, and 0.2 for the one-tap, two-tap, and three-tap RAKE receiver, respectively, while a Λ of 0.1 was used for all three different tap RAKE receivers in a Rician fading channel having a K factor of 12 dB. Since the dynamic range of the fading profile was so small in the Rician fading

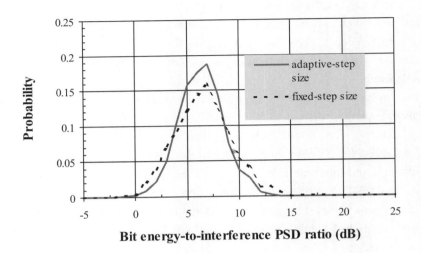

Figure 5.59: Histogram of the received E_b/I_0 for the FS/C/O and AS/C/O schemes having a two-tap RAKE at the BS with a mobile speed of 48 km/h.

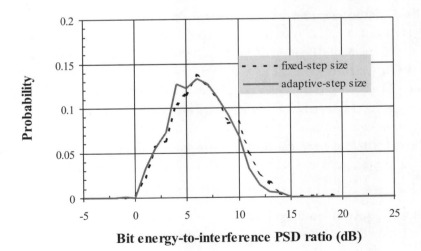

Figure 5.60: Histogram of the received E_b/I_0 for the FS/C/O and AS/C/O schemes having a two-tap RAKE at the BS with a mobile speed of 96 km/h.

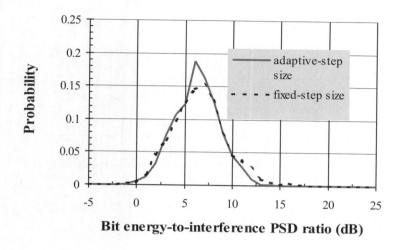

Figure 5.61: Histogram of the received E_b/I_0 for the FS/C and AS/C schemes having a two-tap RAKE at the BS with a mobile speed of 48 km/h.

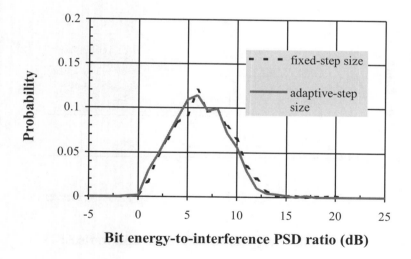

Figure 5.62: Histogram of the received E_b/I_0 for the FS/C and AS/C schemes having a two-tap RAKE at the BS with a mobile speed of 96 km/h.

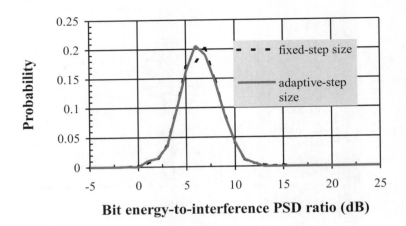

Figure 5.63: Histogram of the received E_b/I_0 for the FS/C/O and AS/C/O schemes having a three-tap RAKE at the BS with a mobile speed of 48 km/h.

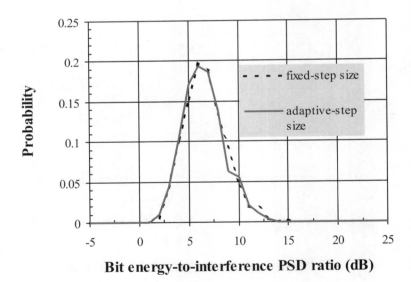

Figure 5.64: Histogram of the received E_b/I_0 for the FS/C/O and AS/C/O schemes having a three-tap RAKE at the BS with a mobile speed of 96 km/h.

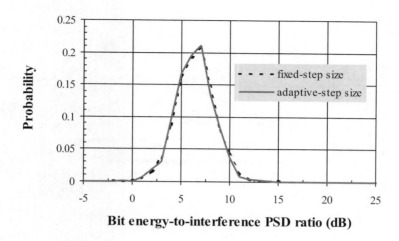

Figure 5.65: Histogram of the received E_b/I_0 for the FS/C and AS/C schemes having a three-tap RAKE at the BS with a mobile speed of 48 km/h.

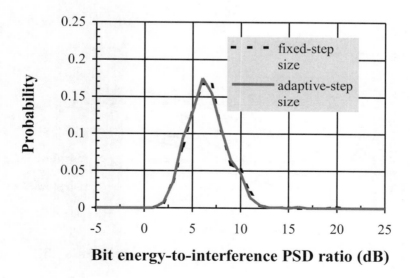

Figure 5.66: Histogram of the received E_b/I_0 for the FS/C and AS/C schemes having a three-tap RAKE at the BS with a mobile speed of 96 km/h.

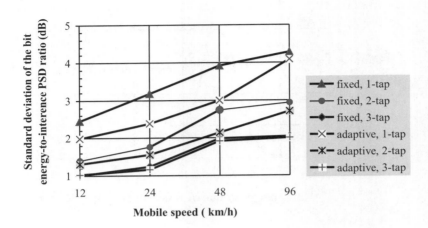

Figure 5.67: The standard deviation of the power control errors in E_b/I_0 versus the mobile speed for the FS/C/O and AS/C/O schemes.

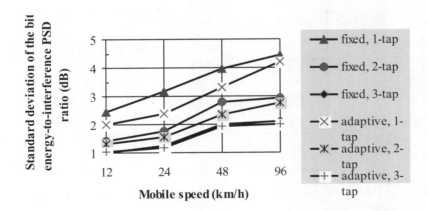

Figure 5.68: The standard deviation of the power control errors in E_b/I_0 versus the mobile speed for the FS/C and AS/C schemes.

channel, the performance of the power control for mobile speeds of 48 km/h and 96 km/h was approximately the same. The results of the histogram of E_b/I_0 for a mobile speed of 48 km/h are shown in Figures 5.69 to 5.80. Since the simulation results for the FS/C/O and AS/C/O schemes and the FS/C and AS/C schemes are approximately the same, we only present here the simulation results for the FS/C and AS/C schemes. The standard deviations of E_b/I_0 as a function of mobile speed are shown in Figures 5.81 and 5.82 for a K factor of 6 dB and 12 dB, respectively.

For a K of 6 dB, the performance of the adaptive-step-size algorithm was only marginally better than that of the fixed-step-size one, while the performance of both schemes was almost the same for a K of 12 dB. The improvement in performance due to a RAKE receiver was still quite significant for a K of 6 dB, while the improvement in performance due to a RAKE receiver became smaller for a K of 12 dB as the dynamic range of the fading was even smaller. For a Rician channel with a minimal K of 6 dB the standard deviation of the power control errors in E_b/I_0 was less than 1.50 dB.

Performance with power control bit errors Since the power control bits are embedded in the down-link traffic, the power control bits are subject to random errors. A power control bit error rate of 10^{-2} was simulated for the FS/C and AS/C power control schemes to investigate the effect of power control bit errors on system performance. From our simulations, we found that the standard deviation of the errors in the received E_b/I_0 was virtually unaffected by the presence of random errors.

5.6.7 Summary

Since the application of RAKE receivers with multi-taps is widely used in cellular CDMA systems for exploiting micro-diversity, the closed-loop power control schemes should make use of multi-tap RAKE receivers for tracking the power variation due to path loss, slow fading, and fast fading. The measurement of the received signal power and the ratio of the bit energy-to-interference power spectral density play an important role in the accuracy of a CDMA power control system, and therefore we proposed a scheme to estimate these two parameters. Both fixed-step-size and adaptive-step-size power control algorithms were simulated.

The performance of the E_b/I_0 estimation scheme was simulated for the situation where the mobile transmitted a constant power through a perfect channel in the presence of interference. The mean value of the estimated E_b/I_0 at the BS for the proposed method was about 7.06 dB, irrespective of the averaging period, while the mean value of E_b/I_0 for the MSVR method varied from 7.87 dB to 7.31 dB as the averaging period increased. For an averaging period of 2.5 ms, the error between the mean value of the estimated E_b/I_0 and the set-up value was 0.06 dB and 0.58 dB for the proposed method and the MSVR method, re-

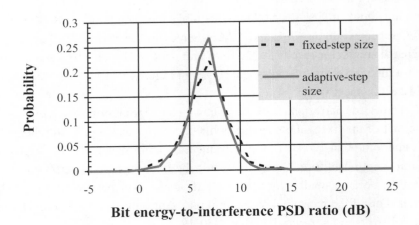

Figure 5.69: Histogram of the received E_b/I_0 for the FS/C/O and AS/C/O schemes having a one-tap RAKE at the BS; $K = 6$ dB; MS speed=48 km/h.

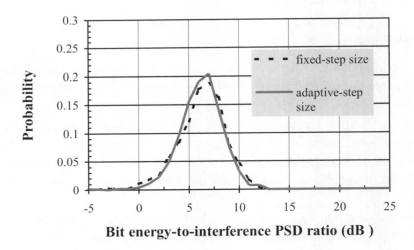

Figure 5.70: Histogram of the received E_b/I_0 for the FS/C and AS/C schemes having a one-tap RAKE at the BS; $K = 6$ dB; MS speed=48 km/h.

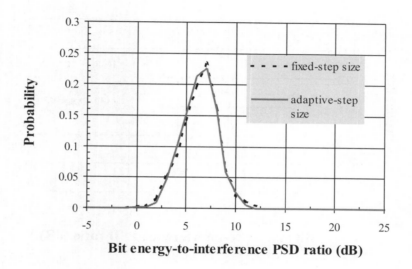

Figure 5.71: Histogram of the received E_b/I_0 for the FS/C/O and AS/C/O schemes having a two-tap RAKE at the BS; $K = 6$ dB; MS speed=48 km/h.

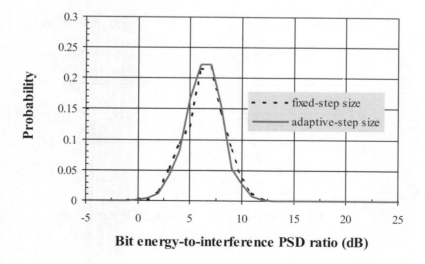

Figure 5.72: Histogram of the received E_b/I_0 for the FS/C and AS/C schemes having a two-tap RAKE at the BS; $K = 6$ dB; MS speed=48 km/h.

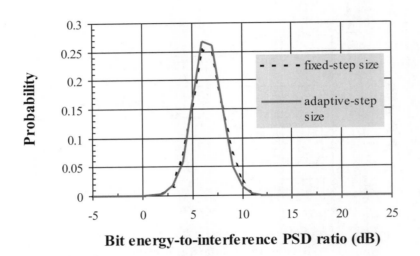

Figure 5.73: Histogram of the received E_b/I_0 for the FS/C/O and AS/C/O schemes having a three-tap RAKE at the BS; $K = 6$ dB; MS speed=48 km/h.

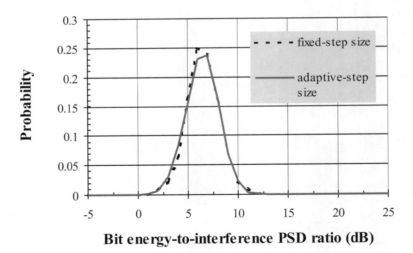

Figure 5.74: Histogram of the received E_b/I_0 for the FS/C and AS/C schemes having a three-tap RAKE at the BS; $K = 6$ dB; MS speed=48 km/h.

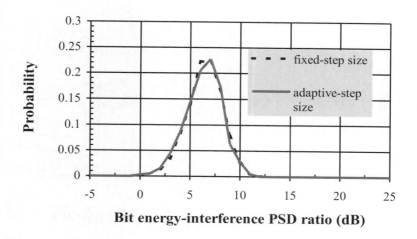

Figure 5.75: Histogram of the received E_b/I_0 for the FS/C/O and AS/C/O schemes having a one-tap RAKE at the BS; $K = 12$ dB; MS speed=48 km/h.

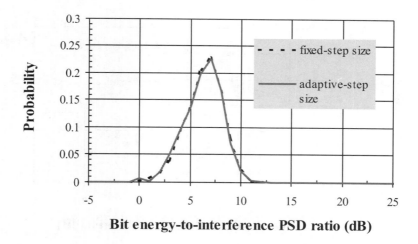

Figure 5.76: Histogram of the received E_b/I_0 for the FS/C and AS/C schemes having a one-tap RAKE at the BS; $K = 12$ dB; MS speed=48 km/h.

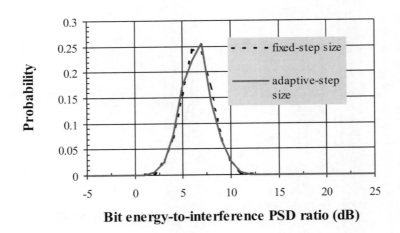

Figure 5.77: Histogram of the received E_b/I_0 for the FS/C/O and AS/C/O schemes having a two-tap RAKE at the BS; $K = 12$ dB; MS speed=48 km/h.

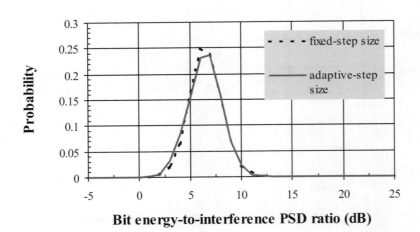

Figure 5.78: Histogram of the received E_b/I_0 for the FS/C and AS/C schemes having a two-tap RAKE at the BS; $K = 12$ dB; MS speed=48 km/h.

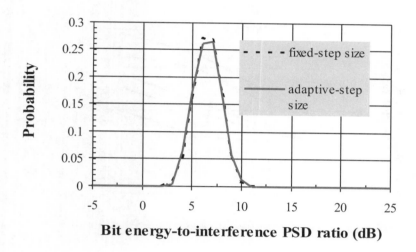

Figure 5.79: Histogram of the received E_b/I_0 for the FS/C/O and AS/C/O schemes having a three-tap RAKE at the BS; $K = 12$ dB; MS speed=48 km/h.

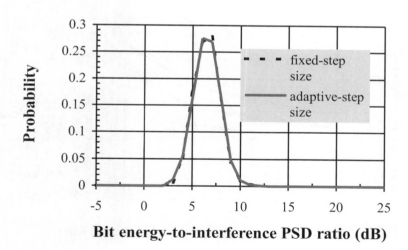

Figure 5.80: Histogram of the received E_b/I_0 for the FS/C and AS/C schemes having a three-tap RAKE at the BS; $K = 12$ dB; MS speed=48 km/h.

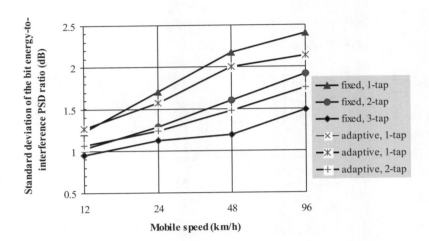

Figure 5.81: The standard deviation of the power control errors in E_b/I_0 versus mobile speed for the FS/C and AS/C schemes for a Rician channel having a K of 6 dB.

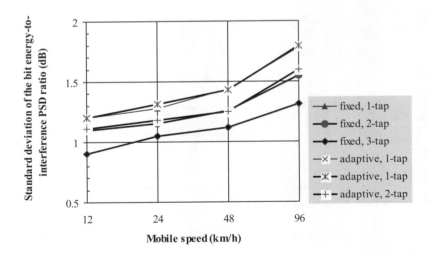

Figure 5.82: The standard deviation of the power control errors in E_b/I_0 versus mobile speed for the FS/C and AS/C schemes for a Rician channel having a K of 12 dB.

spectively. The standard deviation of E_b/I_0 for the proposed method was about half that for the MSVR method. By using the proposed method, the standard deviation of the errors of E_b/I_0 for an averaging period of 1.25 ms and 2.5 ms was 1.28 dB and 0.77 dB, respectively. Consequently, for an averaging period of 2.5 ms, the lower bound of the power control error in E_b/I_0 was limited by this standard deviation of 0.77 dB.

The performance of the closed-loop power control scheme in terms of the standard deviation in E_b/I_0 was simulated for RAKE receivers having a different number of taps. For a mobile speed of 96 km/h and a three-tap RAKE receiver, the standard deviation of power control errors in E_b/I_0 was 2.0 dB and 1.5 dB for Rayleigh and Rician fading channels having a K factor of 6 dB, respectively. For a control bit error rate of 10^{-2}, the performance of the power control scheme in terms of the standard deviation of the errors in E_b/I_0 was essentially unaffected.

The decrease in the standard deviation of the errors in E_b/I_0 due to both the two-tap and three-tap RAKE receivers was larger than 1.0 dB for Rayleigh fading channels. For Rician fading channels, the improvement in performance due to a RAKE receiver was still quite significant for a K of 6 dB, while only marginal for a K of 12 dB. As the number of taps in a RAKE receiver increased, the dynamic range of fading decreased, while the variation in the received power became more random. The reduction in the standard deviation of errors in E_b/I_0 for the adaptive-step-size algorithm compared with the fixed-step-size algorithm decreased as the number of taps increased. For a three-tap RAKE receiver, the decrease in the standard deviation of E_b/I_0 from the adaptive-step-size algorithm was marginal, while the use of a RAKE receiver provided a significant reduction in the standard deviation of the errors in E_b/I_0 for the fixed-step-size power control algorithm. From a power control point of view it was desirable to have a three-tap RAKE receiver.

Bibliography

[1] Salmasi A. and K.S. Gilhousen, On the system design aspects of code division multiple access (CDMA) applied to digital cellular and personal communications networks, *IEEE 41st VTS Conf.*, St Louis, May 1991, 264–275.

[2] Torrieri D.J., Performance of direct-sequence systems with long pseudo noise sequence, *IEEE J. Sel. Area Commun.*, **10**(4), May 1992, 770–781.

[3] Papoulis A., *Probability, Random Variables, and Stochastic Processes*, McGraw-Hill, 1991.

[4] Pickholtz R.L., D.-L. Shilling and L.B. Miltein, Theory of spread spectrum communications–A tutorial, *IEEE Trans. Commun.*, **COM-30**, May 1982, 855–884.

[5] Proakis J.G., *Digital Communications*, McGraw-Hill, 1983.

[6] Lee C.-C. and R. Steele, Closed-loop power control in CDMA systems, *IEE Proc. Commun.*, **143**(4), Aug. 1996, 231–239.

[7] Viterbi A.J. and R. Padovani, Implications of mobile cellular CDMA, *IEEE Commun. Mag.*, **30**(12), Dec. 1992, 38–41.

[8] Cameron R. and B.D. Woerner, An analysis of CDMA with imperfect power control, *IEEE 42nd VTS Conf.*, Denver, May 1992, 97–98.

[9] Gilhousen K.S., I.M. Jacobs, R. Padovani, A.J. Viterbi, L.A. Weaver and C.E. Wheatley, On the capacity of a cellular CDMA system, *IEEE Trans. Veh. Tech.*, **40**(2), May 1991, 303–311.

[10] Lee W.C.Y., Overview of cellular CDMA, *IEEE Trans. Veh. Technol.*, **40**(2), May 1991.

[11] Gejji R.R., Forward-link-power control in CDMA cellular systems, *IEEE Trans. Veh. Technol.*, **41**(4), Nov. 1992, 532–536.

[12] Prasad R., *CDMA for Wireless Personal Communications*, Artech House, 1996.

[13] Viterbi A.J., *CDMA: Principle of Spread Spectrum Communication*, Addison-Wesley, 1995.

[14] Steele R. [Ed.], *Mobile Radio Communications*, Pentech Press, 1992.

[15] Steele R., The cellular environment of lightweight handheld portables, *IEEE Commun. Mag.*, **27**(7), July 1989, 20–29.

[16] Steele R., Towards a high-capacity digital cellular mobile radio system, *IEE Proc.*, **132**, Pt F (5), Aug. 1985, 405–415.

[17] Berg J.-E. et al, Path loss and fading models for microcells at 900MHz, *IEEE 42nd VTS Conf.*, Denver, May 1992, 666–671.

[18] Erceg V., et al, Propagation modelling and measurements in an urban and suburban environment using broadband direct-sequence spread spectrum, *IEEE 42nd VTS Conf.*, Denver, May 1992, 317–320.

[19] Pickholtz R.L. and I.S. Stojanovic, A comparison of TDMA and CDMA in microcellular radio channels, *IEEE 41st VTS Conf.*, Denver, May 1991, 233–235.

[20] Qualcomm, Inc., *CDMA Digital CAI Standard*, October 1990.

[21] Jakes W.C., *Microwave Mobile Radio Communications*, John Wiley, 1974.

[22] Lee W.C.Y., *Mobile Communications Design Fundamentals*, John Wiley 1993.

[23] Steele R., *Delta Modulation Systems*, Pentech Press, 1975.

[24] Lee W.C.Y., Estimate of local average power of a mobile radio signal, *IEEE Trans. Veh. Technol.*, **34**(1), Feb. 1985, 22–27.

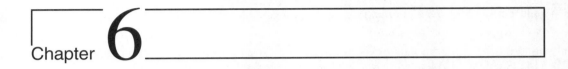

Chapter **6**

Evolution of GSM and cdmaOne to 3G Systems

6.1 Introduction

The previous chapters have concentrated on the two leading second generation (2G) cellular systems: GSM and IS-95. These systems are deployed in many parts of the world and will continue to operate and evolve during the next decade as third generation (3G) systems are rolled out. We may expect that the new 3G systems will be harmonised with their evolved 2G counterparts, and that slowly 2G spectra will be refarmed to provide extra 3G spectra. No 3G systems are currently deployed, although trials are in progress. As a consequence, this chapter, which deals with systems that are about to be deployed, is treated in a qualitative manner, describing how they will work rather than quantifying their performances. Before getting into detail, let us briefly review how cellular communications arrived at today's position.

6.1.1 The generation game

There is no doubt that there was pent-up demand for public mobile telephony networks, and when they arrived in the 1980s as the so-called first generation (1G) analogue cellular networks, they grew at phenomenal rates. These networks initially offered only telephony, but the un-tethering of people from their fixed phones meant that they and businesses could operate in completely new ways. The Europeans identified in the early 1980s the need for a second generation (2G) cellular system that would be totally digital. This 2G system became GSM, and a brief history of GSM has already been provided in Section 2.1. The Europeans have a long view in cellular radio and in 1988 they launched their RACE 1043 project with the aim of identifying the services and technologies for an advanced third generation (3G) system for deployment by the year 2000 [1, 2]. Their 3G system soon became known as the universal mobile telecommunications system (UMTS) [3–5]. The concept was that their 1G, 2G and 3G systems would be independent, and that their deployment would overlap such that the total access communications system (TACS), say, would slowly be replaced by GSM, which in turn would be slowly phased out for UMTS. However, the success of GSM has been so great that evolutionary paths from 2G to 3G needed to be considered. Although the back-haul networks of GSM and UMTS have considerable commonality, their radio interfaces are significantly different.

There were initially great expectations for UMTS [5, 6]. It would not only be cellular, but it would embrace other types of networks from private mobile radio (PMR) (called special mobile radio (SMR) in the United States), to wireless local area networks (WLANs), to mobile satellite systems (MSSs). The cardinal points were that it would operate globally, support high bit rate services and, most importantly, be service orientated. While the Europeans referred to the global 3G network for the turn of the century as UMTS, most of their engineers working on UMTS expected that they would have to yield to international agreements from the ITU to modify UMTS, but that basically UMTS would be accepted as the global standard.

To explain this early expectation we need to point out that the ITU has been in the 3G game from the beginning [6]. Paralleling the European Union (EU) RACE initiative, ITU formed task group TG8/1, originally under the auspices of CCIR. This committee referred to their 3G system as the future public land mobile telecommunications system (FPLMTS). Europeans were, of course, also members of TG8/1, and with commercial and political pressures a long way in the future, FPLMTS and UMTS seemed synonymous in terms of aims and objectives. The important difference between TG8/1 and the happenings in Europe, was that in Europe there was an actual research and development (R&D) 3G programme in process, while TG8/1 was more like a forum.

The Americans did not launch concerted national R&D programmes, neither for 2G nor 3G systems. Their advanced mobile phone service (AMPS) 1G system did evolve into the 2G IS-136, and became dual-mode with IS-95. The United States also introduced the iDEN system with its ability to offer both cellular and dispatch services. It then auctioned a large part of its 3G spectrum for PCS licenses, and allowed GSM to enter the United States in the form of PCS1900. This auctioning of the 3G spectrum meant that there were significant advantages if existing 2G networks could evolve into 3G ones, preferably in a seamless manner.

A big factor, not just in the United States, but in the world, was the advent of IS-95 [7–11]. It arrived late compared with GSM, and some engineers argued that it was a 2.5G system. It had to fight to be born because of the lack of spectrum, and the quasi-religious attitudes of engineers towards methods of multiple access. The meagre spectrum of 1.25 MHz at the top of the AMPS band was just about adequate for cellular CDMA, which was just as well because that was all there was available. CDMA entered the cellular world with a host of technical problems, not made easier as the transceivers from day one had to be dual-mode with AMPS. Its advocates were clear in that CDMA has a high spectral efficiency and is well suited to the 3G multiservice environment. The real significance of IS-95 is that it won the technical argument in that the UMTS and the Japanese Association of Radio Industries and Businesses (ARIB) proposals have CDMA radio interfaces, albeit of wider bandwidth systems as more bandwidth is available for 3G networks. We therefore agree that IS-95 is a 2.5G system and its evolution to 3G should be smooth. This is not so for 2G TDMA systems which will need to migrate to 3G CDMA ones. However, as we will show in Section 6.2, GSM with its TDMA is able to evolve closely to 3G without picking up the CDMA card. Nevertheless there is an evolutionary route from GSM Phase 2+ to UMTS as discussed in Section 6.2

The TG8/1 Committee discarded the unwieldy FPLMTS name for its 3G system, and replaced it with international mobile telecommunications for the year 2000, or simply IMT-2000. It then abandoned all hope of the difficult political objective of a single standard, and has instead opted for a family of standards. Each member of the family had to be able to meet a minimum specification. Sixteen proposals were accepted, ten for terrestrial 3G networks, and six for MSSs. The majority of the proposals advocated CDMA as the multiple access method. A degree of harmonisation between the proposals ensued, and at the time of writing the ITU has agreed that the IMT-2000 family will be composed of the following five technologies.

- IMT DS (Direct Sequence). This is widely known as UTRA FDD and W-CDMA, where UTRA stands for the UMTS Terrestrial Radio Access, and the 'W' in W-CDMA means wideband. We will refer to this system here as UTRA FDD.

- IMT MC (Multicarrier). This system is the 3G version of IS-95 (now called cdmaOne), and is also known as cdma2000. We will use the term cdma2000 as this is its widely used name.

- IMT TC (Time Code). This is the UTRA TDD, namely the UTRA mode that uses time division duplexing.

- IMT SC (Single Carrier). This is essentially a particular manifestation of GSM Phase 2+, known as EDGE, standing for Enhanced Data Rates for GSM Evolution.

- IMT FT (Frequency Time). This is the digitally enhanced cordless telecommunications (DECT) system.

In the authors' opinion, the truly 3G systems are the IMT DS, IMT MC and IMT TC systems.

6.1.2 IMT-2000 spectrum

The World Administration Radio Congress (WARC) in March 1992 assigned 200 MHz in the 2G frequency band to IMT-2000 for world-wide use [3]. The actual frequency bands are 1885–2025 MHz and 2110–2200 MHz. Unfortunately some parts of these bands are already used for other services. Figure 6.1 shows a diagram of the IMT-2000 spectrum, and the current use of this spectrum in Europe, the United States, and Japan.

The IMT-2000 spectrum may be partitioned into seven segments. The frequency of each segment is shown in Table 6.1

Part of Segment 1 is currently used for DECT in Europe, and is also used for PHS, PCS and DECT in other parts of the world. Segment 2 is used at present for PCS and PHS in the United States and Japan, respectively. Segments 3 and 6 form 60 MHz frequency division duplex (FDD) bands. Mobile satellite services (MSS) are in Segments 4 and 7, providing 30 MHz FDD bands. Segment 4 supports the earth-to-space links; while segment 7 provides the space-to-earth links. The 1980–1990 MHz band in Segment 4 is currently used for PCS in the United States. Segments 1, 2 and 5 are unpaired and are suitable for time division duplex (TDD) operation. Segment 5 may be used in the United States for earth-to-space MSS services.

6.2 Evolution of GSM

The GSM system was initially designed to carry speech, as well as low speed data. Much has already been discussed regarding speech, so we will concentrate here on data. The user data rate over the radio interface using a single physical channel, i.e. a single timeslot per

Table 6.1: IMT-2000 spectrum and its segments (MSS stands for mobile satellite services).

Segment number	Frequency band (MHz)	Comment
1	1885–1900	Unpaired
2	1900–1920	Unpaired
3	1920–1980	Paired with 6
4	1980–2010	MSS paired with 7
5	2010–2025	Unpaired
6	2110–2170	Paired with 3
7	2170–2200	MSS paired with 4

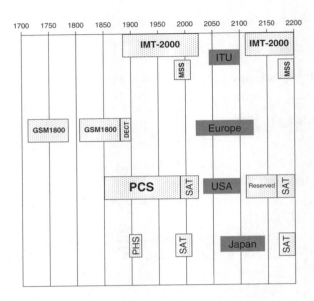

Figure 6.1: IMT-2000 spectrum and the current use of this spectrum in Europe, the United States and Japan. PCS stands for personal communication system, SAT for mobile satellite services, DECT for digitally enhanced cordless telecommunications, and PHS for personal handyphone system.

TDMA frame, was initially 9.6 kb/s. The maximum user data rate available on a single physical channel has since been increased to 14.4 kb/s by reducing the power of the channel coding on the full rate traffic channel by means of code symbol puncturing. Apart from increasing the level of puncturing still further, the other ways to increase the user data rate beyond 14.4 kb/s are either to allow an MS to access more than an one timeslot per TDMA frame or to use a higher level modulation scheme (e.g. quadrature amplitude modulation, QAM) to increase the amount of information that can be transmitted within a single timeslot.

Two new services have been introduced as part of GSM Phase 2+ which allow the user data rate to be increased by permitting an MS to access more than one timeslot per TDMA frame. These new services are the high speed circuit switched data (HSCSD) service and the general packet radio service (GPRS). The HSCSD service allows an MS to be allocated a number of timeslots per TDMA frame on a circuit-switched basis, i.e. the MS has exclusive use of the allocated resources for the duration of a call [12]. In contrast, GPRS uses packet-orientated connections on the radio interface (and within the network) whereby a user is assigned one, or a number of traffic channels only when a transfer of information is required [13]. The channel is relinquished once the transmission is completed. In the following sections we will describe these two services in more detail.

The second approach to increasing the user data rate by employing a higher level modulation scheme is currently being studied under the Enhanced Data Rates for GSM Evolution (EDGE) project [14]. The basic principle behind EDGE is that the modulation scheme used on the GSM radio interface should be chosen on the basis of the quality of the radio link. A higher level modulation scheme is preferred when the link quality is 'good', but the system reverts to a lower level modulation scheme when the link quality becomes 'poor'. At the time of writing it appears that EDGE will use the existing Gaussian minimum shift keying (GMSK) modulation scheme in poor quality channels and eight-level phase shift keying (8-PSK) in good quality channels. EDGE will also include link adaption functions to allow the MS and BS to assess the link quality and switch between the different types of modulation as necessary.

Once developed, the EDGE technology will enhance the range of services offered by GSM. The initial version of the EDGE technology (Phase 1) will be used to enhance the GPRS and HSCSD services, leading to enhanced GPRS (EGPRS) and enhanced circuit-switched data (ECSD). In later releases of EDGE (Phase 2 and beyond) further services will be introduced which utilise the different modulation schemes [14].

In addition to the developments described above, GSM Phase 2+ contains two other important enhancements that have a significant impact on the technology from a radio point of view. In 1993 the European railways, in the form of the Union Internationale des Chemins de Fer (UIC), chose the GSM technology as the basis of all their future mobile radio communication systems [15]. This led to the introduction of a number of advanced speech call items (ASCI) which provide the additional functionality required for railways and other private mobile radio (PMR) environments. The three key elements of ASCI are the voice broadcast service (VBS), the voice group call service (VGCS) and the enhanced multi-level precedence and pre-emption (eMLPP) service. The VBS will allow GSM users to broadcast their voice simultaneously to a number of other users in a chosen talk group. A VBS call will only occupy a single down-link channel in each cell within which the call is broadcast and all 'listening' MSs will monitor this same channel. VBS calls are simplex in that the call originator is the only person who can speak during the call. Many applications require any member of the talk group to become the talker and this functionality is supported in the VGCS. In this case any member of the talk group may become the 'talker' and contention resolution schemes are included to handle situations where more that one user tries to become the talker simultaneously. These PMR systems also support a facility to ensure that important calls are successfully completed even at the expense of less important calls. The eMLPP service allows calls to be prioritised and ensures that the most important calls are completed, regardless of the network loading.

Another important Phase 2+ item from a radio perspective is the adaptive multi-rate (AMR) speech coder [16]. The deployment of the GSM half-rate (HR) codec has been somewhat limited because of concerns relating to the speech quality, but the enhanced full-rate (EFR) codec is more popular amongst GSM operators. The basic concept behind the AMR technology is that the speech coding rate and the degree of channel coding should be chosen according to the channel quality. For example, in 'good' channels a lower rate speech coder can be used in an HR traffic channel thereby increasing the system capacity. However, if the link quality is 'poor' FR then a traffic channel will be used and the level of channel coding increased. At the time of writing the candidate AMR codecs have not yet been chosen.

6.2.1   High speed circuit-switched data

The HSCSD service [12] is a natural extension of the circuit-switched data services that were supported in earlier versions of GSM. No changes to the physical layer interfaces between the different network elements are required for HSCSD. At the higher layers the MS and the network support the additional functionality required to multiplex and demultiplex a user's data onto a number of traffic channels for transmission over both the Abis interface and the radio interface. Additional functionality is also included at the radio resource management level to handle the new situation where a number of different traffic channels are associated with the same connection. For example, when an HSCSD user is handed over between two cells, there must be a mechanism to ensure that sufficient traffic channels are available in the new cell before the handover occurs. An HSCSD connection is, however, limited to a single 64 kb/s circuit on the A interface.

On call set-up the MS provides information to the network which defines the nature of the HSCSD connection. The multislot class of the MS is used by the network to determine the maximum number of timeslots that may be accessed by the MS, and the amount of time that must be allowed between timeslots, e.g. for the purposes of neighbour cell measurements. This information is used to define the MS's capabilities for both the HSCSD and GPRS services. The multislot classes are listed in Table 6.2 along with their associated parameters [17].

Multislot MSs can be either type 1 or type 2 and this information is shown in the right-hand column of Table 6.2. Type 2 MSs are required to be able to transmit and receive simultaneously, whereas type 1 MSs are not. The 'Rx' and 'Tx' columns give the maximum number of receive and transmit timeslots that the MS may occupy per TDMA frame, respectively. The 'Sum' column gives the total number of transmit and receive timeslots the MS may access per TDMA frame. For example, for multislot class 12, 'Sum' is 5 which means that the maximum number of transmit and receive slots cannot exceed 5. So if we have 3 received slots, then we cannot have more than 2 transmit slots in one TDMA frame. The T_{ta} parameter represents the time required for the MS to make a neighbour cell measurement prior to an up-link transmission. This parameter is not applicable to type 2

MSs because they are capable of making measurements and transmitting simultaneously. When the MS is not required to make measurements on neighbouring cells, the T_{tb} parameter defines the minimum number of timeslots that must be allowed between the end of the previous down-link timeslot and the next up-link timeslot, or the time between two consecutive down-link timeslots that are on different frequencies. In other words, T_{tb} is the amount of time the MS needs to prepare to transmit on the up-link after it has received or transmitted on a different frequency. The T_{ra} parameter is the number of timeslots required by the MS to make a neighbour cell measurement prior to the reception of a down-link burst, whereas the T_{rb} parameter is the number of timeslots required between the previous up-link transmission and the next down-link reception, or the time between two consecutive down-link receptions when the frequency is changed in between receptions. In addition to its multislot class, the MS provides a range of additional information on call set-up to allow the network to determine the most appropriate HSCSD configuration. This information includes the fixed network user rate, i.e. the data rate that the MS would like to achieve over the fixed network, the channel coding schemes supported by the MS, and the maximum number of traffic channels to be used during the connection. This final parameter allows the user to control the call cost by limiting the number of traffic channels that will be occupied. The final multislot configuration is chosen by the network based on the MS capabilities and the requirements imposed by the services, e.g. whether neighbour cell measurements are required. The HSCSD service can support both symmetric transmissions, i.e. the same number of up-link and down-link timeslots, or asymmetric transmissions, i.e. more timeslots are allocated in one direction. However, in the case of HSCSD connections, only down-link biased asymmetry is allowed and the up-link timeslot numbers must be a subset of the down-link timeslot numbers.

6.2.2 The general packet radio service

Many services do not require a continuous bi-directional flow of user data across the radio interface. To illustrate this, consider the example of a user browsing the Worldwide Web (WWW) on her lap-top computer using a dial-up connection via a cellular network. Once a page of information has been downloaded, there will be a pause in the information flow between the MS and the network as the user reads the information and before more information is requested. Using circuit-switched connections for 'bursty' services of this nature represents an inefficient use of the radio resources because a user will continue to occupy a radio channel for the duration of a call (or browsing session) even though this channel may only be utilised for a small fraction of the time. Inefficiencies of this type can be overcome by carrying these services using packet-orientated connections.

The GSM system was initially designed to support only circuit-switched connections at the radio interface level with user data rates of up to 9.6 kb/s. However, the Phase 2+

Table 6.2: The MS multislot classes.

Multislot class	Maximum number of slots			Minimum number of slots				Type
	Rx	Tx	Sum	T_{ta}	T_{tb}	T_{ra}	T_{rb}	
1	1	1	2	3	2	4	2	1
2	2	1	3	3	2	3	1	1
3	2	2	3	3	2	3	1	1
4	3	1	4	3	1	3	1	1
5	2	2	4	3	1	3	1	1
6	3	2	4	3	1	3	1	1
7	3	3	4	3	1	3	1	1
8	4	1	5	3	1	2	1	1
9	3	2	5	3	1	2	1	1
10	4	2	5	3	1	2	1	1
11	4	3	5	3	1	2	1	1
12	4	4	5	2	1	2	1	1
13	3	3	NA	NA	a	3	a	2
14	4	4	NA	NA	a	3	a	2
15	5	5	NA	NA	a	3	a	2
16	6	6	NA	NA	a	2	a	2
17	7	7	NA	NA	a	1	0	2
18	8	8	NA	NA	0	0	0	2
19	6	2	NA	3	b	2	c	1
20	6	3	NA	3	b	2	c	1
21	6	4	NA	3	b	2	c	1
22	6	4	NA	2	b	2	c	1
23	6	6	NA	2	b	2	c	1
24	8	2	NA	3	b	2	c	1
25	8	3	NA	3	b	2	c	1
26	8	4	NA	3	b	2	c	1
27	8	4	NA	2	b	2	c	1
28	8	6	NA	2	b	2	c	1
29	8	8	NA	2	b	2	c	1

NA = Not Applicable

$$a = \begin{cases} 1 & \text{if frequency hopping is used} \\ 0 & \text{if frequency hopping is not used} \end{cases}$$

$$b = \begin{cases} 1 & \text{if frequency hopping is used or there is a change from Rx to Tx} \\ 0 & \text{if frequency hopping is not used and there is no change from Rx to Tx} \end{cases}$$

$$c = \begin{cases} 1 & \text{if frequency hopping is used or there is a change from Tx to Rx} \\ 0 & \text{if frequency hopping is not used and there is no change from Tx to Rx} \end{cases}$$

specifications now include provision for the support of a packet-orientated service known as the general packet radio service (GPRS) [13, 18–20]. GPRS attempts to optimise the network and radio resources, and strict separation between the radio subsystem and the network subsystem is maintained, although the network subsystem is compatible with the other GSM radio access procedures. Consequently, the GSM MSC is unaffected. The allocation of a GPRS radio channel is flexible, ranging from one to eight radio interface timeslots in a TDMA frame. Up-link and down-link timeslots are allocated separately. The radio interface resources are able to be shared dynamically between circuit switched and packet services as a function of service load and operator preference. Bit rates vary from 9 kb/s to more than 150 kb/s per user. GPRS can interwork with IP and X.25 networks. Point-to-point and multipoint services are also supported, as well as short message services (SMS).

GPRS is able to accommodate both intermittent, bursty data transfers as well as large continuous data transmissions. Reservation times are typically from 0.5 s to 1 s. Three MS modes are supported, each having a different arrangement with circuit switched GSM services. In this section we provide an overview of the GPRS technology and examine its impact on the GSM radio interface.

6.2.2.1 The GPRS logical architecture

Figure 6.2 is a block diagram showing the architecture of a GSM network that supports GPRS, and the names that have been given to the interfaces that exist between the different network components. GPRS services require two additional network components, the gateway GPRS support node (GGSN), and the serving GPRS support node (SGSN). As its name suggests, the GGSN acts as the gateway between external packet data networks (PDN), and a GSM network that supports GPRS. The GGSN contains sufficient information to route incoming data packets to the SGSN that is serving a particular MS and it is connected to external networks via the Gi reference point. (We note that this point of interconnection is referred to a 'reference point' and not an 'interface' because no GPRS-specific information is exchanged at this point.) The SGSN is connected by the Gn interface to GGSNs belonging to its own public land mobile network (PLMN) and it is connected by the Gp interface to GGSNs belonging to other PLMNs. These two interfaces are very similar, but the Gp supports additional security functions that are necessary for inter-PLMN communications. The GGSN may also interface directly with the home location register (HLR) over the Gc interface, but this is not mandatory.

A SGSN keeps track of the location information and the security information associated with the MSs that are within its service area. A SGSN communicates with GGSNs and SGSNs in its own PLMN using the Gn interface and GGSNs in other PLMNs via the Gp interface. Interfaces also exist between an SGSN and an MSC/VLR (Gs interface),

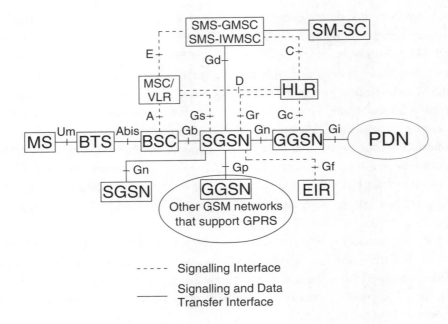

Figure 6.2: The GPRS network architecture.

an HLR (Gr interface), an EIR (Gf interface) and a short message service gateway MSC (SMS-GMSC) and interworking MSC (SMS-IWMSC) (Gd interface). The SMS-GMSC and SMS-IWMSC allow the GSM short message service (SMS) to be carried on the GPRS channels instead of by the SDCCH and the SACCH. The GPRS support nodes (i.e. the GGSNs and SGSNs) of a PLMN are interconnected using an Internet protocol (IP) based backbone network.

6.2.2.2 GPRS transmission plane

This is a layered protocol structure enabling user information transfer and associated control procedures such as flow control, error detection, error correction and error recovery. The GPRS transmission plane is shown in Figure 6.3 [18, 19]. We observe in Figures 6.2 and 6.3 the entities MS, BSS (namely BTS connected via the Abis to the BSC), SGSN and the GGSN; together with the interfaces Um, Gb, Gn and Gi. From the transmission plane we observe that at the highest level the application is leaving the network via Gi. The GPRS tunnelling protocol (GTP) tunnels user data and signalling between GPRS support nodes, SGSN and GGSN, in the backbone network. The transport control protocol (TCP) carries the GPRS tunnelling protocol (GTP) data units (PDUs) in the GPRS backbone network for protocols that need a reliable data link, such as X.25. The user datagram protocol (UDP) conveys GTP PDUs for protocols, such as Internet protocol (IP) that do not require reliable links. The TCP provides flow control and protection against lost and corrupted GTP PDUs. The GPRS backbone network protocol is IP, and is used for routing user data and signalling. The subnetwork dependent convergence protocol (SNDCP) maps network-level character-istics onto the lower layer network. The provision of a highly reliable ciphered logical link is achieved by the logical link control (LLC). The LLC is independent of the radio interface protocols.

In the BSS the relay function relays LLC PDUs between the radio interface Um and the Gb interface. In the SGSN the relay function relays packet data protocol (PDP) PDUs between the Gb and the Gn interfaces. The BSS GPRS protocol (BSSGP) conveys routing and quality-of-service (QoS) information between a BSS and an SGSN. It does not perform error control. The network service (NS) layer transports BSSGP PDUs. The radio link control (RLC) provides a radio dependent reliable link, while the medium access control (MAC) controls the access signallings, both request and grant, for the radio channel; and the mapping of the LLC frames onto the GSM physical channel. GSM RF is the GSM physical layer.

6.2.2.3   High-level functions required for GPRS

Network access control function   An access protocol has a set of procedures, such as registration, authentication and authorisation, admission control, charging and so on, that allow a user to use the services and facilities the network provides. A user may make an access attempt, or the user may be paged. The fixed network interface, Gi, may support, at the discretion of the PLMN operator, multiple access protocols to external networks such as X.25 and IP.

Anonymous access   The idea of anonymous access has also been introduced into GPRS whereby an MS can access the network without being authenticated and without air-interface encryption being required. There is also no requirement for the MS to supply its IMSI or IMEI, although there is provision for the network to request these. One example of an ap-plication that could make use of this facility is automatic road-tolling, whereby a road user could use a pre-paid card, inserted into a GSM terminal, automatically to make payment as she approaches a toll booth.

Packet routing and transfer functions   A route consists of an originating network node, and if required, relay nodes, and finally a destination node. Routing is the transmission of messages within and between PLMNs. A node forwards data received to the next node using the relay function. The routing function determines the network GPRS support node

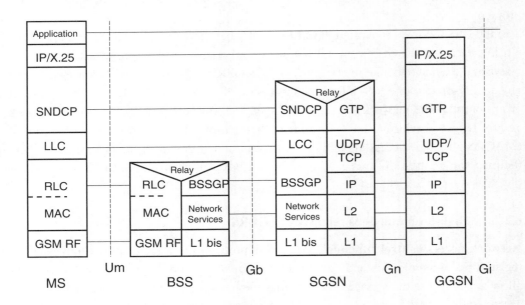

Figure 6.3: GPRS transmission plane [18–20].

(GSN) where the message is sent using the destination address in the message. The routing function also selects the transmission path to the next GSN along the route. Address translation is needed to convert one address to another when packets are routed between PLMNs. Encapsulation is the addition of address and control information to the PDU for routing within and between PLMNs. Encapsulation, and its reverse, are performed between the GGSN nodes of PLMNs, and between the SGSN and an MS. The tunnelling function (see the transmission plane) is the transfer of encapsulated PDUs within and between PLMNs from where they are encapsulated to where they are decapsulated. There is a compression function that removes as much overhead information as possible, prior to radio transmission. The ciphering function provides confidentiality of a user's data, while the domain name server function is a standard IP function that resolves any name for GSNs and other

GPRS nodes within the GPRS backbone networks.

GPRS mobility management and the MS states Mobility management describes the processes in the mobile radio system that are associated with tracking the movements of subscribers as they travel around a network. The non-GPRS GSM system uses location areas and location update procedures to ensure that it always knows the whereabouts of its MSs. The GPRS uses a similar approach, but instead of using the existing location areas (LAs) it uses routing areas (RAs). If an MS detects that it has entered a new RA it will submit a routing area update request message to the network, which if successful, will be acknowledged with a routing area update accept message. The decision as to whether a GPRS-equipped MS should perform the RA update procedure as it moves between RAs will depend on the GPRS state of the MS.

A GPRS-equipped MS may be in any one of three different states for the purposes of mobility management. In the *idle* state the MS is 'not reachable' as far as the GPRS is concerned. If an MS is in the idle state, the network does not hold any information regarding the location of the MS, and hence the MS cannot be paged. It also means that the MS does not need to perform any RA updates as it moves around the network. MSs in the idle state cannot access the packet data services without first performing a procedure known as *GPRS attach* where the MS announces its identity.

In the *standby* state the user is attached to the GPRS mobility management (MM). The MS performs RA and cell selection, and may receive point-to-multipoint multicast (PTM-M) and point-to-point multipoint group call (PTM-G) data. The SGSN may send data or signalling information to an MS. The MM state in the MS changes to *ready* when an MS responds to a page, as does the state in the SGSN when the response from the MS is received. The MM state in the MS goes to ready when data or signalling information is sent by the MS to the network, hence the MM in the SGSN also changes to the ready state. Either the MS or the SGSN may initiate the GPRS detach procedure to move to the idle state.

In the ready state, the MS informs the network of the selected cell by means of an identifier included in the BSSGP header of the data packet from the MS. The MS is able to send and receive PDP PDUs. The network initiates no pages for an MS in the ready state, but pages for other services may be executed via the SGSN. The ready state is supervised by a timer, and when the timer expires the MM moves from ready to standby. To move from ready to idle states, the MS initiates the GPRS detach procedure.

Logical link management functions These are involved with link establishment, maintenance and release procedures between an MS and the PLMN over the radio interface. These functions involve the co-ordination of link state information and the supervision of data activity over the logical link.

Radio resource management functions   Allocation and maintenance of radio communication channels are provided by these functions. The GSM radio resources are dynamically shared between the circuit mode and GPRS. The GPRS radio resource management is concerned with the allocation and release of timeslots for a GPRS channel; monitoring GPRS channel utilisation; congestion control; and the distribution of GPRS channel configuration information that is broadcast on the common control channels.

Network management functions   These support operation and management functions for GPRS.

6.2.2.4   Low-level functions

The radio interface functions for GPRS are the medium access control (MAC) and radio link control (RLC) that operate above the physical layer. The MAC function arbitrates between MSs attempting to transmit at the same time. It is therefore concerned with collision avoidance, detection, and recovery following a collision. The MAC function may let a single MS use several physical channels simultaneously. The multiplexing of data and control signalling on both links is affected by the MAC function, as well as by priority scheduling.

The GPRS RLC function supports the transfer of logical link control layer PDUs (LLC-PDU) between the LLC and MAC entities, the segmentation and reassembly of LLC-PDUs into RLC data blocks, and backward error correction for the retransmission of uncorrectable code words.

Packet data logical channels   Although circuit switched and GPRS services use the same physical channels, they have different logical channels. The physical channel dedicated to packet data is called a packet data channel (PDCH). The logical channels for common control signalling for packet data are carried by the packet common control channel (PCCCH), and there is also the packet random access channel (PRACH), an up-link-only channel used by MSs to initiate data or signalling packet transmission. The other PCCCHs are all down-link ones. There is the packet paging channel (PPCH) that uses paging groups of MSs to enable discontinuous reception (DRX) to be used. PPCH can be used for both circuit switched and packet services. The packet access grant channel (PAGCH) identifies the resource assignment to be used by an MS prior to packet transfer, while the packet notification channel (PNCH) provides notification to a group of MSs that a PTM-M packet transfer is imminent. The packet broadcast control channel (PBCCH) informs the MSs of packet data specific information. This information may also be transmitted on the BCCH if a PBCCH has not been allocated.

The packet data traffic channel (PDTCH) is allocated for data transmissions. It is temporarily allocated to an MS (or a group of MSs in the PTM-M case). An MS may be assigned multiple PDTCHs.

There are also packet dedicated control channels. The packet associated control channel (PACCH) conveys signalling information to a specific MS. The type of information includes acknowledgements, power control, resource assignment and reassignment. The PACCH and PDTCH share the resources allocated to an MS. An MS that is transferring a packet can be paged for circuit switched services on the PACCH. Another dedicated control channel is the packet timing advance control channel, up-link (PTCCH/U) that conveys in random access bursts the information to allow the network to deduce the necessary timing advance for the MS packet transmissions. There is also the packet timing advance control channel, down-link (PTCCH/D) that transmits timing advance information updates to MSs. One PTCCH/D is paired with several PTCCH/Us.

A number of PDCHs can share the same physical channel. They are mapped dynamically onto a 52-multiframe consisting of 12 blocks of four consecutive frames, two idle frames and two frames for the PTCCH, as shown in Figure 6.4. The 52-multiframe has a duration of 240 ms, namely two 26-multiframes of the GSM TCH. The first of the 12 blocks in the 52-multiframe is B0 and its PDCH contains PBCCH. On any PDCH with PCCH, up to 12 blocks can be used for PAGCH, PNCH, PDTCH or PACCH on the down-link. On the up-link PDCH that contains PCCCH, the blocks in the multiframe can carry the PRACH, PDTCH or PACCH. The mapping of channels onto multiframes is controlled by parameters broadcast on the PBCCH.

6.2.2.5 The GPRS MS classes

There are three classes of MS as far as the GPRS is concerned and these are based on the ability of the MS to support the simultaneous use of packet-based and circuit-switched services. The *Class A* MSs are able to support the simultaneous transfer of both packet-based and circuit-switched traffic using different timeslots within the GSM TDMA frame structure. The *Class B* MSs can be simultaneously 'attached' to both the circuit switched and packet-based services, e.g. they can receive pages for either service; however, they cannot transfer packet-based traffic and circuit-switched traffic at the same time. If, for example, a Class B MS is engaged in a packet-based data transfer and a circuit-switched connection is established, the transfer of packet data will be suspended for the duration of

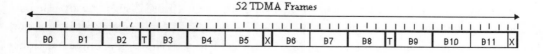

Figure 6.4: Multiframe structure for PDCH.

the circuit-switched connection. However, in this example there is no requirement for the GPRS to be deactivated for the particular MS. *Class C* MSs do not support any simultaneous use of circuit-switched and packet-based services. A Class C MS can be viewed as operating in two distinct modes, the GPRS mode and the circuit-switched mode, and it can only be in one mode at a particular time. For example, if the class C MS is in GPRS mode (i.e. it is engaged in a packet-based data transfer) and a circuit-switched call arrives at the PLMN for this particular MS, then the MS will be considered 'not reachable' as far as the circuit-switched call is concerned. It should be noted that the GPRS class of an MS and its multislot class (see Table 6.2) are separate parameters and there is no direct correlation between the two.

6.2.2.6 The GPRS quality-of-service

There are four different quality-of-service (QoS) profiles supported in GPRS and these are dependent on the type of service. For example, an email transaction can tolerate greater delays than, say, an interactive video service, and the QoS profile will be chosen appropriately in each case. The MS and the network will agree on a particular QoS profile during the initial service negotiation stages and, as far as possible, the network will attempt to deliver this QoS to the MS. The QoS service profile is made up of a number of factors, including the delay class (i.e. the average packet transmission delay), the precedence class (i.e. the priority value attached to the packets in the event of packet erasure being required as a result of network congestion), the reliability class (i.e. the probability of errors in the received data packet), and the peak and average throughput class (i.e. the peak and average rates at which data are transferred through the network, respectively).

6.2.3 The enhanced data rates for GSM evolution (EDGE)

The driving force behind EDGE is to improve the data rates of GSM by means of enhancing the modulation methods [14, 21]; specifically, to increase the data rate transmission per radio timeslot compared with GMSK modulation. Different types of enhanced modulation methods have been considered starting with quaternary offset quadrature amplitude modulation (Q-O-QAM) and binary offset quadrature amplitude modulation (B-O-QAM), and ending with 8-level phase shift keying (8-PSK). Although the initial drivers were to increase the user bit rates and thereby increase the range of services, EDGE has been gilded as a 3G system and is now a member of the IMT-2000 family. EDGE's new heady role has a lot to do with the evolutionary strategy of IS-136, the US TDMA system that is itself an evolution of the former analogue system, AMPS. The Universe Wireless Communications (UWC) Consortium advocated a family of mutually compatible TDMA standards known as UWC-136 that would be developed from the second generation IS-136. The concept was that voice services would be provided by the existing IS-136 network and by a new variant called 136+ which would employ enhanced modulation methods, such as $\pi/4$ differential quadrature phase shift keying ($\pi/4$ DQPSK), coherent QPSK, and coherent 8-PSK offering a maximum user rate of 43.2 kb/s. Higher rate services would be provided by 136 HS, which supports user rates up to 521 kb/s for pedestrian mobiles and for vehicular mobiles at speeds up to 100 km/h. For speeds in the range of 100–500 km/h, rates of 182 kb/s would be accommodated. In indoor environments, the maximum bit rate would be 4.7 Mb/s. For 136 HS the carrier spacing is 200 kHz (as in GSM) for outdoor/vehicular environments, and 1600 kHz for offices.

The 136 HS (outdoor/vehicular) with its 200 kHz carrier spacing, eight slots per frame, FDD, with multilevel modulation was part of the UWC-136 IMT EDGE proposal. The evolution of GSM to EDGE and IS-136 to EDGE is now undertaken jointly by ETSI and the UWC Consortium. Consequently, EDGE will be compatible with both GSM and IS-136. The plan [14] is to deploy GPRS, then enhanced GPRS (EDPRS) and enhanced circuit switch data (ECSD). Then the high level of modulation will be deployed to realise 3G EDGE services.

EDGE radio interface The radio interface will continue to have GMSK available, but will be able to use 8-PSK which has three bits per symbol instead of the one-bit type symbol of GMSK. Since the symbol rate is 271 ksymbols/s then the gross bit rates per slot (includes overhead) is 22.8 and 69.2 kb/s for GMSK and 8-PSK, respectively. The pulse shape for 8-PSK is such that the 8-PSK spectrum fits within the GMSK spectrum mask. The normal burst format for EDGE is the same as for GSM, except the two sets of 58 symbols now have three bits per symbol.

Packet switched transmission The EDGE concept has both packet switched and circuit switched modes. Indeed, EDGE is more like the grand evolutionary plan of GSM that includes both GPRS and HSCSD. The enhanced GPRS (EGPRS) differs from GPRS because with multilevel modulation the channel coding must be improved because it is more vulnerable to interference and noise. Accordingly a link adaption scheme regularly estimates link quality and selects GMSK or 8-PSK and the appropriate channel coding to provide the highest user bit rate. At the time of writing various schemes are being considered for standardisation. The coding rate is determined by the amount of puncturing. The rate per time slot for GMSK is 11.2, 14.5, 16.7 and 22.8 kb/s for code rates of 0.49, 0.64, 0.73, and 1.0, respectively. For 8-PSK the rate per time slot is 22.8. 34.3, 41.25, 51.6, 57.35 and 69.2 kb/s for code rates of 0.33, 0.50, 0.60, 0.75, 0.83 and 1.0, respectively.

The enhanced circuit switched (ECSD) mode has the data interleaved over 22 TDMA frames. For GMSK modulation the rate per time slot is 3.6, 6, 12 and 14.5 kb/s for a code rate of 0.16, 0.26, 0.53 and 0.64, respectively; while for 8-PSK the bit rates have the higher values of 14.5, 29, 32 and 38.8 kb/s for code rates of 0.42, 0.46, and 0.56, respectively.

For EGPRS when QoS issues are addressed where different factors, such as priority of packet transmissions, packet delay, packet throughput, maximum and minimum bit rates that must be handled, and so on, need to be considered, then the maximum bearer rate should be able to accommodate data rates of 384 kb/s for MS speeds up to 100 km/h and 144 kb/s for an MS travelling at 250 km/h. These rates require a user to make use of multiple slots per frame if these high bit rates are to be achieved. Fewer slots are required if 8-PSK is used. Similar comments can be made regarding bearer rates for ECSD. For low bit error rate transmissions the maximum user bit rate is 57.6 kb/s.

Because of different user requirements there is one class of MSs where 8-PSK is used in the down-link and only GMSK in the up-link. This provides higher bit rate down-link transmissions than the up-link ones while at the same time decreasing the complexity of the MS. Another class of MS will support 8-PSK transmissions on both links.

As with all the 3G standards, we await with interest to see how well they will perform when they are deployed in operational networks.

6.3   The Universal Mobile Telecommunication System

The universal mobile telecommunication system (UMTS) is, at the time of writing, being shaped within the Third Generation Partnership Project (3GPP) [22]. The participants have come together for the specific task of specifying a 3G system based on an evolved GSM core network and the UTRA FDD and TDD radio interfaces. The 3GPP is composed of organisational partners, market representation partners and observers. The organisation partners, i.e. standards organisations, are: ARIB (Japan), CWTS (China), ETSI (Europe), TI (USA), TTA (Korea) and TTC (Japan). The market representation partners are: Global Mobile Suppliers Association (GSA), the GSM Association, the UMTS Forum, the Universal Wireless Communications Consortium (UWCC), and the IPv6 Forum. The observers are TIA (USA) and TSACC (Canada). The 3GPP activity is overseen by a project coordination group (PCG). The specifications are developed by four technical specification groups (TSGs) responsible for the core network, the radio access networks, services and system aspects, and terminals. Each TSG has a number of working groups. There will be a roll-out of the specifications; the initial release of the specifications is Phase 1 Release 99. New capabilities and services will be introduced according to annual specification releases.

The UMTS terminology introduces a number of new terms, and re-names some familiar ones. Many of these new terms will be defined as they appear in the text, but to assist the reader give a list of UMTS abbreviations in Table 6.3. We also draw the reader's attention to some familiar GSM terms that are different in UMTS. These are presented in Table 6.4

Table 6.3: UMTS abbreviations.

ACLR	Adjacent channel leakage power ratio
AI	Acquisition indicator
AICH	Acquisition indication channel
BCH	Broadcast control channel
CCPCH	Common control physical channels
CPCH	Common packet channel
CPICH	Common pilot channel
DPCCH	Dedicated physical control channel
DPDCH	Dedicated physical data channel
DCH	Dedicated channel
FACH	Forward access channel
FBI	Feedback information
FDD	Frequency division duplex
GMSC	Gateway MSC
GGSN	Gateway GPRS support node
I_{uCS}	Interface between an RNC and an MSC
I_{uPS}	Interface between an RNC or BSC and an SGSN
I_{ur}	Interface between RNCs
MAC	Medium access control
MSC	Mobile switching centre
MUD	Multiuser detection
Node B	Base station transceiver
OVSF	Orthogonal variable spreading factor
P-CCPCH	Primary common physical channel
PCH	Paging channel
PCPCH	Physical common packet channel
P-CPICH	Primary CPICH
PI	Paging indicator
PICH	Pilot channel
PRACH	Physical random access channel
PSC	Primary synchronisation code
QPSK	Quadrature phase shift keying
RACH	Random access channel
RNC	Radio network controller (like BSC in GSM)
RNS	Radio network subsystem
S-CCPCH	Secondary common control physical channel

S-CPICH	Secondary CPICH
SCH	Synchronisation channel
SF	Spreading factor
SGSN	Serving GPRS (generalised packet rate service)
SSC	Secondary synchronisation code
TFCI	Transport format combination indicator
TPC	Transmit power control
UARFCN	UTRA absolute radio frequency channel number
UE	User equipment
UMTS	Universal mobile telecommunication system
USIM	Universal subscriber identity module
UTRA	UMTS terrestrial radio interface

The UMTS network architecture block diagram is displayed in Figure 6.5. The core network is encased by a dotted line. The mobile switching centre (MSC) and gateway MSC (GMSC) are for circuit-switched GSM networks. Because GSM Phase 2+ will also accommodate GPRS, and therefore handle packet data, there is a serving GPRS support node (SGSN) and a gateway GPRS support node (GGSN). The other core network elements to do with authentication, home and visitor location registers and equipment identity registers are essential to support both circuit-switched and packet data networks. Thus, the core network is architecturally a GSM Phase 2+ core network that is powered up so that it can also handle the higher volume, higher bit rate, UMTS traffic.

Shown below the core network in Figure 6.5 are two GSM base station subsystems (BSSs) and two UMTS radio network subsystems (RNSs). The A-interface is between a base station controller (BSC) and a mobile switching centre (MSC), and there is an I_{uPS} between a BSC and SGSN, where the subscript uPS signifies a packet switch interface. The A_{bis} interface between a BTS and a BSC is also shown.

The UMTS network uses the same core network as GSM, and has interfaces between the RNC and MSC, SGSN and RNC of I_{uCS}, I_{uPS} and I_{ur}, respectively. The subscript uCS

Table 6.4: GSM and UMTS terminologies of some key entities.

GSM	UMTS
Mobile station (MS)	User equipment (UE)
Base station transceiver (BTS)	Node B
Base station controller (BSC)	Radio network controller (RNC)
Base station subsystem (BSS)	Radio network subsystem (RNS)
Subscriber identity module (SINM)	Universal subscriber identity module (USIM)

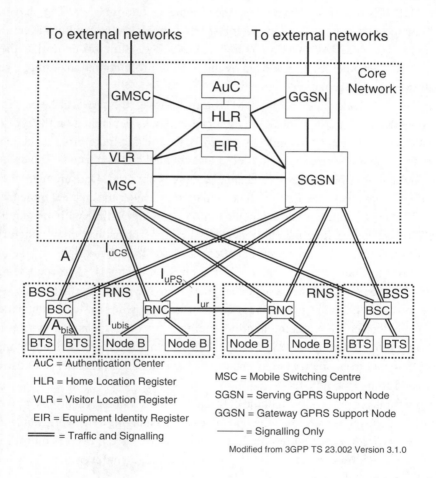

Figure 6.5: UMTS network architecture.

represents circuit-switched. The equivalent of the A_{bis} in UMTS is I_{ubis}.

This figure is important because it illustrates the evolutionary path from GSM Phase2+ to UMTS. We see the different radio interfaces of GSM and UMTS plugging into a common backbone network.

The physical channels in UMTS transfer information across the radio interface. In the FDD version of UMTS, a physical channel is defined by its code and carrier frequency, while in TDD it is in terms of its code, carrier frequency and timeslot. The physical layer (layer 1) of the protocol stack supports transport channels to the medium access control (MAC) in layer 2. The MAC layer offers logical layers, e.g. radio link control, to the higher layers. A logical channel is characterised by the type of information transferred, e.g. it may be handling control information of data traffic.

Figure 6.6 is a block diagram of a UMTS transmitter at the physical layer. Transport channel data from layer 2 and above are arranged in blocks depending on the type of data. These blocks are cyclically redundancy coded (CRC) to facilitate error detection at the receiver. The data are segmented into blocks and channel coding ensues. The coding may be convolutional or turbo. Sometimes channel coding is not used. Data are then interleaved to decrease the memory of the radio channel and thereby render the channel more Gaussian-like. The interleaved data are then segmented into frames compatible with the requirements of the UTRA interface. Rate matching is performed next. This uses code-puncturing and data repetition, where appropriate, so that after transport channel multiplexing the data rate is matched to the channel rate of the dedicated physical channels. A second stage of bit interleaving is executed, and the data are then mapped to the radio interface frame structure.

Suffice to say at this point is that there are different types of physical channels, namely pilot channels that provide a demodulation reference for other channels; synchronisation channels that provide synchronisation to all UEs within a cell; common channels that carry information to and from any user equipment (UE); and dedicated channels that carry information to and from specific UEs.

The physical layer procedures include cell search for the initial synchronisation of a UE with a nearby cell; cell reselection which involves a UE changing cells while not engaged in a call; access procedure that allows a UE to initially access a cell; power control to ensure that a UE and a BS transmit at optimum power levels; and handover, the mechanism that switches a serving cell to another cell during a call.

6.3.1   The UTRA FDD mode

The UMTS terrestrial radio interface (UTRA) frequency duplex (FDD) mode is the W-CDMA radio interface of the UMTS, and is designated by the ITU as IMT DS. Referring to Table 6.1, the UTRA FDD mode uses segment 3 for up-link transmission, and segment 6 for down-link transmission, i.e. from node B to UE communications. These two segments are

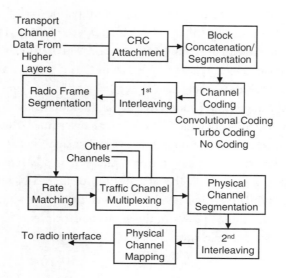

Figure 6.6: Block diagram of a UMTS transmitter at the physical layer.

called band A. (There is a band B for PCS services in the United States, where 1850–1910 and 1930–1990 MHz are for the up-link and down-link transmission, respectively).

For the UTRA FDD the duplex spacing, i.e. the frequency separation between paired channels, is in the range 134.8 MHz to 245.2 MHz, and all UE must support a duplex spacing of 190 MHz. The nominal spacing between radio carriers is 5 MHz, with a channel raster of 0.2 MHz. This means that the carrier separation may be adjusted in steps of 0.2 MHz, e.g. the carrier spacing may be 4.8 MHz.

The carrier frequency is defined by the UTRA absolute radio frequency channel number (UARFCN). This number is defined over a frequency band from 0 to 3.7 GHz, and it is the transmission frequency multiplexed by five. Consequently, the UARFCN, which we denote as N_u and N_d for the up-link and down-link, respectively, will always be an integer because of the raster frequency of 0.2 MHz. Note that the radio channels in the UTRA FDD are not necessarily paired as they are in GSM.

Transmitted spectrum The CDMA chip rate is 3.84 Mchips/s, and the transmitted spectrum must conform to, i.e. be within, a spectrum mask [23]. As an example, the mask for a BS with a maximum output power of ≥ 43 dBm is shown in Figure 6.7. The abscissa is the frequency offset, $\triangle f$, from the carrier. The carrier is at $\triangle f = 0$, not shown in the figure, and the adjacent carrier would be positioned at $\triangle f = 5$ MHz. The lightly shaded part of the figure corresponds to the left-hand ordinate, which is the power measured by a spectral analyser using a bandwidth of 30 KHz. The maximum output power as a function of $\triangle f$ between 2.7 and 3.5 MHz must be less than $-14 - 15(\triangle f - 2.7)$ dBm. The darker shaded area of the figure is associated with the right-hand ordinate, which is the power measured in a 1 MHz bandwidth. The value of $\triangle f_{max}$ is 12.5 MHz, or the edge of the allotted frequency band for UMTS, whichever is the greater.

The adjacent channel leakage power ratio (ACLR) is the ratio of the transmitted power to the power measured in an adjacent band following a receiver filter. Both the transmitted and received power are measured at the output of a raised cosine filter with a 0.22 roll-off factor and a noise power bandwidth equal to the chip rate. For an adjacent channel offset of 5 MHz. i.e. $\triangle f = 5$ MHz in Figure 6.7, the ACLR limit for the BS is 45 dB less than the transmitted signal power, while for a UE it is 33 dB down on the transmitted power of the adjacent carrier, or if the transmitted power of the adjacent carrier is -50 dBm, whichever is the higher. The corresponding figures for $\triangle f = 10$ MHz are 50 dB for the BS, and 43 dB or -50 dBm, whichever is the greater, for the MS.

Physical channels   We have already introduced the concept of UTRA physical channels, and that for UTRA FDD they are defined by a specific carrier frequency and code. There are two basic types of physical channels, called *dedicated channels* and *common channels*. The former are used by UEs for the duration of a call, while the latter carry information to all UEs within a cell and are used by the UEs to access the network.

There are two types of dedicated physical channels. The dedicated physical control channel (DPCCH) carries physical layer (i.e. layer 1) control information and the dedicated physical data channel (DPDCH) transports the user traffic, as well as control information from layer 2 and from higher layers.

The DPCCH contains pilot symbols, transmit power control (TPC) symbols, and a transport format combination indicator (TFCI). The pilot symbols enable the receiver to estimate the impulse response of the radio channel and to perform coherent detection. They are also necessary when adaptive antennae are used that have narrow beams. The pilot symbols constitute a pilot word having a duration of 0.667 ms.

The TPC commands the fast closed-loop power control, and are used on both the up-link and down-link. TPC symbols are included in every transmitted packet, and they convey a binary instruction, namely to increase or decrease the transmitted power by a specific amount.

The TFCI informs the receiver of the instantaneous parameters of the different transport channels; that is, it tells the receiver of the data rates currently in use. The TFCI also contains feedback information (FBI) on the up-link which is used to provide a feedback loop for transmit diversity and selection diversity.

The frame structure, multiplexing arrangement and spreading schemes are different for

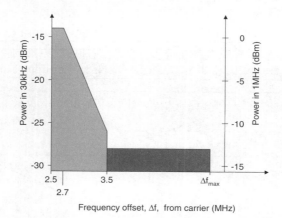

Figure 6.7: Transmitted spectrum mask for a UMTS BS.

the up-link and the down-link. The DPCCH and DPDCH are time multiplexed i.e. the DPCCH and DPDCH are multiplexed within a slot prior to spreading and transmission at a BS. By contrast the up-link transmissions from a UE have the DPDCH and DPCCH transmitted in parallel on the quadrature components of the dual-channel QPSK modulator.

6.3.1.1   The UTRA FDD down-link

The down-link frame structure is shown in Figure 6.8. The duration of a frame is 10 ms, and contains 15 slots. Each slot has a duration of 0.667 ms and contains 2560 chips as the chip rate is 3.838 Mchips/s. The actual data per slot is 10×2^k, $k = 0, 1, \ldots, 7$, namely 10 to 1280 bits, where k is a system parameter. The spreading factor is

$$SF = \frac{512}{2^k} \qquad (6.1)$$

and since k is an integer with a value from 0 to 7, the SF has a value of either 4, 8, 16, 32, 64, 128, or 256.

We observe in Figure 6.8 that the ith slot contains sub-slots. First there is the DPDCH containing N_{data1} bits of data 1, then comes a DPCCH having N_{TFCI} bits of TFCI, followed by the next DPCCH having N_{TPC}, followed by a second DPDCH having N_{data2} bits of data 2, and finally the last DPCCH with N_{pilot} bits of pilot signal. This means that when two adjacent frames are considered, the pilot bits are placed between data 2 bits and data 1 bits. We also note that, unlike GSM where sounding is executed every 4.6 ms, UTRA sounds the radio channel every 0.667 ms. The greater sounding rate of UTRA is necessary because of the higher bit rate services supported by UTRA compared with those in GSM.

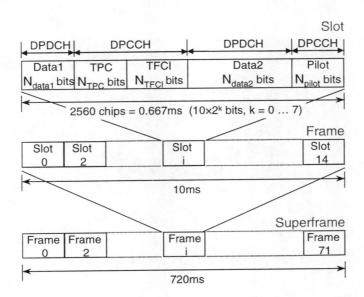

Figure 6.8: Down-link slot and frame structure for dedicated channels.

There are 17 different down-link slot formats that relate to the channel bit rate. For example, if the bit rate is 15 kb/s then $N_{data1} = 2, N_{data2} = 2, N_{TFCI} = 0, N_{TPC} = 2$, and $N_{pilot} = 4$, totalling 10 bits in a slot, and corresponding to $k = 0$, $SF = 512$. We see that in this situation there are only four bits out of 10 bits carrying data, corresponding to an overhead of 60%! For a channel bit rate of 1920 kb/s (slot format 16), $N_{data1} = 240, N_{data2} = 1008, N_{TFCI} = 8, N_{TPC} = 8$ and $N_{pilot} = 16$, giving a total of 1280 bits per slot and $SF = 4$. The overhead now is only 2.5%.

All the down-link slots contain pilot symbols, and the pilot bit patterns are defined for N_{pilot} values of 2, 4, 8 and 16 bits. For a particular N_{pilot}, the logical values of the bits change from slot to slot creating a pilot pattern that repeats for each 15-slot frame. By this procedure the pilot symbols affect both frame synchronisation as well as channel estimation.

The TPC field has $N_{TPC} = 2, 4$ or 8 bits, and these N_{TPC} bits are either all logical 1s or all logical 0s, depending on whether the power control command is to have the UE increase or decrease its transmitter power by the step size, respectively.

The TFCI provides information about the data rates currently being used on the data channels, e.g. in the case of multiple simultaneous services. The TFCI may be omitted, e.g. in the presence of fixed rate services. The UE will know the down-link slot format, and hence that $N_{TFCI} = 0$. It will also know the value of the SF, the pilot pattern, and N_{data1}, N_{data2}, and N_{TPC} (but not their logical values). The UE must apply blind detection in which it must try different decoding strategies that are likely to be the inverse of those used in conditioning the data. The decoding strategies that provide the lowest bit error rate (BER) is deemed to be the correct one.

We have seen that the DPCCH carries the TFCI, TPC and pilot bits on a time division basis within each slot, while the DPDCH carries data 1 and data 2 also on in a time division mode within a slot. Further more, both the DPCCH and DPDCH are multiplexed together. The DPCCH and DPDCH signals are converted from a serial stream to two parallel streams, one called the inphase component and the other the quadrature component. At this point although they are designated as inphase and quadrature components, they are, in fact, of the same phase. Both bit streams are then spread to 3.84 Mchips/s using the same code, known as the *channelisation code*, $C_{ch,SF,n}$, and $SF = 2^n$. Note that if a bit from the converter is a logical 0, then the code replaces the bit, and if the bit is a logical 1, then the code is inverted. The SF must be chosen so that the chip rate is always 3.84 Mchips/s. As an example, if the combined DPDCH/DPCCH is 960 kb/s, then the I and Q branches will operate at 480 kb/s, and the SF is 3.84 Mchips/s divided by 480 kb/s, namely 8.

If the required data rate exceeds the capacity of a single DPDCH channel, then additional DPDCHs can be added. Suppose N DPDCHs are required, then we still need only one DPCCH, but each DPDCH requires its own channelisation code. The arrangement is shown in Figure 6.9 [24], where the DPCCH is multiplexed with DPDCH$_1$. The absence of any control data on DPDCH$_i$, $i = 2,3,\ldots,N$, means that transmissions are suspended during those portions of the slots when DPCCH transmission would normally occur. The down-link slot format for these multicode transmissions is displayed in Figure 6.10 [25].

The so-called inphase and quadrature spread signals from each set of multipliers in Figure 6.9 are applied to adders to create the so-called inphase signal I and quadrature signal Q for multicode transmissions. The Q signal is shifted through 90° degrees so that the I and Q signals are now really in quadrature, and on adding these signals we have $I + jQ$. This resulting signal is scrambled by a code, C_{scramb}, but before discussing this code we will describe the channelisation codes.

Channelisation codes Channelisation codes are orthogonal variable spreading factor (OVSF) codes that identify the down-link channel. They are basically Walsh codes of different lengths that are able to preserve orthogonality between channels even when they are operating at different data rates. The OVSF codes are arranged in a tree structure for code allocation purposes. The OVSF code tree is shown in Figure 6.11. For a spreading factor of $SF = 1$ there is a channelisation code $C_{ch,1,0} = (1)$, i.e. a word of one bit that is a logical 1. For $SF = 2$, there are two codes $C_{ch,1,0} = (1,1)$ and $C_{ch,2,1} = (1,-1)$. Doubling the SF to 4 gives four codes: $C_{ch,4,0} = (1,1,1,1); C_{ch,4,1} = (1,1,-1,-1); C_{ch,4,2} = (1,-1,1,-1)$; and $C_{ch,4,3} = (1,-1,-1,1)$. We observe that $C_{ch,4,0}$ is $C_{ch,2,0}$ followed by $C_{ch,2,0}$, i.e. the $C_{ch,2,0}$ and its repeat, whereas $C_{ch,4,1}$ is $C_{ch,2,0}$ followed by its $C_{ch,2,0}$ but with its bits inverted. In

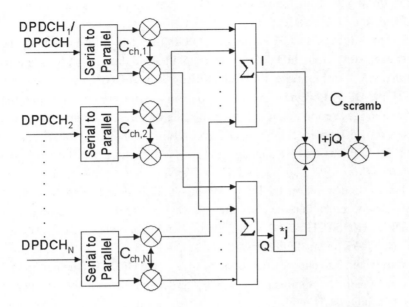

Figure 6.9: Down-link spreading arrangement for multicode transmission.

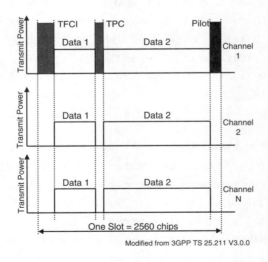

Figure 6.10: Down-link slot format for multicode transmission.

general

$$C_{ch,2^{(n+1)},2i} = (C_{ch,2^n,i}C_{ch,2^n,i}) \tag{6.2}$$

and

$$C_{ch,2^{(n+1)},2i+1} = (C_{ch,2^n,i}, -C_{ch,2^n,i}). \tag{6.3}$$

For an additional code to be used in a cell the following criterion is applied: none of the codes in the path from the target code to the root of the tree are already in use, nor are any of the codes in use in the branches above the target code. Figure 6.12 is an example showing which codes are available for selection if currently only one code is in use. As an illustration in support of this criterion, suppose a cell is using $C_{8,4}$. If it then decides to use the longer code $C_{16,8}$ the cross-correlation between two concatenated $C_{8,4}$ codes and the $C_{16,8}$ code is non-zero and hence there will be interference if these codes are used. By selecting $C_{8,3}$ and $C_{16,8}$ on different branches the codes are uncorrelated. This situation is illustrated in Figure 6.13.

Observe that UMTS has codes of different lengths simultaneously in use in a cell, depending on the services being handled. Good code allocation is essential if a large number of codes are to be available. For example, if all the short codes are used, then no long codes are available, thus restricting capacity. Care must also be exercised in selecting codes to prevent the crest factor, i.e. the ratio of the peak-to-mean transmitted power, from becoming too high. The crest factor is allowed to be up to 18 dB. We recall that in the presence of multiple users the amplitude values of the I and Q signals are dependent upon the sum of the user codes. If during any chip interval the codes sum to a large number, then the peak power becomes large. This makes it necessary for the transmitter equipment, i.e. the power amplifier and filters, to be specified for these surges in power. A low crest factor eases equipment design and cost.

Scrambling code Walsh codes have poor autocorrelation functions and, in the presence of fading, non-zero cross-correlation functions. This renders them unsuitable for multiple access codes. Since the channelisation codes are Walsh codes of differing lengths it follows that scrambling these codes by another code having good autocorrelation and cross-correlation codes is essential. This arrangement is used in cdmaOne. The scrambling code employed in UTRA FDD is a 38 400-chip segment of a $2^{18} - 1$ length Gold code. The scrambling codes have inphase and quadrature components, and there are a total of 8192 codes. For the 3.84 Mchips/s transmitted chip rate, the 38 400 chip code lasts 10 ms, i.e. the duration of the scrambling codes lasts for one 15-slot frame.

The 8192 codes are divided into 512 sets, with each set having 16 codes. These 16 codes are composed of a primary scrambling code, and 15 secondary scrambling codes. Eight sets

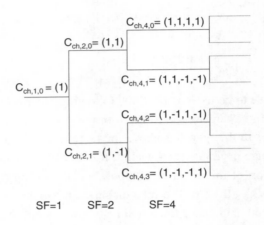

SF=1   SF=2   SF=4

Figure 6.11: OVSF code tree.

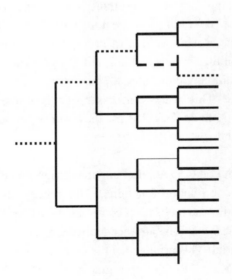

Figure 6.12: Channelisation codes. – – – code in use, ······ unavailable codes, ———— available codes.

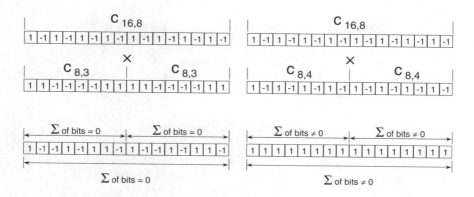

Figure 6.13: Code allocation is branch sensitive. (a) Codes $C_{16,8}$ and $C_{8,4}$ are on the same branch and must not be used; (b) codes $C_{16,8}$ and $C_{8,3}$ are permissible.

are formed (having $8 \times 16 = 128$ codes) to give a group of codes, and there are 64 groups. Each cell is allocated only one primary scrambling code. The scrambling code hierarchy is shown in Figure 6.14.

The block diagram for the down-link spreading and modulation is shown in Figure 6.15. The two outputs from the serial-to-parallel converter are multiplied by the channelisation code, C_{ch}, followed by multiplication by the complex scrambling code C_{scramb}. The output of each multiplier is low pass filtered (LPF) by filters having a raised cosine impulse response with a frequency roll-off of 0.22. They are then multiplied by the quadrature carriers $\cos(\omega_c t)$ and $\sin(\omega_c t)$ and the quadrature signals added to give the transmitted carrier signal, where ω_c is the angular carrier frequency. The arrangement is quadrature phase shift keying (QPSK) modulation.

Down-link common physical channels   These are: the common pilot channel (CPICH) that provides a common demodulation reference over all or part of a cell; the primary common control physical channel (P-CCPCH) that carries general network information; the secondary common control physical channel (S-CCPCH) for paging and packet data; the

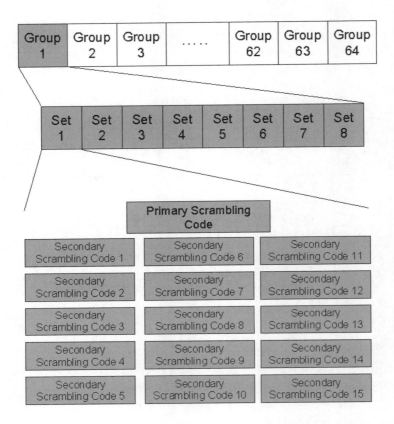

Figure 6.14: Scrambling code hierarchy.

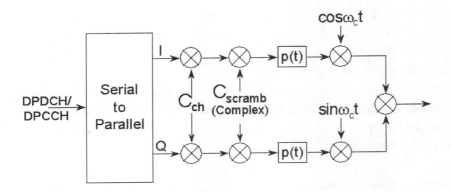

Figure 6.15: Down-link spreading and modulation.

synchronisation channel (SCH) that a UE uses in its initial cell search; and the acquisition indication channel (AICH) that controls the use of common up-link channels.

Common pilot channel   The CPICH operates at 30 kb/s with a spreading factor of 256. It has a predefined bit sequence, which for a single transmit antenna is an all logical 1 sequence. There are two types of CPICH: primary and secondary. The P-CPICH provides a coherent reference to obtain the SCH, P-CCPCH, AICH and PICH at the UEs, as these channels do not carry their own pilot information. The channelisation code used by the P-CPICH is $C_{ch,256,0}$, an all logical 1 code, while its scrambling code is the cell's primary scrambling code. In the case of a single transmit antenna, the CPICH is the unmodulated primary scrambling code. There is only one P-CPICH in each cell, and is broadcast over the entire cell.

A UE estimates the channel impulse response from the received pilot signal, and armed with this response the data may be recovered. The procedure for achieving this is described in Section 4.2.2.6. Thus the pilot and data must be transmitted over the same radio channel (which includes the transmitter and receiver antennae). Consequently, since the CPICH is transmitted over the entire cell or sector, it cannot be used to recover data from a narrow beam of a smart antenna because the radio channels for the data and CPICH may be very different. A smart antenna with its narrow beams will create radio channels with few or no significant multipath components, unlike a wide angle beam.

The secondary common pilot channel (S-CPICH) provides a common coherent reference within part of a cell or sector. The antennae have narrow beams, e.g. from a smart antenna, and may be used to target individual UEs or groups of UEs in close proximity to one another. The node B (a BS) may use any channelisation code having a length of 256 chips. The S-CPICH may be used as reference for the S-CCPCH (which transmits paging messages) and the down-link dedicated channels.

Common control physical channels   Having described the CPICH, let us now consider the next down-link common physical channels on our list, namely the common control physical channels (CCPCH). The P-CCPCH is a 30 kb/s down-link channel that carries the broadcast control channel (BCH). It is transmitted continuously over the entire cell. The spreading factor of the P-CCPCH is 256, and it is always scrambled by the cell's primary scrambling code. The P-CCPCH does not have any TPC, TFCI or pilot bits. It occupies 90% of a slot as shown in Figure 6.16. (The first 256 chips, marked no transmission of the P-CCPCH, carry the synchronisation channel, a channel that occupies only 10% of a slot.) Observe that the P-CCPCH is transmitted in every slot, although there will generally be different BCH data in each slot.

The S-CCPCH transmits paging messages as and when required. Data rates range from

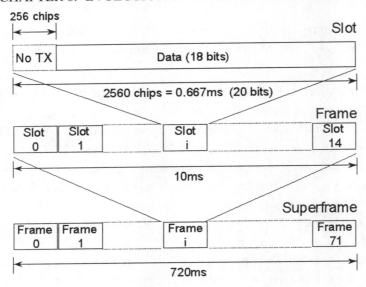

Figure 6.16: Slot and frame structure of the primary common control physical channel.

30 kb/s to 1920 kb/s, corresponding to spreading factors of 256 to 4, respectively. The data rate can be varied as required using the TFCI. The S-CCPCH is scrambled by either the primary or secondary scrambling code, and can be transmitted with or without its own pilot bits. Packet data may be sent to particular UEs via the S-CCPCH. The slot and frame structures for the S-CCPCH are displayed in Figure 6.17. The S-CCPCH occupies a complete slot, and is transmitted in every slot. However, the S-CCPCH is only transmitted when there is information to send, unlike the continuous transmission of the P-CCPCH.

Page indication channel   The page indicator channel (PICH) is always associated with a paging channel (PCH) on an S-CCPCH. The PICH carries page indicators (PIs), where a PI indicates to a subset of UEs within a cell whether they should examine the next S-CCPCH frame for paging messages. Each UE is associated with a particular PI. The PICH uses a spreading factor of 256, and its 10 ms frame has 15 slots (i.e. the basic structure). Since each slot contains 20 data bits, there are 300 bits per frame. The PICH uses only 288 bits, leaving 12 bits unused. In one frame the PICH may carry 18, 36, 72 or 144 PIs, which means that each PI consists of 16, 8, 4 or 2 bits, respectively. The PIs occupy a fixed position within a frame, depending on the number of PIs per frame. Figure 6.18 is an example when the PICH carries 18 PIs per frame, with 16-bit PIs. All 16 bits will be set to a logical 1 if the PI is set, i.e. a paging message is being sent on the PCH, otherwise all the 16-bit PIs contain logical 0s.

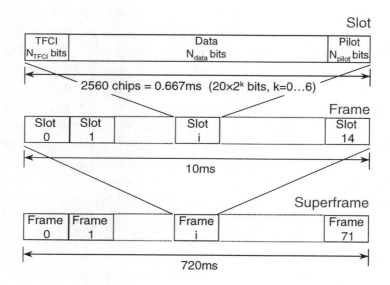

Figure 6.17: Slot and frame structure of the secondary common control channel.

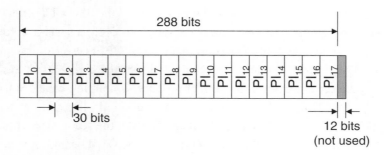

Figure 6.18: Page indication channel frame showing an example of 18 page indicators.

The synchronisation channel   A cardinal function in any communication system is synchronisation. In UTRA the synchronisation channel (SCH) is used by UEs in the initial cell search process. In Figure 6.16 the primary CCPCH occupies 90% of each slot. The SCH is transmitted in the first 10% of these slots, during 66.67 μs by a 256-chip word. There are two sub-channels of the SCH, the primary and secondary synchronisation channels. Figure 6.19 shows the primary and secondary SCH over a 15-slot frame. The primary synchronisation code (PSC) (shown as C_p in the figure) is the same in every slot, and is the same in every node B, although the node Bs are not synchronised. The PSC is a generalised hierarchical Golay sequence, chosen to reduce the complexity of the cell search procedure. It has good autocorrelation properties as can be seen with reference to Figure 6.20.

The PSC is generated by modulating a 16-chip code running at 3.84 Mchips/s by another 16-chip code generated at 240 kchips/s. The result is a 256-chip sequence at 3.84 Mchips/s whose autocorrelation function can be rapidly found, as we will describe in the cell search procedure. A UE receiver utilises the PSC initially to detect the presence of a nearby BS and then to identify the start of each timeslot.

The secondary SCH (shown as $C_s^{i,1}$, $j = 0, 1, \ldots$, 14 in Figure 6.19) is transmitted simultaneously with the primary SCH during the first 256 chips of every slot. The BS transmitter arrangement is shown in Figure 6.21. The basic function of the secondary SCH is to enable the identity of the scrambling code group used by the BS to be determined by a UE. We observe in Figure 6.19 that there are 15 different 256-chip secondary synchronisation codes (SSCs), i.e. the SSC changes from slot to slot such that the SSCs in a frame constitute a predefined sequence that is associated with the scrambling code *group* used by the cell; see Figure 6.14. Figure 6.22 shows an example of the SSC sequence for scrambling code group 1.

There are 64 scrambling code groups, and with the aid of the SSC a UE knows the actual group number. The group identified has eight sets, and each set has a unique primary scrambling code. This code is also the pilot code. The UE receiver cross-correlates the pilot code with all the eight primary scrambling codes of the eight sets in the group. By this means the receiver determines the correct primary scrambling code. Since this code also scrambled the BCH data, this data can now be recovered. The main goal of the SSC is thereby achieved. The BCH data, include, if appropriate, the secondary scrambling code being used. The secondary scrambling codes are used when spot beams are employed.

The PSC, Cp, and the SSC sequence $C_s^{i,j}$, $i = 1, \ldots, 512$ and $j = 0, 1, \ldots, 14$, are added to the other down-link channels. The Cp and $C_s^{i,j}$ are not subjected to multiplication by either channelisation or scrambling codes.

The acquisition indication channel   The acquisition indication channel (AICH) carries acquisition indicators (AIs). An AI is part of the random access procedure which indicates

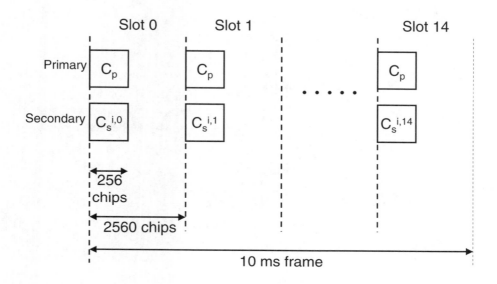

Figure 6.19: The synchronisation channel slot and frame structure.

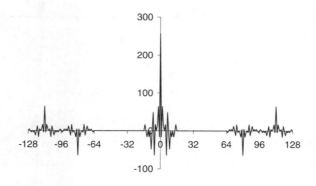

Figure 6.20: Primary synchronisation code autocorrelation function.

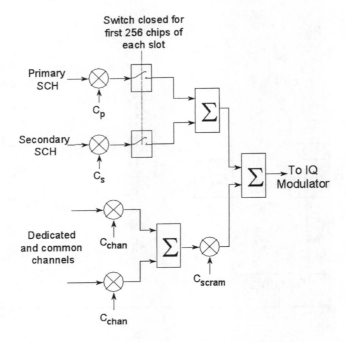

Figure 6.21: BS transmitter block diagram.

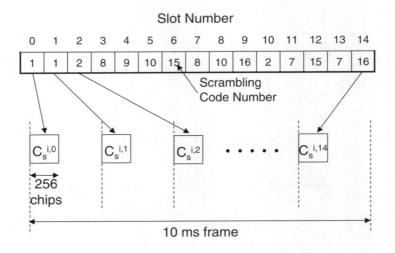

Figure 6.22: The SSC sequence for scrambling code group 1.

to a UE that it should proceed with the transmission of its access request. The AI can take three different values: a 0, which causes the UE to transmit with more power until it gets a -1 or $+1$ response; a -1, which instructs the UE to discontinue with the access attempt; and a $+1$ which enables the UE to go to the next stage in the access procedure. Observe that the three states of 0, -1, and $+1$ result in no code being sent, the code sent but inverted, and the code sent, respectively.

The AICH has a 15-slot frame, whose duration is 20 ms rather than the usual 10 ms. Accordingly each slot has a duration of 1.33 ms, equivalent to 5120 chips. The first 4096 chips correspond to 16 AI symbols as the spreading factor is 256. The last 1024 chips are four zero AI symbols. Figure 6.23 shows the slot and framing arrangement.

The AI part of the slot has 16 orthogonal code words which have been multiplied by the AI. Each code word corresponds to one of 16 different Walsh codes, each of 256 chips. These codes are called *preamble signature codes*. (We will discuss these codes later when we come to the access procedures.) Figure 6.24 shows the 16 preamble signature codes $W_1, W_2, \ldots, W_{16}$. Each code will be associated with a particular UE, and this code is multiplied by the appropriate AI, e.g. $W_2 \times AI_2$, resulting in either no output, or all the chips inverted in W_2, or just W_2, depending on whether AI_2 is 0, -1 or $+1$, respectively. Up to 16 $W_i AI_i, i = 1, 2, \ldots, 16$, are summed and loaded into the AICH slot. The switch setting in Figure 6.24 for the last four AI symbols introduces logical 0s, which means that no codes are transmitted. Consequently, the AICH slot has data corresponding to 16 symbols arranged over the first 80% of the slot.

Transport channels   Having discussed a range of physical channels, we now describe the down-link transport channels that define the manner in which information is mapped onto the physical channels. There are two categories of transport channels: the dedicated channel (DCH), and the common transport channels.

There is only one DCH, and it has the ability rapidly to change the transmitted data rate, if necessary, every 10 ms. It also supports fast power control, and provides addressing of UEs. The DCH is always mapped onto the DPDCH, and may be transmitted over the entire cell, or to zones illuminated by narrow beam antennae.

The situation is more complex for common transport channels. The broadcast channel (BCH) is a point-to-multipoint channel that is used to broadcast network and cell-specific information to all UEs within a cell. It is mapped onto the P-CCPCH. The forward access channel (FACH) carries small amounts of information such as short packets of user data. The FACH does not need to be transmitted over the entire cell, e.g. if smart antennae are used, and it is mapped onto the S-CCPCH. The paging channel (PCH) is transmitted over the entire cell and conveys pages to the UEs. It uses the S-CCPCH.

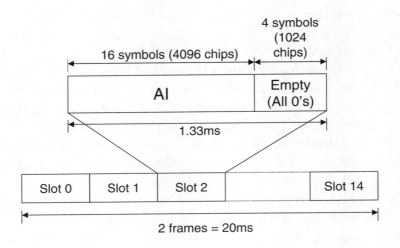

Figure 6.23: The acquisition indication channel slot and framing structure.

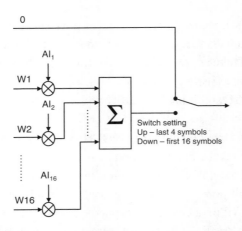

Figure 6.24: The acquisition indication channel arrangement.

6.3.1.2 The UTRA FDD up-link

Communications from a UE to a BS, i.e. a node B, employ a radically different slot structure compared with that used on the down-link. On the up-link the DPDCH and the DPCCH are carried on the inphase and quadrature channels of the carrier, respectively. At baseband, therefore, the DPDCH and DPCCH are packetised, as shown in Figure 6.25. We observe that during a slot period there are 10 ms of N_{data} bits gathered in a packet, and over the same 10 ms the DPCCH packet is assembled having pilot, TFCI, FBI, and TPC bits in a TDMA arrangement.

Again the parameter k determines the number of bits in each slot and is related to the SF, as previously described for the down-link, except that now the values of k are 0, 1, 2, 3, 4, 5, and 6. This means that the SF varies from 256, 128, 64, 32, 16, 8, and 4, and the number of bits per slot is 10×2^k or 10, 20, 40, 80, 160, 320 and 640 bits, respectively. The SF may be different for the DPDCH and the DPCCH transmitted via the same carrier. These different bits per slot, N_{data}, correspond to seven slot formats 0, 1, 2, 3, 4, 5 and 6, with slot format 6 having $N_{data} = 640$ bits per slot. As an example, slot format 3 has $N_{data} = 80$ bits per slot, a channel rate of $80/(0.667 \times 10^{-3}) = 120$ kb/s and there is one bit per symbol, an SF of 32, and 1200 bits per frame.

There are six different DPCCH slot formats, but all use an SF of 256 resulting in 10 bits per slot and a corresponding channel bit rate of 15 kb/s. Table 6.5 displays the number of N_{pilot}, N_{TFCI}, N_{FBI}, and N_{TPC} bits for each slot format. Different pilot bit patterns of a frame duration are specified for $N_{pilot} = 5, 6, 7$, and 8, and therefore the pilot bits change from slot to slot over a frame. As a consequence, the BS receiver on inspecting the received pilot pattern is able to identify the frame boundaries and the current position within a frame.

The TPC bits from a UE to its BS request either step increases or decreases in the BS's transmit power. The power control step size, $\triangle_{TPC}$, is controlled by the network to be either 1 or 2 dB. No channel coding of the TPC bits is employed as the delay involved is unacceptable. As shown in Table 6.5, N_{TPC} may be either one or two bits. When $N_{TPC} = 1$

Table 6.5: Different slot format assignments. A spreading factor of 256 results in 10 bits per slot, giving a rate of 15 kb/s.

Slot format	N_{pilot}	N_{TFCI}	N_{FBI}	N_{TPC}
0	6	2	0	2
1	8	0	0	2
2	5	2	1	2
3	7	0	1	2
4	6	0	2	2
5	5	2	2	1

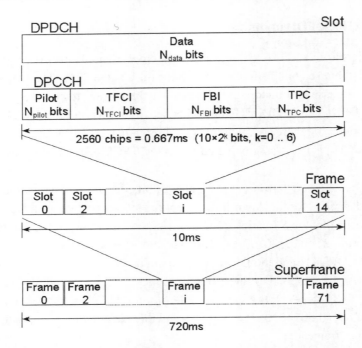

Figure 6.25: Slot and frame structure for the up-link transmissions

bit, the command to increase the transmit power is a logical 1, while to decrease power it is a logical 0. The corresponding logical values when $N_{TPC} = 2$ is 11 and 00.

The TFCI bits provide the same functionality as they do in the down-link, namely indicating the bit rates that are currently in use on each of the dedicated channels. In the absence of TFCI bits, i.e. $N_{TFCI} = 0$, blind detection is used by the BS. However, we recall that there are six slot format structures for the up-link, and that N_{TFCI} is either zero or two bits, whereas in the down-link there are 17 different slot format structures and that N_{TFCI} can be much larger than two bits.

Table 6.5 shows the presence of up-link-only feedback information (FBI) bits, N_{FBI}, that have values 0, 1, or 2 depending on the slot format. These bits are up-link specific. Their role is to provide a layer 1 feedback path between a UE and its BS. The FBI bits are used for closed-loop mode transmit diversity and site selection diversity (SSDT).

Note in Table 6.5 that the system is able to trade the 10 bits per slot for DPCCH between the pilot, TFCI, FBI and TPC requirements for a particular service or network requirement. Power control bits and pilot bits are essential, but if TFCI and FBI are not required, then more pilot bits can be transmitted with advantage.

Up-link modulation and spreading   Dual-channel quadrature phase shift keying (QPSK) modulation is used with the inphase channel carrying the DPDCH, while the quadrature channel conveys the DPCCH. This is the situation when a UE has only one DPDCH. The baseband DPCCH and DPDCH are spread, i.e. multiplied, by a channelisation code. The channelisation code is generated at the chip rate of 3.84 Mchips/s, and as with the down-link, they are OVSF codes. For multimedia services, multichannel operation may be required. The additional DPDCHs may be transmitted on either the inphase channel or the quadrature channel. Each additional DPDCH has its own channelisation code. The channelisation code allocation rules apply only within each UE. Two different DPDCHs on different quadrature branches may use the same channelisation code because the branches are orthogonal. One DPCCH and up to six DPDCHs may be transmitted simultaneously. The DPCCH is always spread by the channelisation code $C_{ch,256,0}$. If there is only one DPDCH, it is spread by $C_{ch,SF,k}$ where $k = SF/4$. For example, if an SF of 16 is required, then $C_{ch,16,4}$ is used. Should there be more than one DPDCH code, then all of them have an $SF = 4$. So for DPDCH$_i$, $i = 1, 2, \ldots, 6$, the channelisation codes are $C_{ch,4,i}$, respectively. Figure 6.26 shows the block diagram of the up-link spreading arrangement.

The DPDCH$_i$ and DPCCH signals after spreading are weighted by a gain factor β_d. We observe in Figure 6.26 that DPDCH$_1$, DPDCH$_2$ and DPDCH$_3$ when spread and weighted are added together to form the baseband inphase signal I which is multilevelled with changes occurring at the rate of 3.84 Mchips/s. The other three DPDCH signals, and the DPCCH duly spread and weighted, are summed and on phase shifting by 90° the quadrature component Q is obtained. The complex $I + jQ$ signal is scrambled by a scrambling code.

It is the scrambling code that provides the multiple access on the up-link, and consequently each UE must have a different scrambling code. There are 2^{24} up-link scrambling codes, and each is required to have good autocorrelation and cross-correlation properties. A UE can use either a short or a long scrambling code. A short scrambling code is primarily designed for BSs effecting multi-user detection (MUD) techniques [26–29]. When MUD is not used the long scrambling code is preferred. Both scrambling codes are complex, which means that the scrambling of the I and Q signals is done by different codes.

The long scrambling codes are 38 400 chip segments of Gold codes and last one frame of 10 ms. The codes are formed by a bit-wise addition of two m-sequences. By contrast the short scrambling codes have only 256 chips. They are based on the periodic extension of the S(2) codes [30]. The basic codes are 255 chips that are extended to 256 chips by repeating the first chip at the end of the code.

After scrambling, the data are filtered and modulate an RF carrier. The filters have the same characteristics as those used on the down-link.

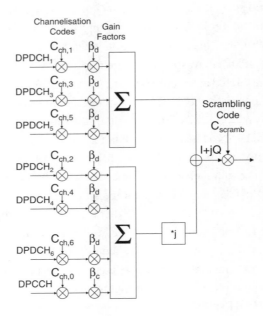

Figure 6.26: The up-link spreading arrangement at the UE.

Physical random access channel   There are two common physical up-link channels: the physical random access channel (PRACH) and the physical common packet channel (PC-PCH). Let us first deal with the PRACH. Its format is different from the dedicated up-link channels described previously. There are 15 access slots in two 10 ms up-link frames, as shown in Figure 6.27. The PRACH access slots are 5120 chips of 1.33 ms duration. An access transmission can last for 10 ms, i.e. for one frame.

A PRACH transmission consists of a preamble, one or a number of them, followed by a message. Each preamble lasts for 4096 chips. The preamble used for the UE during an access attempt to the network is initially transmitted at low power. If the access is unsuccessful, then the preamble is again transmitted, but at a higher power. This process continues until the preamble is successfully received. The preamble(s) is used by the UE to inform the BS that it is about to send an access request, namely the access message, and the second part of the PRACH. Although the preamble(s) does not carry information data (that is in the message) it can be used by the BS to synchronise with the UE. The 4096 chips of the preamble are composed of 256 repetitions of a 16-chip Walsh code. The term for the preamble having a repetitive Walsh code is a *preamble signature*. There are 16 different preamble signatures available. A UE randomly selects one of them prior to each access attempt. Prior to transmission by the UE, the preamble is spread using the up-link long

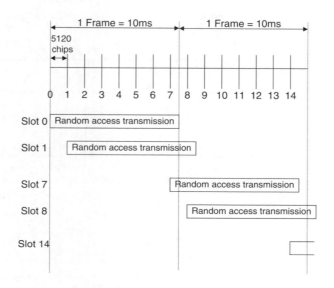

Figure 6.27: The random access frame structure.

scrambling codes, although only the first 4096 chips are used.

Once the preamble has been received and synchronisation established, there are a number of procedures to be adopted. These will be described in detail when we come to access procedures. At this juncture we are concentrating on the PRACH, and so we need to discuss the message part of the PRACH. The message length is 10 ms, composed of 15 slots, each of 2560 chips. Each slot contains a data component which carries the access information to the higher layers, and a control part containing pilot and TFCI bits. The data and control parts are transmitted on the inphase and quadrature components, respectively. Figure 6.28 shows the PRACH message part of the slot and frame structure.

Table 6.6 displays the four different message data slot formats, and the only message control slot format of the PRACH message part. We observe that the channel bit rate for the message data varies from 15 to 120 kb/s by varying the *SF*, while the message control slot channel bit rate is always 15 kb/s with 10 bits per slot and $N_{pilot} = 8$ bits and $N_{data} = 2$ bits. The eight-bit pilot pattern changes from slot to slot and repeats during each PRACH frame.

Physical common packet channel The physical common packet channel (PCPCH) is an up-link only channel that carries the common packet channel (CPCH). This transport CPCH allows bursts of data to be transferred from a UE to the network without recourse to a dedicated channel. The CPCH is typically used for short, infrequent bursts of data, whereas the dedicated channels are more suited to long bursts or short frequent bursts of

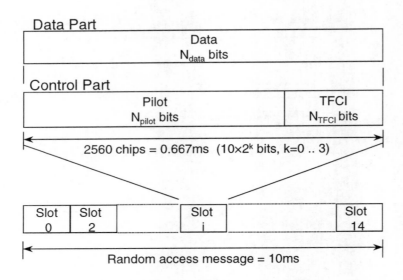

Figure 6.28: The slot and frame structure of the PRACH message part.

Table 6.6: The slot formats of the PRACH message part.

Slot format	Channel bit rate (kb/s)	Spreading factor	Bits per slot
0	15	256	10
1	30	128	20
2	60	64	40
3	120	32	80

Message data slot formats

Channel bit rate (kb/s)	Spreading factor	Bits per slot	N_{pilot}	N_{data}
15	256	10	8	2

Message control slot format

traffic.

The PCPCH transmissions are closed-loop power controlled. This is achieved by utilising the PCPCH to carry power control information during the packet transmission to control the transmit power of the BS, and the down-link DPCCH to adjust the transmit power of the UE.

The PCPCH is similar to the PRACH, but offers a higher data capacity. The PCPCH has four parts: an access preamble, a collision detection preamble, a power control preamble, and the message data. The access preamble, like that for the PRACH, indicates to the BS that the UE seeks to commence packet transmission on the PCPCH. The preamble consists of a preamble signature (one of a possible 16) modulated by a preamble scrambling code. These preamble signatures are identical to those used on the PRACH, while the preamble scrambling codes are 4096-chip codes that are similar to those used by the PRACH codes. A UE transmits an access preamble, and if this is successfully received it will transmit no more. However, it may be necessary to transmit a number of access preambles, each at a higher power level, until one is correctly received, or the access attempt is aborted.

Once an access preamble attempt is completed, the UE transmits the collision detection preamble. This is only transmitted once, and its function is to inform other UEs not to transmit their packets on the same PCPCH at this time. The collision detection preamble is similar to the access preamble, consisting of a signature and a scrambling code.

After the collision detection preamble has been sent the UE transmits a power control preamble. This is a DPCCH that contains only the pilot, FBI and TPC bits. The BS uses this PDPCCH to measure the power received from the UE, and in addition, the DPCCH enables the UE to control the power of the associated down-link DPCCH. The power control preamble lasts for 10 ms.

The last part of the PCPCH is the message part. This may consist of one or a number of 10 ms frames, each consisting of 15 slots. Each slot carries a DPCCH and a DPDCH, transmitted on quadrature arms of the carrier. The DPDCH can use any spreading factor between 4 and 256, but the DPCCH always uses a spreading factor of 256.

The four parts of PCPCH are displayed in Figure 6.29.

6.3.1.3 System procedures

Cell search procedure   The cell search is initiated each time a UE is switched on, or if a UE loses contact with the network. The process is divided into three steps. The first is for the UE to achieve slot synchronisation. The UE searches for the primary synchronisation code (PSC) using a correlator. This code is the same for all cells in the network. Referring to Figure 6.19, we observe that the PSC (shown as C_p) is transmitted at the beginning of every slot (for 10% of the slot). A peak at the output of the PSC correlator enables the commencement of the next slot to be known. We recall that the PSC is produced by taking

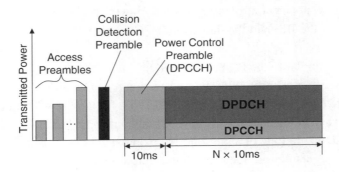

Figure 6.29: The physical common packet channel.

a 16-chip code at 240 kchips/s and multiplying it by another 16-chip code at 3.84 Mchips/s. Because of the way the PSC is formed the correlation process is done in steps. Firstly, the received signal is correlated with the 16-chip component at 3.84 Mchips/s, and the results are stored in a buffer. Next, every 16th entry in the buffer is correlated with the 16-chip component at 240 kchips/s. This procedure is equivalent to correlating over the entire 256-chip sequence. The complexity is considerably decreased compared with the conventional approach of doing a 256-chip correlation, particularly as this must be done over an entire slot period of 2560 chips, i.e. for each chip offset of 2560 chips. If the data are oversampled by two to give 5120 samples, then 5120×256 correlations would be required per time slot. Even by producing the PSC from two 16-chip words, the number of correlations is still substantial, although reduced by a factor of eight. Figure 6.30 shows the arrangement whereby the sliding correlator moves along the received signal making correlations with a 16-chip code, storing the results in a buffer, followed by another 16-chip sliding correlation to give the peak and hence a knowledge of the slot boundary.

The BS selected depends on the size of the correlator peaks. Figure 6.31 shows the output of the correlator in the presence of three nearby BSs. BS_A would be the preferred choice for the UE.

Having achieved slot synchronisation the next step is to obtain frame synchronisation and code-group identification. Let us return again to Figure 6.19 and recall that the secondary synchronisation codes (SSCs) also occur at the same time as the PSC, but the SSCs provide a code sequence over each frame. It is this sequence which enables the scrambling code to be identified. Further more, recall from Figure 6.14 that there are 64 code groups. The

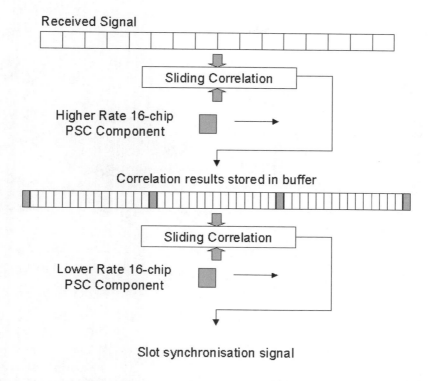

Figure 6.30: Correlation procedures for slot synchronisation.

received signal is correlated with all the 64 possible SSC sequences (over a frame). The sequence that yields the maximum correlation identifies the group used. As a consequence of identifying the SSC sequence, which repeats every frame, the receiver knows the frame boundaries and hence frame synchronisation is obtained. The arrangement to identify the scrambling code group is shown in Figure 6.32. Observe that the process is a parallel one to provide a rapid decision.

The third step is scrambling code identification. Once the receiver knows the code group, it knows the eight sets of codes associated with this code group. It now has to identify the set that was used by the transmitter. Each set is associated with a particular primary scrambling code. Accordingly it cross-correlates the common pilot channel (CPICH) with each of the eight scrambling codes and thereby deems which code is the most probable. It does this because it is the CPICH that is also the primary scrambling code. Armed with the primary scrambling code, the UE decodes the CCPCH and obtains the system and cell information from the BCH.

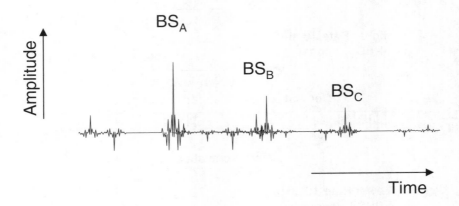

Figure 6.31: Output from the primary synchronisation code output in the presence of three nearby BSs.

The access procedure   Before a UE attempts an access to the network it must obtain the following information from the BCH.

- The preamble and message spreading codes used by the BS.

- The available preamble signatures and access slots.

- The available message part spreading factors.

- The up-link interference level at the BS.

- The transmit power level on the primary CCPCH.

- The AICH transmission timing parameter.

- The power offset between the preamble and the message part, $\triangle P_{p-m}$.

- The power step-size $\triangle P_o$ to be used during access.

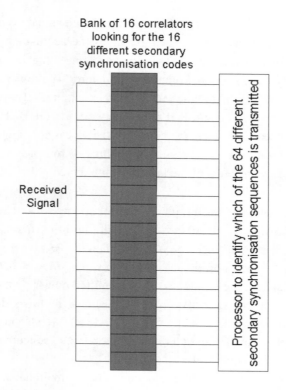

Figure 6.32: Identification of the scrambling code group.

The UE randomly selects one of the preamble codes specified by the BCH, and sets the transmit power for its PRACH transmission according to

$$P_{PRACH}(\text{dBm}) = L_p(\text{dB}) + I_{BS}(\text{dBm}) + C, \tag{6.4}$$

where L_p is the path loss between the UE and the BS, I_{BS} is the interference power at the BS, and C is a constant. The parameter L_p is deduced from the transmit power on the primary CCPCH, I_{BS} is part of the BCH message, and C is a fine adjustment parameter conveyed by the BCH.

Next the UE employs a dynamic persistence algorithm to determine whether it should proceed with the access attempt. This algorithm regulates the access load to the network. The persistence factor N is downloaded from the BCH. If it is zero, the UE is allowed to proceed with its access attempt, but should $N \neq 0$, the UE generates a random integer R between 0 and $2^N - 1$. When $R \neq 0$, the UE waits for one frame and repeats the persistence test. If $R = 0$, the UE continues with the access attempt. When the BS deems the RACH to be too high, it increases the value of N on the BCH.

The UE randomly selects an access slot and a preamble signature from those available, and transmits its access preamble. It then examines the acquisition indicator (AI) it receives on the BCH. However, if the UE does not detect an AI that corresponds to its access preamble it randomly selects another preamble signature. The transmission power is now increased by the step size $\triangle P_0$ and the preamble is transmitted. The number of times this procedure may be repeated is controlled by the network via the BCH. Should the UE receive a negative AI it ceases its access attempt, i.e. the access attempt has failed. A positive AI causes the UE to transmit its access message three or four access slots later, with the transmit power $\triangle P_{p-m}$ above that of the last transmitted preamble.

Power control   Closed-loop up-link power control attempts to maintain the up-link *SIR* at a target value of SIR_{Target}. The BS estimates the up-link *SIR*, namely SIR_{Est}, on the dedicated physical channel. If $SIR_{Est} > SIR_{Target}$, then the BS sets the down-link TPC bits to decrease the UE's transmit power, and vice versa when $SIR_{Est} \leq SIR_{Target}$. On receiving a power control command, the UE adjusts its transmitted power by the power control step size, $\triangle_{TPC}$, which is either 1 or 2 dB. Figure 6.33 shows the block diagram of the power control elements between the UE and the BS (or Node B). We also note in Figure 6.33 the role of the RNC which provides an outer control loop that is concerned with ensuring that the bit error rate (BER) is satisfactory.

In soft handover (SHO) a UE is able to communicate simultaneously with two or more BSs. As a consequence, the UE can get conflicting power control commands from the BSs. In this situation the UE examines the quality of the TPC bits. If one TPC commands a power decrease, then this is executed, irrespective of the TPC from the other BSs. This results in the UE always being power controlled to the satisfaction of the BS receiving the best signal level.

Fast *down-link* power control is also supported. The UE compares the received *SIR* with a target *SIR* and sets its transmitted TPC bits accordingly. This way the transmit power of the BS(s) is regulated. The range of down-link power control is, however, smaller than that of the up-link range.

6.3.1.4 Handover

The handover procedure is the key factor in supporting roaming. In UMTS there are five different types of handover. Softer handover involves the UE communicating with two sectors on the same cell site using a common carrier frequency. Eventually the UE changes from its original sector to the new sector. When two or more BSs forming different cells communicate simultaneously via a common carrier frequency with a UE as it roams between cells, then a soft handover is said to be in process. The process ends when the UE is communicating with a single BS. Sometimes a UE entering a different cell has to have its carrier

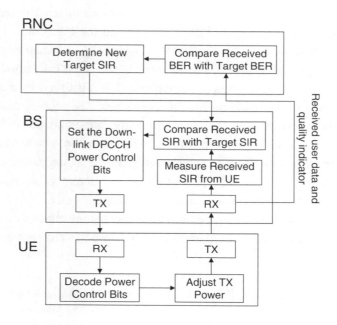

Figure 6.33: Up-link power control system.

frequency changed. This handover procedure is called a hard handover. A hard handover can also occur at the same BS when a UE is forced to switch carrier frequencies in order to enhance radio link or network performance. The fourth type of handover is between a cell operating in a UTRA FDD mode and a cell supporting a UTRA TDD communications (the latter is dealt with in Section 6.3.2). The last type of handover in UTRA FDD is between a UTRA FDD BS and a GSM BS. Observe that the last two types of handovers are hard because the frequency bands of UTRA FDD, UTRA TDD and GSM are different.

The management of the handover process at the UE is facilitated using three sets of cells. First, there is the handover monitoring set that contains a list of all the cells the UE should *monitor* while it is in an active connection. This list is provided to the UE by the network via the BCCH and may contain details of nearby UTRA TDD and GSM cells. The second set is the *active set* listing the cell or cells that are currently being used in the active connection. Should the UE be in a softer or soft handover, the active set will contain the identities of the sectors or cells, respectively The active set can only contain UTRA cells. The third set is the *handover candidate set*. This set contains the identities of cells that are not in the active set, but which measurement information suggests that they could support an active connection if required. The names of these cells are reported to the network by the UE. These cells can be on the same or different frequencies, and may contain UTRA TDD cells or GSM cells.

The set contents can be changed by using different procedures. The radio link addition procedure is where a cell is added to the active set and the UE commences transmission with that cell. The radio link removal procedure is where a cell is removed from the active set and the UE ceases communications with that cell. The combined radio link addition and removal procedure results in a cell in the active set being substituted by a different cell.

A prerequisite to forming sets is making measurements. The primary measurement quantity is the E_c/I_0 of the primary CPICH. These measurements are made in every frame on the CPICH from cells in the active set. Other measurements that may be evoked are the down-link path loss, the down-link transport channel BER and the block erasure rate (BLER), and the time difference between the signals from the target cell and the serving cell. Measurements on cells not in the active set are made less frequently.

An intra-frequency measurement report can be triggered for a number of reasons. A report will be sent when the primary CPICH from a candidate BS is received with an (E_c/I_0) of M_{new} such that

$$10\log_{10}M_{New} \geq S \cdot 10\log_{10}\left(\sum_{i=1}^{N_A} M_i\right) + (1-S)10\log_{10}M_{Best} - (R+H_{1A}) , \qquad (6.5)$$

where M_i and M_{Best} are the E_c/I_0 measurements for the ith cell in the active set and the strongest cell in the active set, respectively, R is the reporting range parameter, H_{1A} is a hysteresis parameter, and S is a network-defined parameter. For $S = 0$,

$$10\log_{10}M_{New} \geq 10\log_{10}M_{Best} - (R+H_{1A}), \qquad (6.6)$$

which means a report is made by the UE to the network whenever M_{New} (dB) from the candidate BS is within $(R+H_{1A})$ of the E_c/I_0 dB of the best cell. Observe that E_c is the energy per chip since the CPICH does not carry data. Figure 6.34 shows that the report is triggered at the moment when the candidate cell E_c/I_0(dB) is $< (R+H_{1A})$ from the (E_c/I_0)dB of the best cell.

Figure 6.34 shows the situation when a UE issues a report to the network that another cell should be added to active set. By contrast, Figure 6.35 shows the condition when a UE reports that a cell has a received (E_c/I_0)dB that renders it inappropriate for inclusion in the candidate set. The equation employed in making this report is

$$10\log M_{Old} \leq S \cdot 10\log_{10}\left(\sum_{i=1}^{N_A} M_i\right) + (1-S)10\log_{10}M_{Best} - (R+H_{1B}), \qquad (6.7)$$

where M_{Old} is the (E_c/I_0) dB measurement for the old cell and H_{1B} is a hysteresis parameter for reporting when a cell should leave the active set.

Having considered the situation when a received primary CPICH (E_c/I_0) is sufficiently large to enter, and also sufficiently small to leave the reporting range, we now consider

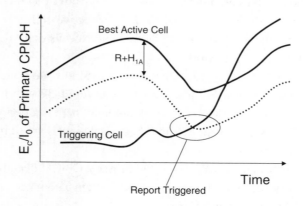

Figure 6.34: (E_c/I_0) dB versus time showing curves of: the best active cell; a candidate entering cell; a dotted curve of $R + H_{1A}$ below the best active cell, with $S = 0$. The report is triggered when the two curves intersect.

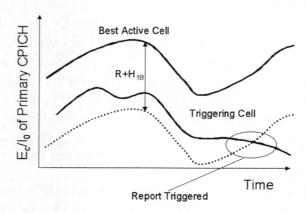

Figure 6.35: (E_c/I_0) dB versus time showing curves of: the best active cell, a cell that triggers a report when its (E_c/I_0)dB falls below the curve of the best active cell minus $R + H_{1B}$, with $S = 0$.

reporting when a non-active primary CPICH (E_c/I_0) becomes greater than that of an active primary CPICH. Initially, say, there are three active cells, and a non-active cell. As the UE roams towards the non-active cell, the (E_c/I_0) of the non-active cell increases, as shown in Figure 6.36, until eventually it exceeds the E_c/I_0 of an active cell. A report is duly filed with the network. Later the E_c/I_0 of the non-active cell exceeds that of another active cell, and again a report is sent to the network.

From the last two equations it is evident that the network needs to be kept informed of the current BS that provides the best E_c/I_0 at the UE. In Figure 6.37 we observe that initially the best cell is the cell associated with curve (a), but later there is a period when the cell of curve (c) is best, until curve (a) again becomes best. The reports are triggered when curves (a) and (c) intersect.

Reports are also sent when the (E_c/I_0) of the primary CPICH becomes either better or worse than an absolute threshold. This threshold is a system parameter, and sets a minimum E_c/I_0 that can be used in the handover process.

The UTRA specifications define the format of the measurement information delivered by a UE to the network. It is up to the equipment manufacturers to decide how the network will use the reports to determine whether a soft handover should be performed, i.e. the soft handover algorithm is not in the UTRA specifications. However, these specifications do offer the following suggestions for a soft handover algorithm. There are four parameters: an active set threshold ($Thresh_{AS}$) and an active set hysteresis ($Hyst_{AS}$) level for entry to, or removal from, the active set; an active set replacement hysteresis ($Hyst_{AS-REP}$) value required for one cell to replace another in the active set; and a time to trigger ($\triangle T$) that sets the time between a trigger event (a report) occurring and the resulting action.

The algorithm utilises these parameters as follows. A cell is removed from the active set if its E_c/I_0 falls ($Thresh_{AS} + Hyst_{AS}$) below that of the best active cell. Alternatively, a cell is added to the active set when its E_c/I_0 exceeds the trigger level which is set at ($Thresh_{AS} - Hyst_{AS}$) below the E_c/I_0 of the active cell. Should the active set be full and the best non-active cell is better than $Hyst_{AS-REP}$ above the worst active cell, then the latter is replaced by the former cell. Figure 6.38 shows the variation of E_c/I_0 as a function of time for three cells. Observe that actions follow $\triangle T$ after reports are made to the network. All the handover algorithm changes are shown in Figure 6.38.

Having described the types of handover, handover sets and a soft handover algorithm, the handover execution is done by the network. After the UE provides down-link measurements information to the network, and the BS provides the network with measurements of the received signal from the UE, the network considers the overall impact of a handover before it is executed, e.g. the network will need to assess whether there is sufficient capacity in the target cell.

So far we have concentrated on soft handovers. With these type of handovers it is rela-

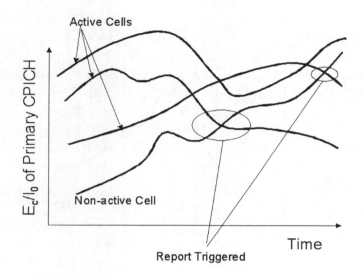

Figure 6.36: Non-active cell is better than active cells and reports are triggered when this occurs.

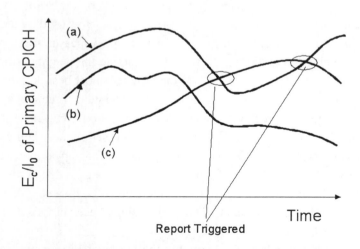

Figure 6.37: Reports are issued when there are changes in the best cell.

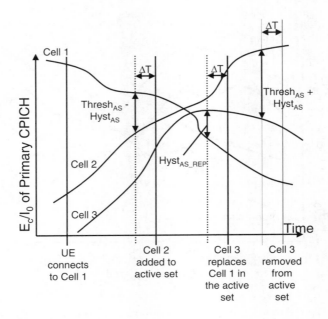

Figure 6.38: Example of the soft handover algorithm to illustrate changes in the active set.

tively straightforward for a UE to make E_c/I_0 measurements on the primary CCPCH from neighbouring cell sites because the same radio carrier frequency is used. Attempting a hard handover is a more complex problem. For a UE to make a measurement from a signal in a neighbouring cell using a different carrier, for example the neighbouring BS may be a UTRA FDD, a UTRA TDD or a GSM BS, then the technical best option would be for the UE to have a dual receiver. One receiver is used to maintain communications with the serving cell, while the other receiver implements the neighbour cell measurements. An alternative is for a single receiver to operate in a compressed mode when making neighbouring cell measurements. In compressed mode, transmission gaps or periods are formed in the down-link FDD transmissions, i.e. transmissions are suspended in these gaps. This allows the UE to retune its single receiver to the frequency used by a neighbouring cell and make the required E_c/I_0 measurements. UTRA FDD single receiver UEs must support the down-link compressed mode.

In compressed mode a number of slots within the normal 10 ms frame are not used to transmit data. Because of this, the data rate in the remaining slots in the frame is increased. This is achieved either by puncturing the data, where some FEC bits are deleted and therefore not transmitted, or by decreasing the spreading factor by two. The transmitted data are now more vulnerable, and so to compensate, the data are transmitted at a higher power. Figure 6.39 shows the variation in the transmitted power with time. The first and third frames are when the compressed mode is not invoked. The second frame shows what happens to the down-link transmission in compressed mode. We observe the transmission gap to allow the UE to make E_c/I_0 measurements on adjacent BS transmissions, and the increase in transmitted power. The position and duration of the transmission gap can be varied depending on the measurement requirements. For example, if a UE is trying to locate the SCH on a neighbouring GSM cell, then the transmission gap changes its position in the frame to align with the control channel multiframe on the GSM BS.

In UTRA FDD the up-link and down-link transmissions occur simultaneously, but in different frequency bands. Interference between the relatively higher power transmissions and the low level received signals in negligible because of the wide frequency separation between the up-link and down-link bands. So when there is a hard handover between two BSs operating at different frequencies there is no need to suppress transmissions from the UE when it is making measurements via the down-link compressed mode. However, when the hard handover is from a UTRA FDD to a UTRA TDD or to a (GSM 1800 or GSM 1900) the frequency spacing of the TDD BS carrier from the UTRA FDD UE's transmit carrier may be insufficient to prevent the UE's transmission interfering with the measurements made by its own receiver. As a consequence, the UE must exercise the up-link compressed mode at the same time as it utilises the down-link compressed mode. All UEs must be able to support the up-link compressed mode of operation, even if they have dual receivers. In the up-link compressed mode only the spreading factor decrease can be used to increase the throughput on the remaining timeslots in the frame.

The UTRA FDD up-link uses switched diversity and down-link combining, as previously described in connection with soft and softer handovers in cdmaOne; see Section 4.3.4.1.

6.3.1.5 System performance

There is a limited amount of performance data available at the time of writing. Link level simulations have been made using two tapped delay line channel models, A and B, for each of three different environments [31]. The characteristics of models A and B for the office environment are displayed in Table 6.7, and relative average power versus the relative delay of each model are shown in Figure 6.40. For the office environment, the Doppler spectrum is considered to be flat.

The corresponding tables and figures for the outdoor-to-indoor/pedestrian and the vehicular channel models are shown in Table 6.8 and Figure 6.41, and Table 6.9 and Figure 6.42, respectively. The Doppler spectrum for these models is the classic Doppler spectrum which is proportional to $\left\{1 - (f/f_m)^2\right\}^{1/2}$, where $\pm f_m$ is the maximum Doppler frequency shift relative to the carrier, and f is the frequency. Outside of $f_c \pm f_m$, the spectrum is deemed to

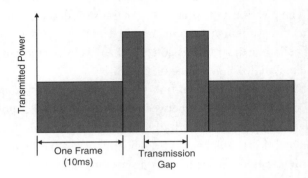

Figure 6.39: Transmitted power versus time for the down-link transmission showing a frame in compressed mode.

Table 6.7: Characteristics of indoor office channel models A and B.

	Channel A		Channel B		
Tap	Relative Delay (ns)	Relative Average Power (dB)	Relative Delay (ns)	Relative Average Power (dB)	Doppler Spectrum
1	0	0	0	0	Flat
2	50	−3.0	100	−3.6	Flat
3	110	−10.0	200	−7.2	Flat
4	170	−18.0	300	−10.8	Flat
5	290	−26.0	500	−18.0	Flat
6	310	−32.0	700	−25.2	Flat

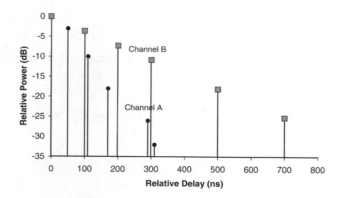

Figure 6.40: Relative power versus delay for models A and B: indoor environment.

be zero.

The indoor office environment used in the simulation consists of three floors, with UEs moving between the offices and the corridors at 3 km/h. Sixty BSs are deployed, with 20 on each floor, and they have omnidirectional antennae. Figure 6.43 shows the layout of one floor, with one BS every second office.

The indoor-to-outdoor and pedestrian environment is a Manhattan-like grid of roads with square city blocks, as shown in Figure 6.44. The block size is 200×200 m, and the width of the street is 30 m. There are 72 BSs spaced as shown in the figure with their omnidirectional antennae located 10 m above street level. The UEs' speed is 3 km/h. We note that this environment consists of clusters of street microcells.

The vehicular environment assumes three-sectored hexagonal cells, with the 'cell radii' being 2 km and 500 m for bit rates of <144 kb/s and ≥ 144 kb/s, respectively. The standard deviation of the log-normal shadow fading is set at 10 dB, and the UEs move at a speed of 120 km/h.

The cell capacity and spectral efficiency for the 8 kb/s speech service is shown in Table 6.10 for the A models in offices, street microcells, and vehicular environments. UL and DL refer to up-link and down-link, respectively. The capacity and spectral efficiency are not the same for both links, with the up-link having the better performance. (We wonder if subsequent experiments will concur with these finding?) Long constrained data (LCD) have a less stringent time delay and time variation compared with speech services, and there is also unconstrained delay data (UDD). The performances of these two types of services are also shown in Table 6.10. There are the LCD pedestrian services at 64 and 384 kb/s, and the UDD indoor services at 64 and 2048 kb/s. UDD is used for packet-based radio transmissions, where the unrestrained delay feature can be exploited for increased data throughput. The LCD and UDD services have a high spectral efficiency and a lower cell capacity com-

Table 6.8: Characteristics of outdoor-to-indoor/pedestrian channel models A and B.

	Channel A		Channel B	
Tap	Relative Delay (ns)	Relative Average Power (dB)	Relative Delay (ns)	Relative Average Power (dB)
1	0	0	0	0
2	110	−9.7	200	−0.9
3	190	−19.2	800	−4.9
4	410	−22.8	1200	−8.0
5	-	-	2300	−7.8
6	-	-	3700	−23.9

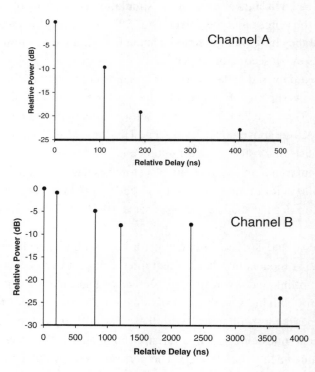

Figure 6.41: Relative power versus delay for models A and B: outdoor-to-indoor/pedestrian environments.

Table 6.9: Characteristics of vehicular channel models A and B; classic Doppler spectrum.

Tap	Channel A		Channel B	
	Relative Delay (ns)	Relative Average Power (dB)	Relative Delay (ns)	Relative Average Power (dB)
1	0	0	0	−2.5
2	310	−1.0	300	0
3	710	−9.0	8900	−12.8
4	1090	−10.0	12900	−10.0
5	1730	−15.0	17100	−25.2
6	2510	−20.0	20000	−16.0

pared with speech services which operate at the much lower bit rate of 8 kb/s.

We now include an example of the link budget for the 8 kb/s speech service in an indoor environment when channel A is applicable. The various parameters that determine the link budget are displayed in Table 6.11 for both the down-link and up-link. The path loss in dBs for these links and the maximum range in meters are 122.7 dB and 121.5 dB, and 717 m and 654 m, respectively.

Technology enhancements There are a number of techniques for enhancing the performance of UTRA FDD. Dedicated pilots allow beam-forming antennae to be deployed at the BSs. Short scrambling codes are available to allow multi-user detection, a method of decreasing multiple access interference (MAI), to be implemented [26–29]. Both open-loop and closed-loop transmit diversity can be employed. This type of diversity requires the transmission of the CDMA signal from two antennae. In the open-loop mode, space–time block coding transmit diversity (STTD) is used whereby the information is sent at different times on each antenna. With closed-loop transmit diversity, the same information is transmitted on each antenna. The UE feeds back information of the received signal power using the FBI bits. The BS then adjusts the phase and amplitude of the signals transmitted from its two antennae, and the UE reports the change in received power. This closed-loop process continues until the received power at the UE is maximised [30]. Figure 6.45 shows the arrangement at the BS for the closed-loop transmit diversity.

6.3.2 UTRA TDD system

The UMTS terrestrial radio interface (UTRA) that uses time division duplex (TDD) will be deployed in the unpaired IMT-2000 frequency bands. Band A is the 3G unpaired frequency

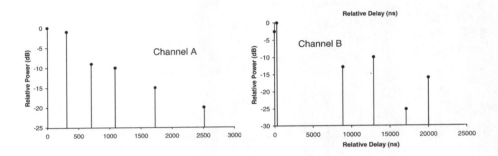

Figure 6.42: Relative power versus delay for models A and B: vehicular environment.

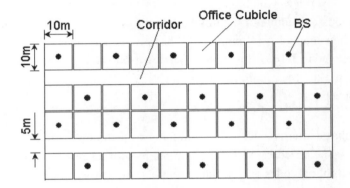

Figure 6.43: Office environment used in the simulation. A black dot represents a BS.

allocation in Europe: 1900–1920 MHz and 2010–2025 MHz. In the United States there is band B, the PCS spectrum allocation: 1850–1910 MHz and 1930–1990 MHz. The United States also has band C, an unlicensed band from 1910 MHz to 1930 MHz. The nominal channel spacing in UTRA TDD is 5 MHz, with a channel raster of 200 kHz, which means that the carrier frequency is a multiple of 200 kHz. The UTRA absolute radio frequency channel number (UARFCN) in the lower IMT-2000 band is

$$N_t = 5(F - 1885.2), \quad 1885.2 \leq F \leq 2024.8 \text{ MHz}, \tag{6.8}$$

and so N_t varies from zero to 698.

The transmitted spectrum is defined by a number of spectrum masks. As an example, the mask for a UE transmitting a maximum output power of 21 dBm is shown in Figure 6.46.

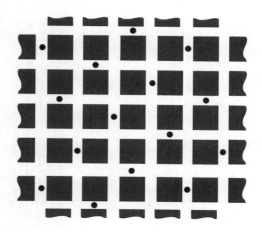

Figure 6.44: The indoor-outdoor and pedestrian environment. A black dot represents a BS.

Table 6.10: Capacity and spectral efficiency of UTRA FDD services in different environments.

Service	Scenario	Bit rate (kb/s)	E_b/N_0 (dB) [UL/DL]	Cell Capacity [Erlangs per carrier per cell] (UL/DL)	Spectral Efficiency [kb/s per MHz per cell] (UL/DL)
Speech	Indoor A, 3 km/h	8	3.2/6.1	192/92	135/74
	Pedestrian A, 3 km/h	8	3.3/6.1	154/157	123/125
	Vehicular A, 120 km/h	8	5.4/7.9	112/89	90/71
LCD	Pedestrian A, 3 km/h	64	2.4/1.9	16.4/33.7	210/431
	Pedestrian A, 3 km/h	384	1.3/1.1	3.5/6.0	269/461
UDD	Indoor A, 3 km/h	64	1.5/1.2	96/154 (parallel sessions)	224/372
	Indoor A, 3 km/h	2048	0.6/0.1	50/82 (parallel sessions)	273/453

Table 6.11: Link budget for the speech service in an indoor environment; model A used.

Parameter (units)	Down-link	Up-link
TX power (dBm)	10	4
Cable/combiner losses (dB)	2	0
TX antenna gain (dBi)	2	0
TX EIRP (dB)	10	4
RX antenna gain (dBi)	0	2
Cable/combiner losses (dB)	0	2
RX noise figure (dB)	5	5
Thermal noise density (dB/Hz)	-174	-174
RX noise floor (dBm/Hz)	-169	-169
Data rate (dB/Hz)	39	39
Required E_b/I_0 (dB)	6.0	3.2
RX sensitivity (dBm)	-124.0	-126.8
Power control TX power increase (dB)	2.0	2.0
Handoff gain (dB)	6.1	6.1
Log-normal fading margin (dB)	15.4	15.4
Maximum path loss (dB)	122.7	121.5
Maximum range (m)	717.2	654.1

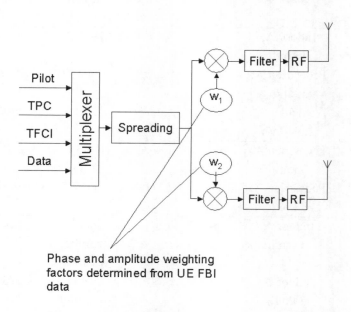

Figure 6.45: The closed-loop transmit diversity system at the BS.

The power in the 30 kHz measurement band is flat for frequency offset $\triangle f$ between zero (the carrier) and 2.5 MHz, and then decreases until $\triangle f = 3.5$ MHz. This latter part of the mask is shown by the dark shading. The mask then has the form shown in the figure for $\triangle f$ values from 3.5 to 12.5 MHz when the measurement bandwidth of the spectrum analyser is 1 MHz. The adjacent channel leakage power ratio (ACLR) is the same as discussed for UTRA FDD.

The physical layer   The UTRA TDD has a similar frame structure to the UTRA FDD. There are 15 slots in a frame which has a period of 10 ms. Each slot has 2560 chips and lasts for 0.667 ms. A superframe consists of 72 frames and lasts for 720 ms.

A physical channel consists of bursts that are transmitted in the same slot in each frame. It is also defined by its repetition period, repetition length and superframe offset. The number of frames between slots belonging to the same physical channel is the repetition period. This must be a sub-multiple of 72, i.e. 1, 2, 3, 4, 6, 8, 9, 12, 18, 24, 36, and 72. As an example, if the physical channel occupies slot 0 and the repetition period is 12, then the physical channel occupies slot 0 in every 12th frame. The superframe offset defines the repetition period offset with the superframe. So, in our example, if the superframe offset is 3, the physical channel will occupy slot 0 in frames $3, 15, 27, 39, \ldots$, namely offset by 3 frames, but 12 frames apart. The repetition length defines the number of slots associated with each repetition, and may have values of 1, 2, 4, or 8. For the example where the physical channel occupies slot 0, the repetition period is 12, the superframe offset is 3, and, say, the repetition length is 4, then the physical channel will occupy slot 0 in frames (3, 4, 5, 6), (15, 16, 17, 18), (27, 28, 29, 30), etc. The superframe arrangement for this physical channel is shown in Figure 6.47.

Time division duplex   The UTRA TDD employs time division duplex to achieve bi-directional transmission links. Each slot in a frame can be used to carry either up-link or down-link information. The switching point or points between up-link and down-link slots is variable, as is the number of slots allocated on each link. At least one slot must be allocated in each direction. Figure 6.48 shows examples of the time duplex slot structure per frame.

Burst structure   An important feature of the UTRA TDD is that multiple users can be accommodated per slot by means of CDMA. Each user within a slot is separate at the receiver due to the uniqueness of their spreading codes. The chip rate within each burst is always 3.84 Mchips/s. Thus, whether a UTRA UE is in FDD or TDD mode it will transmit at the same chip rate. There are two different types of bursts, both with the format of a midamble training sequence, with two data blocks on either side. Burst type 1 has the

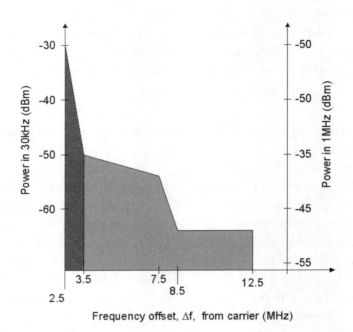

Figure 6.46: Transmitted spectrum mask for a UE transmitted power of 21 dBm.

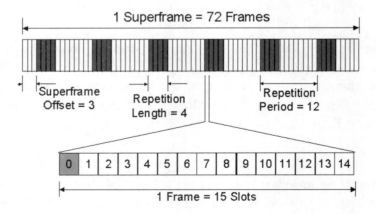

Figure 6.47: UTRA TDD physical channel defined by slot =0, repetition period =12, superframe offset = 3 and repetition length =4.

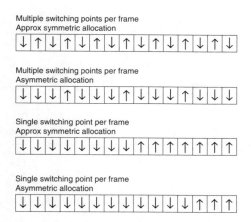

Figure 6.48: Multiple switching points per frame for different slot per frame allocations.

longer training sequence of 512 chips, and is used on the down-link with up to 16 users per slot. The other burst type 2 has a shorter training sequence of 256 chips. When used on the down-link it can support 16 users per slot, but on the up-link only four users per slot are allowed. Figure 6.49 shows how the data and midambles are constituted for both burst types. Observe the wide variation of bits per data field depending on the spreading code employed.

The bursts can also carry a transmission format combination indicator (TFCI). The actual use of a TFCI is negotiated at call set-up, along with the number of bits to be used. The TFCI is located either side of the midamble. The power control (TPC) bits can also be included in the bursts, and a guard period is added at the end of each burst. Figure 6.50 shows the burst within one slot when both TFCI and TPC are carried. Note that TPC immediately follows the midamble. The modulation used is quadrature phase shift keying (QPSK).

The midamble codes are training sequences that enable the receiver to estimate the impulse response of the radio channel. The sequences used by different users in the same slot are time-shifted versions of the same code. Consequently, the channel impulse response of all the users transmitting in the same slot can be estimated using a single cyclic correlation. The specific time-shift between the different sequences is chosen depending on the maximum delay spread in the cell. Nearby cells with the same carrier frequency must use midamble codes that are different to avoid corrupting the channel estimates. All midamble codes are complex, alternating between real and imaginary values.

The spreading of the data bursts at 3.84 Mchips/s is done by OVSF codes, as in the FDD mode. This is to allow the users to have orthogonal codes within a burst, regardless of the spreading factor (*SF*). The codes must be of length 1, 2, 4, 6 and 16, and the *i*th chip of the

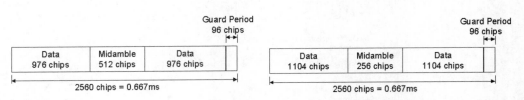

Spreading Factor	Number of Symbols per Data Field	Number of Bits per Data Field
1	976	1952
2	488	976
4	244	488
8	122	244
16	61	122

(a)

Spreading Factor	Number of Symbols per Data Field	Number of Bits per Data Field
1	1104	2208
2	552	1104
4	276	552
8	138	276
16	69	138

(b)

Figure 6.49: The UTRA TDD burst types: (a) type 1 and (b) type 2.

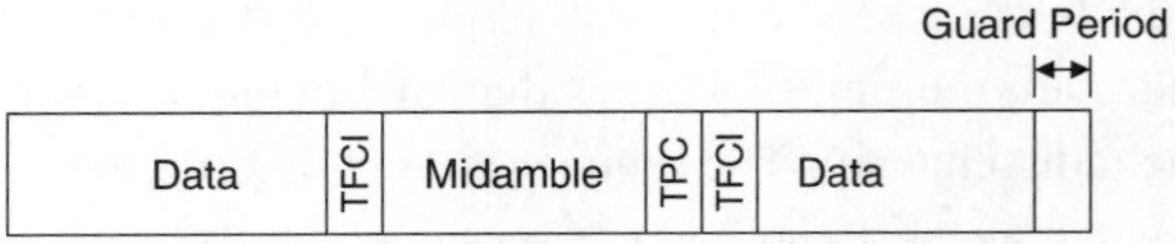

Figure 6.50: The burst arrangement when TFCI and TPC data are included.

*k*th complex code is given by

$$c_i^{(k)} = (j)^i a_i^{(k)} \begin{cases} i & = 1, \ldots, SF, \\ k & = 1, \ldots, \text{number of users}, \end{cases} \tag{6.9}$$

where $a_i^{(k)}$ is the *i*th chip of the *k*th OSVF code, and j causes a 90° phase shift for incrementing i.

After the spreading has been performed, the data are scrambled by a cell-specific 16-chip code. There are 128 of these 16-chip codes. They are formed into 32 code groups, each having four scrambling codes. Each scrambling code has a direct association with one long and one short basic midamble sequence (from which the actual midamble sequences are generated). Figure 6.51 shows the 32 code groups, where the particular group 16 has scrambling codes 64 to 67.

The dedicated physical channels (DPCH) convey user data and signalling information between a BS and a UE. The transport channel, the dedicated channel (DCH), is transmitted on the DPCH. To accommodate variable rate transmissions there are two methodologies: the multicode with a fixed spreading factor, or a single code using variable spreading. In the former, each timeslot can carry a number of bursts having their unique codes. Some or all of the bursts can be assigned to a single user or to a number of users, as indicated in Figure 6.52, where six users are sharing the 16 CDMA signals within one timeslot. In the other option each user transmitting in a timeslot has a single code, but the spreading factor of the code, i.e. the code length, is varied according to the required data rate. Figure 6.53 is an example of five users all transmitting in the same slot, where each user has a code length such that for the data to be transmitted, the chip rate is always 3.84 Mchips/s. So user 1 has 16 chips per word (chips/w) while user 2 only has 2 chips/w. This means that user 1 is transmitting fewer bits than user 2. By avoiding transmitting multiple codes, a UE does not increase the crest factor, namely the peak-to-average transmitted power ratio, of its transmitted signal.

Common control physical channels The down-link common control physical channels (CCPCH) transmit cell and system-specific information and paging signals to the UEs. The physical random access channel (PRACH) is an up-link channel that UEs use initially to access the network.

The down-link CCPCH is similar to the dedicated channels. However, it always uses burst type 1, a spreading factor of 16 and does not carry TFCI bits. One of the down-link CCPCHs is the primary CCPCH (PCCPCH). This is transmitted as a reference power over the entire cell, i.e. no beam forming can be used on this channel. It occupies a known position relative to the synchronisation channel, and it carries the broadcast channel (BCH).

When a UE initially accesses the network, it does not know the propagation delay to

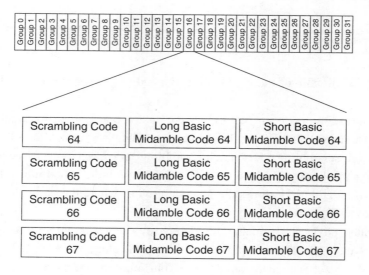

Figure 6.51: The 32 code groups.

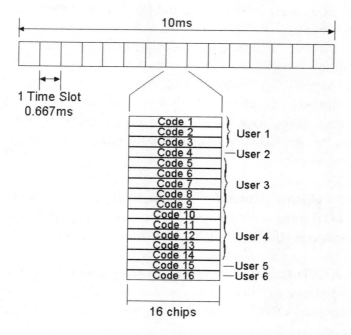

Figure 6.52: Multicode transmission with fixed spreading codes.

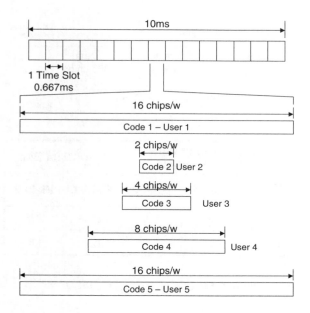

Figure 6.53: Each user transmitting in a slot has a single code whose length is selected according to the required data rate.

the chosen cell site and is, therefore, unable to apply timing advance. Consequently, the PRACH bursts have a large guard time to avoid burst collisions at the BS. The guard time is 50 μs, corresponding to 192 chips. This sets the UE-BS separation to be less than 7.5 km. Figure 6.54 shows the burst formed and the guard period when no data are transmitted. Beneath the burst is a table giving the number of data symbols and data bits for the spreading factors of 8 and 16.

The physical synchronisation channel (PSCH) enables a UE initially to synchronise with a BS. The PSCH consists of a primary synchronisation code (PSC), and three secondary synchronisation codes. From the PSCH a UE is able to determine the code group used by the BS, the position of the frame within the 20 ms interleaving period, the position of the slot within the frame, and the position of the PCCPCH.

There are three different ways in which the PSCH and the PCCPCH can be mapped onto a down-link frame. Option 1 is to map them both onto the same single timeslot; in option 2 they are both mapped onto the same two timeslots which must be separated by eight timeslots; while for option 3 the PSCH is mapped onto two timeslots that are separated by eight slots, and the PCCPCH is mapped onto any timeslot.

The PSC in the TDD mode is identical to the PSC in the FDD mode, namely it is a 256-

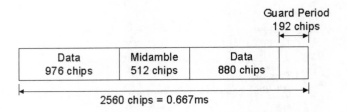

	No of Data Symbols		No of Data Bits	
Spreading Factor	Field 1	Field 2	Field 1	Field 2
8	122	110	244	220
16	61	55	122	110

Figure 6.54: Burst format of the PRACH.

chip length code produced by modulating one 16-chip code with a second 16-chip code, as previously described. All systems use the same PSC. Again, the secondary synchronisation codes (SSCs) are the same as those used in the FDD mode. They are generated from 256-chip Walsh–Hadamard sequences, and there is a total of 16 available secondary scrambling codes. The SSCs are modulated to indicate the code group and the slot within the frame.

In public TDD systems the BSs must maintain synchronisation to prevent interference between UEs, e.g. the situation where one UE is receiving at the same time when another nearby UE is transmitting. Given that BS synchronisation is achieved, there is a danger that the PSC transmitting from the synchronised BSs could overlap, preventing the UEs from distinguishing the different BSs. To overcome this problem the start of the PSCH is offset from the start of the timeslot by a factor

$$t_{offset, n} = n T_c \left[\frac{2560 - 96 - 256}{31} \right]$$
$$= 71 \, n \, T_c \, ; \, n = 0, \, 1, \, \ldots, \, 31, \tag{6.10}$$

where T_c is the chip period of $1/(3.84 \times 10^6)$, and n is chosen such that no two neighbouring or nearby BSs use the same offset. Figure 6.55 shows the position of the PSCH, represented by the PSC of C_p, offset by $t_{offset, n}$.

The three secondary synchronisation codes, $C_{s,a}$, $C_{s,b}$ and $C_{s,c}$, are selected from a set of 16 available codes. Bits b_a, b_b and b_c in Figure 6.55 carry the information that is spread by the codes $C_{s,a}$, $C_{s,b}$ and $C_{s,c}$, respectively. These secondary synchronisation codes are chosen based on the code group, and the combination of the PSCH and the PCCPCH used

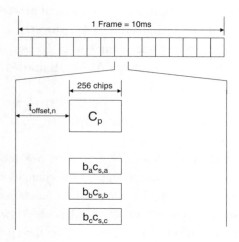

Figure 6.55: Physical synchronisation channel.

in the cell. As an example, if option 1 is used, then the PSCH and the PCCPCH are mapped onto one timeslot. Code groups 0–15 and 16–31 are formed into code sets 1 and 2, respectively; see Table 6.12. The identities of the secondary scrambling codes are chosen based on the code set used by the cell.

Note that the value of n for the offset is the same as the code group number; that the codes may be transmitted on the inphase I channel, or on the Q channel (denoted by j); and that for a given code group the synchronisation codes change periodically with the two frame periods.

Transport channels   There are seven different transport channels. There is a broadcast channel (BCH) that broadcasts cell and system-specific information from a BS to its UEs. The paging channel (PCH) pages UEs, while the forward channel (FACH) transports information such as packet data to the UEs. The UE accesses a BS using the random access channel (RACH). There is a synchronisation channel (SCH), which we have just discussed. The remaining channels are the up-link shared channel (USCH), and the down-link shared channel (DSCH).

Table 6.12: Secondary synchronisation codes $C_{s,a}$, $C_{s,b}$ and $C_{s,c}$.

Code set	Code group	$C_{s,a}$	$C_{s,b}$	$C_{s,c}$
1	0–15	C_0	C_1	C_2
2	16–31	C_3	C_4	C_5

Power control   The UE adjusts its initial up-link transmit power of the PRACH as in the case of the FDD mode according to Equation (6.4), where L_p is the path loss measured using the PCCPCH. The value of the interference level I_{BS} at the BS is broadcast on the BCH. Once a dedicated connection is established, the UE utilises open-loop power control where the transmitted power of the UE is

$$P_{UE} = \alpha L_p + (1 - \alpha)L_0 + I_{BS} + SIR_{target} + C \tag{6.11}$$

and L_0 is the average long-term path loss, SIR_{target} is the SIR the UE attempts to provide at the BS receiver, α is a weighting factor to smooth the changes in the UE transmit power and C is a constant. The value of α can be selected as a function of the delay between the current up-link timeslot and the most recent down-link access timeslot.

In addition to the open-loop power control exercised on the up-link, there is fast closed-loop power control on the down-link which is implemented using the TPC bits in the up-link burst. The UE periodically measures the down-link SIR and compares this with a target value. If the target SIR value is less than the received SIR, then the UE sets its up-link TPC bit to a logical 0. The TPC bit is made a logical 1 if the target SIR exceeds the received SIR, causing the BS to increase its transmit power. The target SIR is set by a higher level outer loop.

Timing advance   The UTRA TDD system can be used in wide area cells, but to achieve this a timing advance mechanism is necessary in order to prevent up-link burst collisions at the BS receiver. The timing advance operates to a resolution of four chips or 1.04 μs. The BS measures the time offset on the UE's PRACH transmissions and calculates the UE's initial timing advance. The timing advance parameter is transmitted as an 8-bit number giving a maximum timing advance of 256×1.04 μs corresponding to a range of 40 km. During a call the BS continuously measures the arrival time of the up-link transmissions from the UE and transmits the appropriate timing advance parameter to the UE so that slot collisions with other UEs are avoided.

There are proposals to have an enhanced timing advance mechanism with a resolution of one-eighth of a chip period. This opens up the prospect of all the up-link transmissions within a slot arriving in synchronisation, and this will decrease the multiple access interference since all the transmitted codes from the UEs are orthogonal.

Handover   The UE uses the same cell set classification as employed in the FDD mode, namely an active set, a handover monitoring set and a handover candidate set. An advantage of UTRA TDD over the FDD version is that the slotted mode of operation for neighbour BS measurements is not required because the TDMA structure provides this inherently.

System performance Little information is available on system performance at the time of writing. System level simulation results from UTRA Evaluation Document submitted to the ITU in September 1988 are collated in Table 6.13 [31]. The speech service is at 8 kb/s; LCD means long constrained data bearer services and the numbers following LCD signify the bit rate in kb/s; and UDD stands for unconstrained delay data bearer services which are at either 144, 384 or 2048 kb/s. LCD and UDD services are described in Section 6.3.1.5. The spectral efficiency for three different services in three different environments is shown in Table 6.13. We observe that for speech the spectral efficiency is the same for the vehicular and indoor environments, but is double for pedestrians in microcells.

An example of the link budget for the speech service in an indoor environment for UEs moving at 3 km/h is provided in Table 6.14. We observe that the maximum range is greater on the down-link compared with the up-link.

Multiuser detection In Chapter 4 we described the single-user detection method used in cdmaOne, where at a BS there is a separate receiver for each MS signal. This is a suboptimal approach because the multiple access interference (MAI) between the users is not removed. Instead the presence of MAI is compensated for by a number of measures, such as accurate power control, FEC coding, space diversity, variable bit rate source encoders, sectorisation, and so forth. The presence of MAI is due to radio channels that significantly degrade the orthogonality of the spreading codes used by the MSs when they arrive at their BS. Instead the cross-correlations between the received codes from the MSs at the BS are non-zero and this creates the MAI. Failure to remove the MAI results in an irreducible bit error rate (BER).

Multiuser detection (MUD) CDMA receivers may be classified as optimal when using maximum likelihood sequence detection, and suboptimal when employing either joint detection (JD) or subtractive interference cancellation (SIC). JD receivers apply a linear mapping to the soft outputs of the conventional matched filter detectors (MFDs), whereas an SIC receiver creates estimates of the interference that are then cancelled from the wanted signals.

MUD receivers can be implemented at both the BS and MS. However, to avoid added complexity at an MS the MUD can be used only at the BS receiver. Because the MAI is largely removed at the BS the MS can use less spreading (or less powerful FEC codes) and therefore less bandwidth. Consequently, more bandwidth can now be assigned to the down-link, enabling high spreading codes to be deployed and thus raising the processing gain at the MS receivers. The QoS on both links is subsequently enhanced.

The subject of MUD is vast and the reader is advised to consult References [26]– [29]. All we will do here is to provide a brief introduction to this complex subject. The starting point is to appreciate that if there are K mobiles transmitting to a conventional CDMA BS

Table 6.13: Simulation results of spectral efficiency from UTRA TDD for different environments.

Environment	Service	Spectral Efficiency kb/s per MHz per cell
Vehicular	Speech	70
	LCD144	201
	UDD144	320
Pedestrian	Speech	148
	LCD384	330
	UDD384	642
Indoor	Speech	73
	LCD2048	62
	UDD2048	400

Table 6.14: UTRA TDD: link budget for speech service in indoor environment.

Parameter (units)	Down-link	Up-link
TX power (dBm)	22	16.0
Cable/combiner losses (dB)	2	0
TX antenna (dBi)	2	0
TX EIRP (dB)	22	16
RX antenna gain	0	2
Cable/combiner losses	0	2
RX noise figure (dB)	5	5
Thermal noise density (dB/Hz)	-174	-174
RX noise floor (dBm/Hz)	-169	-169
Data rate (dB/Hz)	52.1	52.1
Required E_b/I_0 (dB)	6.0	3.6
RX sensitivity (dBm)	-110.9	-113.3
Handoff gain (dB)	5.9	5.9
Log-normal fade margin (dB)	15.4	15.4
Maximum path loss (dB)	123.4	119.8
Maximum range (m)	761.1	577.3

(we will ignore for simplicity intercellular interference) there are K single-user MFDs. Just prior to the hard-limiting detections we have the soft outputs of the MFDs. Since there are K of these soft outputs $(y_k, k = 1, 2, \ldots, K)$, then we may view the K outputs of the MSs transmitting to a BS as a vector $\mathbf{y}$ having K elements. A JD acts on this vector by applying a linear mapping to either remove the MAI or drastically to decrease it. We may represent the vector of the soft outputs at a symbol epoch as

$$\mathbf{y} = \mathbf{RAd} + \mathbf{z} \tag{6.12}$$

where $\mathbf{R}$ is the matrix containing the partial cross-correlation functions between the received signals from the MSs, $\mathbf{A}$ is a matrix containing the amplitude components of the channel impulse response, $\mathbf{d}$ is the data vector and $\mathbf{z}$ is the noise vector. In the decorrelation detector versions of the JD, the BS performs a subsequent operation on the $\mathbf{y}$ vector, namely

$$
\begin{aligned}
\mathbf{R}^{-1}\mathbf{y} &= \mathbf{R}^{-1}(\mathbf{RAd} + \mathbf{z}) \\
&= \mathbf{Ad} + \mathbf{R}^{-1}\mathbf{z} \\
&= \mathbf{d_{det}}.
\end{aligned} \tag{6.13}
$$

All the MAI has been removed to yield $\mathbf{Ad}$, but the noise is increased to $\mathbf{R}^{-1}\mathbf{z}$ by the matrix inversion. Computations of $\mathbf{R}^{-1}$ are difficult in real-time processing, particularly as K and the multipath delays are varying with time. The noise-enhancing effect is acceptable, provided K is not excessive, and that is why in UTRA TDD, K is restricted to a maximum value of four users per slot on the up-link when burst type 2 is used.

Another JD method is the minimum mean-squared error (MMSE) detector, which attempts to minimise the error between the true data $\mathbf{d}$ that is transmitted and the estimated data $\hat{\mathbf{d}}$ at the output of the JD. The error vector $(\mathbf{d} - \hat{\mathbf{d}})$ is set to be orthogonal to the soft output vector $\mathbf{y}$. Rather than performing $\mathbf{R}^{-1}\mathbf{y}$, the MMSE detector performs

$$\left[\mathbf{R} + (\mathrm{No}/2)\mathbf{A}^{-2}\right]^{-1}, \tag{6.14}$$

where $\mathrm{No}/2$ is the power spectral density (PSD) of the additive white Gaussian noise (AWGN). (We are conveniently avoiding channel fading here.) Using the MMSE detector the MAI is not completely removed, but the noise enhancement is less than that for the decorrelation detector.

In our discourse we have described how a conventional bank of MFDs yields a vector $\mathbf{y}$ which is subsequently processed by a second receiver called a joint detection (JD) receiver. This latter receiver produces all K output symbols per system epoch, but from a joint process rather than from a set of parallel detectors as used in the first stage. It is by using JD that the MAI can be eliminated or greatly decreased. JD is independent of the modulation used (which is important in adaptive radio interfaces) and provides both MAI reduction, equalisation to mitigate intersymbol interference (ISI) and, when used with coherent receiver diversity, it combats the effects of multipath fading. The receiver must know all the codes used by the MSs; channel estimation is necessary for every MS, and estimation of the received signal levels may be required.

We turn now to the other form of multiuser detectors, namely SIC. These multiuser interference cancellation detectors are impressive, particularly when the y vector is processed first by a serial canceller followed by a cascade of parallel cancellers [32]. The single-user bound is attainable, which means that with K users the performance is as good as if only one user is present. The operation of a serial canceller will now be described with reference to Figure 6.56. The received signal is a composite of K CDMA signals from the K MSs. Single-user MFD is done as previously described and the $y_k, k = 1, 2, \ldots, K$, soft outputs are obtained as previously discussed in connection with JD. This time these soft outputs are not processed as a vector. Instead the soft outputs are ranked in terms of amplitude. For simplicity we will assume that y_1 is the largest amplitude, y_2 the next largest amplitude, etc., with y_K the smallest amplitude. In order to perform interference cancellation the received composite signal must be decomposed into its separate CDMA signals. A delay is introduced into the received signal to allow for the MFD and the ranking process, and the delayed received signal $r(t)$ is applied to the serial canceller.

Because $y_1(t)$ has been identified as the largest amplitude we are very confident that a hard decision of $y_1(t)$ will not yield an error, and so the MFD of $y_1(t)$ is used to realise the recovered bit $\hat{b}_1(t + \tau_0)$, where τ_0 is the time for a signal to pass through a multiplier and an estimate of the channel. However, the other users will experience considerable interference from MS$_1$ as $y_1(t)$ is the largest soft output from the K MFDs. So $\hat{b}_1(t + \tau_0)$ is multiplied by its code c_1 (known to the BS) and the resulting signal is passed through an estimate of the baseband radio channel for MS$_1$ (which is computed by the BS, for example, with the aid of a sounding sequence transmitted by MS$_1$). This yields an estimate of the received CDMA signal $\hat{s}_1(t)$ from MS$_1$, i.e. we have separated $\hat{s}_1(t)$ from $r(t)$. This CDMA component is subtracted from $r(t)$ and MFD ensues using c_2 to give $\hat{b}_2(t - \tau + \tau_0)$. This MFD is performed in the presence of a lower MAI than originally experienced when the MAI from the strongest interferer MS$_1$ was present. The next strongest component in $r(t)$ is now obtained by multiplying $\hat{b}_2(t - \tau + \tau_0)$ by c_2 and convolving it with the estimated impulse response $\hat{h}_2(t)$ of MS$_2$ to yield $\hat{s}_2(t - \tau)$, where τ is the added processing time. We now delay $s_1(t)$ by τ to give $s_1(t - \tau)$ which is now co-phased with $\hat{s}_2(t - \tau)$, delay $r(t)$ by τ to give $r(t - \tau)$, and then subtract these two large sources of MAI from $r(t - \tau)$. Following MFD we obtain a more accurate estimate of $\hat{b}_3(t - 2\tau + \tau_0)$. The CDMA signal associated with MS$_3$ is formed as $\hat{s}_3(t - 2\tau)$, i.e. delayed by 2τ relative to $\hat{s}_1(t)$. To generate bit $\hat{b}_4(t - 3\tau + \tau_0)$ we proceed in a similar fashion. This time signals $r(t)$, $s_1(t)$, $s_2(t - \tau)$ and

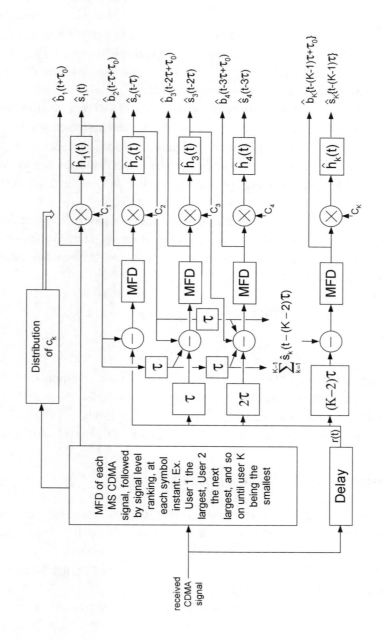

Figure 6.56: Serial canceller of multiple access interference.

$s_3(t - 2\tau)$ are involved. They must all be co-phased such that all are delayed by 2τ relative to $\hat{s}_1(t)$ by introducing additional delays shown by the blocks marked τ in Figure 6.56. This process of removing the interference by progressively using the separated CDMA signals as they become available continues until we arrive at the MS that has the lowest soft output, namely MS_K. For this MS, $r(t)$ is delayed by $(K - 2)\tau$ and all the interferring CDMA signals are co-phased with this delay relative to $s_1(t)$. MFD on the difference signal between $r(t - (K - 2)\tau)$ and all the co-phase interferring signals yields $\hat{b}_K(t - (K - 1)\tau + \tau_0)$. The new sequence of bits $\{\hat{b}_k\}$ is less likely to have errors compared with the sequence $\{b_k\}$ obtained from the outputs of the original set of MFD.

During the next symbol interval the ranking of the set of soft outputs from the single-user bank of MFDs may change. MS_K may now not be the smallest user, nor MS_1 the largest, say, and distribution of c_k for each row of Figure 6.56 may change. The generation of $\hat{b}_k$ differs, but the time difference between the first symbol and the last is approximately $(K - 1)\tau$ plus the delay in the MFD and ranking process. In general, this is a small fraction of a bit period.

We could continue with this serial cancellation technique using $\{\hat{b}_i(t)\}$, but it is better to employ a parallel canceller [33]. The received signal $r(t)$ is delayed some more (to allow for the processing delay) to be $r_d(t)$. To create improved estimates of the data bits for MS_i (the ith ranked MS), we form

$$s_i(t - \tau) = r_d(t - \tau) - \sum_{\substack{k=1 \\ k \neq i}}^{K} s_k(t - \tau), \quad i = 1, 2, \ldots, K, \tag{6.15}$$

and these subtractions are implemented in parallel. Next, $s_i(t - \tau)$ for all i are MFD to yield a detected symbol for each MS. These symbols may be used as inputs to the next parallel cancellation stage. This process can continue until the single-user bound is approached.

Multiuser detection systems are likely to be introduced into both FDD and TDD CDMA systems, but because of their complexity we may expect their arrival to be delayed.

6.4 Evolution of IS-95 to cdma2000

The pioneering of CDMA in cellular radio is due to Qualcomm Inc. Starting in the late 1980s, Qualcomm embarked on a series of experiments that culminated in demonstrating that CDMA had the potential to provide an efficient radio interface for use in cellular networks. By July 1993, the Qualcomm proposals were adopted by the Telecommunications Industry Association as Interim Standard 95 (IS-95) [34].

From its dual-mode beginnings with AMPS in the 800 MHz band, the auctioning of spectrum in the 1900 MHz band (a 3G band) for PCS resulted in IS-95 operating in this higher band as a system in its own right. This enabled IS-95, in conjunction with IS-41-C [35], to go US-wide and, as we have previously stated, IS-95 has now become a *de facto* world-wide standard. The original message rates, now known as rate set 1 (RS1), have been augmented by a second rate set, RS2, and there have been other changes to the original IS-95 specification. What we now discuss is the evolutionary path that IS-95 is following that will take it to wideband CDMA and the multimedia 3G environment. We emphasise that at the time of writing much standardisation work is in progress, and by the time the book is published substantial additions to the specifications will have been completed.

The IS-95 CDMA system, now called cdmaOne (the evolution of a name), is destined to evolve to the 3G version known as cdma2000. The basic services that cdma2000 will provide are traditional mobile telephony plus enhanced voice services, such as audio-conferencing and voice mail. In addition to the low data rate services, there will be medium data rate services at 64 kb/s to 144 kb/s for applications such as Internet, and a high data rate, up to 2 Mb/s, for high speed packet and circuit-switched services. cdma2000 will enable MSs to communicate multimedia services where combinations of audio, video and data signals will be handled simultaneously. The 2 Mb/s services are likely to be restricted to indoor environments, while the 144 kb/s will be supported in all environments. There is complete compatibility between cdmaOne and cdma2000 in that both can work with each other, although only for the low bit rate services supported within the 1.25 MHz carrier occupancy. Like cdmaOne, the cdma2000 BSs are synchronised to each other and to the cdmaOne BSs. As a consequence, fast handovers between the two systems are supported. The chip rates of cdma2000 are multiples of the cdmaOne 1.2288 Mchips/s rate, and the carrier spacing in cdma2000 is $1.25N$ MHz, $N = 1, 3, 6, 9$, and 12. The minimum bandwidth allocation in FDD mode (we will later see that a TDD mode has been added) is the IS-95 bandwidth of 2×1.25 MHz, or $2 \times (1.25 + 0.625)$ MHz should a guard band be required, as in the case of an unco-ordinated spectrum. The carrier spacings of 1.25 MHz (IS-95), 3.75 MHz, 7.5 MHz, 11.25 MHz and 15.00 MHz mean that both narrowband and wideband CDMA can be supported. A high data rate message can therefore be handled by demultiplexing the message into N parallel low rate data streams, and each stream spread and modulated onto separate 1.25 MHz carriers at 1.2288 Mchips/s. Alternatively, the high data rate can be transmitted on a single wideband carrier at a rate of $1.2288N$ Mchips/s. These methods are known as multicarrier and direct spread, respectively. (There is another method where each stream is modulated onto the same carrier as if there are N low bit rate users.)

Figure 6.57(a) shows the channel occupancy of three contiguous cdma2000 1.25 MHz carriers in a 5 MHz band where frequency guards of $(1.25/2) = 0.625$ MHz have been left at the edges of the band. The arrangement is an $N = 3$ multicarrier deployment. The direct spread for $N = 3$ is displayed in Figure 6.57(b). Although the two methods have similar link performances the multicarrier approach when used in a cdmaOne overlaid by cdma2000 environment enables both carriers to be dynamically assigned to either system as required. The multicarrier method also enables forward link diversity to be incorporated without any increase in complexity at the MS. With this form of diversity, the different N carriers are transmitted from spatially separated antennae. For example, in Figure 6.57(a) instead of transmitting carriers f_1, f_2 and f_3 from a single wideband antenna, each carrier could be transmitted from its own antenna. The fading channel associated with each antenna is essentially uncorrelated. The MS receiver performs maximal ratio combining (MRC) on the received multicarrier signal.

It is interesting to note that the cdma2000 system with $N = 1$ might be thought to support the same number of voice users as a cdmaOne carrier. In fact cdma2000 can support twice the number of voice users because it uses quaternary phase shift keying (QPSK) which doubles the number of Walsh codes, i.e. channels, available. It also employs transmit diversity, and employs fast power control on the forward link which cdmaOne does not. So even in the $N = 1$ state, cdma2000 has a higher capacity than cdmaOne.

6.4.1 Forward link

6.4.1.1 Forward common channels

Because of the compatibility with IS-95, cdma2000 has the same forward common control channels as IS-95. There is the forward pilot channel (F-PICH), but cdma2000 may also use auxiliary pilots when adaptive multibeam antenna arrays are used. The forward common area pilot channel (F-CAPICH) is used by all mobiles with the geographic area of the beam, e.g. shopping malls, where the propagation loss is high and the teletraffic is high. By using the spot beam the transmit power of both the BS and MSs is decreased. There is also a forward dedicated area pilot channel (F-DAPICH) that is employed when an antenna spot beam follows, and is used exclusively by, a particular mobile. F-DAPICH is used by high bit rate MSs, or an individual MS experiencing high propagation losses. An MS will use the IS-95 pilot F-PICH that covers the sector to obtain cell identification, phase reference and timing information, but if an adaptive antenna is used, then the MS must have an individual pilot that is sent via the narrow beam so that the radio channel for that beam can be uniquely estimated for RAKE receiver operation. By using multiple pilots the capacity and link performance can be enhanced.

Since each MS may have a unique Walsh code we must take care that a sufficient number of Walsh codes remain to handle users' traffic. Accordingly, the auxiliary pilot codes are obtained from a set of expanded Walsh codes. Starting with a Walsh code W_i^m, a code having length m, sequence i, we can form longer codes as shown in Figure 6.58. In each branch we double the length of the code, and we observe that if $m = 64$, then from one 64 code, UEs, W_i^m, we have produced four 256 Walsh codes. Hence the use of one Walsh code (W_i^m) has yielded four unique pilot codes, each of four times the length, i.e. $4m$. We can

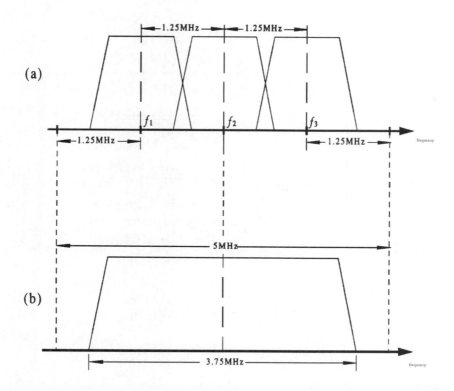

Figure 6.57: (a) Multicarrier, and (b) direct spreading spectra for $N = 3$.

continue with the tree structure, and for N branches we generate Walsh codes of length Nm. The correlation period at the receivers must be over the period associated with Nm, and the limit on N is that the channel must remain essentially stationary over the correlation period. We note that these codes are the same as the orthogonal spreading codes (OVSFs) described in Section 6.3 dealing with UTRA; see Figure 6.11.

The forward synchronisation channel (F-SYNC), the forward paging channel (F-PCH) and the forward common control channel (F-CCCH) are the same as in IS-95; see Chapter 4.

6.4.1.2   Forward dedicated channels

The forward dedicated channels are the forward fundamental channels (F-FCH) and the forward supplemental channels (F-SCH), as well as the forward dedicated control channel (F-DCCH) for call-specific control messages. A basic voice service requires one F-FCH;

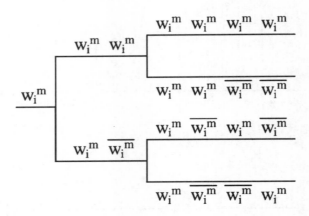

Figure 6.58: The pilot code tree structure.

control messages are multiplexed onto the F-FCH using either 'blank and burst', where a complete traffic frame is replaced by control data, or by using 'dim and burst', where the control information utilises part of a traffic frame. As another example, if a user requires to transmit voice, packet data and circuit data, then one F-FCH, two F-SCH and an F-DCCH may be assigned. The media access control (MAC) and higher layer signalling for the packet or circuit data is either carried on the F-FCH or on the F-DCCH. Dual frame sizes (5 ms and 20 ms) are used on these channels to support the mixture of data and signalling.

The generation of coded data is achieved using the arrangement shown in Figure 6.59. Reserved bits whose function is undecided (all logical 0s) are added to the traffic and control data, followed by CRC bits and tail bits. FEC coding ensues, and the symbols are repeated and puncturing performed as required. This is followed by block interleaving. The FEC code is a convolutional code of constraint length 9, but for the F-SCH operating rates above 14.4 kb/s turbo codes with a constraint length of 4 may be used. The structure in Figure 6.59 allows the output rate to be varied in an easy way, e.g. by changing the repetition rate or the amount of code puncturing.

Table 6.15 shows the F-FCH parameters for one ($N = 1$) carrier and RS1. Note that the Walsh code length is 128 chips, whereas in our descriptions of IS-95 in Chapter 4 the code length is 64 chips. The difference lies in the type of modulation used in cdma2000 where different data are conveyed on the I and Q channels, i.e. the data rate per channel

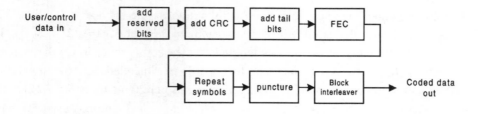

Figure 6.59: Structure for generating coded data in F-FCH and F-SCH.

is halved so the length of the Walsh code is doubled to maintain the same chip rate of 1.2288 Mchips/s. We also observe in Table 6.15 that there is a 5 ms frame having 96 bits after FEC coding corresponding to a rate of 96 bits/5 ms = 19.2 kb/s, or 9.6 kb/s on each quadrature component. The short frame data are coded by a Walsh code of 128 chips to give the rate of 1.2288 Mchips/s on each quadrature arm. Consider the row in the table for an uncoded rate of 2.7 kb/s. In a 20 ms frame there are $(40+6+8)2$ coded bits that become $54 \times 2 \times (8/9) \times 4 = 384$ bits after puncturing and a bit repetition of four. The chip rate is 128×384 bits/20 ms/2 (the divisor of 2 is because of the quadrature arms) to give 1.2288 Mchips/s. By means of adjusting the coding parameters the chip rate is constant for a range of uncoded bit rates.

When the RS is changed to RS2 with its values of 14.4, 7.2, 3.6 and 1.8 kb/s, then F-FCH, $N = 1$, has a set of parameters that is similar to those in Table 6.15 except that the coded symbol rate is 19.2 kb/s instead of 9.6 kb/s for RS1. As a consequence, we must spread with Walsh code of half the length, namely 64 chips.

Table 6.15: F-FCH parameters for $N = 1$.

Uncoded rate (kb/s)	Frame duration (ms)	Bits per frame	CRC bits	Tail bits	FEC rate	Repet- ition	Punct- uring	Walsh length (chips)
9.6	5	24	16	8	1/2			128
	20	172	12	8	1/2			128
4.8	20	80	8	8	1/2	×2		128
2.7	20	40	6	8	1/2	×4	1/9	128
1.5	20	16	6	8	1/2	×8	1/5	128

The F-SCH is used to support high bit rate services. The rates are derived from RS1 and RS2. The 20 ms frames are used. In Table 6.15 the highest rate service is 9.6 kb/s. The F-SCH has a set of rates that starts with 9.6 kb/s, and are 9.6α kb/s, α=1, 2, 4, 8, 16, 32. To keep the same chip rate, the Walsh code length is decreased in proportion to the increase in α. Table 6.16 shows the Walsh code lengths for different uncoded rates based on RS1. We observe that for an uncoded rate of 307.2 kb/s, the bits per 20 ms frame are 12288, the bit rate is 12288 bits/20 ms = 614.20 kb/s, and as quadrature modulation is used, we have 307.20 kb/s per quadrature component. For a chip rate of 1.2288 Mchips/s, each bit must be spread by only a four-chip code.

The uncoded rates based on RS2 are $14.4\,\alpha$ kb/s, $\alpha = 1, 2, 4, 8$ and 16, corresponding to Walsh codes of length 64, 32, 16, 8 and 4, respectively. The FEC rate is 1/3 and the puncturing is 1/9, making an equivalent rate of 3/8, which is a more powerful code than the 1/2 used with the rates based on RS1.

When more than one carrier is used (i.e. $N > 1$), F-FCH maintains RS1 and RS2. However, F-FCH is able to support a wide range of data rates.

6.4.1.3   The transmitter

The block diagram showing the essential components of a single carrier transmitter is presented in Figure 6.60. The figure does not explicitly show the pilot channels, nor the sync channel information coded to 4.8 kb/s, nor the paging channel coded to 19.2 kb/s, while the F-FCH and F-SCH structure is shown in Figure 6.59. The figure does show a set of 'cards', one for each user, as well as the cards for the forward common channels. The visible card shows the long code generator, decimator, the power control bits, the multiplexer (MUX) and the IQ signal point mapping. It is in the MUX/IQ map that the data is divided equally into inphase (I) and quadrature (Q) components. This is not done in IS-95 where the same data occur on I and Q.

The I and Q data are multiplied by the Walsh channelisation code W_m for the mth user. The length of W_m is varied in accordance with the input data rate, the FEC rate and puncturing rate, as well as the symbol repetition rate employed, to ensure that the chip rate on I and Q is 1.2288 Mchips/s. This orthogonal coding of the data is done by a different code on each card, and all the I components are added together, and so are all the Q components. The resulting I and Q signals are now spread by the short PN codes of length 2^{15} chips, i.e. by the forward pilot channel (F-PICH) that is used throughout the sector (or cell if no sectorisation is employed). This is the same arrangement as used in IS-95 where the phase offset of the code identifies the sector (or cell); see Chapter 4.

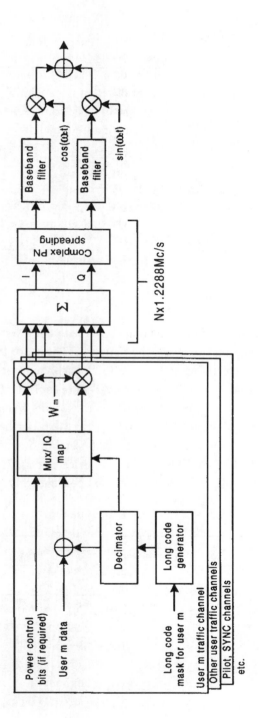

Figure 6.60: $N = 1$ and $N > 1$ (direct spread mode) spreading and modulation.

Table 6.16: F-SCH parameters for $N = 1$.

	Uncoded rate (kb/s)	Bits per frame	CRC bits	Tail bits	FEC rate	Punct- uring	Rate out (ks/s)	Walsh length
	9.6	168	16	8	1/2		9.6	128
	19.2	360	16	8	1/2		19.2	64
Rate set	38.4	744	16	8	1/2		38.4	32
1 (RS1)	76.8	1512	16	8	1/2		76.8	16
	153.6	3048	16	8	1/2		153.6	8
	307.2	6120	16	8	1/2		307.2	4
	14.4	264	16	8	1/3	1/9	19.2	64
Rate set	28.8	552	16	8	1/3	1/9	38.4	32
2 (RS2)	57.6	1128	16	8	1/3	1/9	76.8	16
	115.2	2288	16	8	1/3	1/9	153.6	8
	230.4	4584	16	8	1/3	1/9	307.2	4

6.4.2 Reverse link

The reverse access channel (R-ACH) is used in conjunction with the reverse pilot channel (R-PICH) when an MS is attempting to access the network. These are the only reverse common control channels. When operating at 9.6 kb/s, 172 bits containing the request have 12 CRC and eight tail bits appended in a 20 ms frame. This is equivalent to 9.6 kb/s for each quadrature signal. The R-ACH rate may be switched to 4.8 kb/s when the number of information bits per packet decreases from 172 to 80. Figure 6.61 shows the block diagram of the R-ACH coding parameters for the 9.6 kb/s and 4.8 kb/s conditions.

The R-PICH that accompanies the R-ACH is a code that carries no data. Both the R-PICH and the R-ACH are the same for all MSs within a cell, and consequently MSs make their

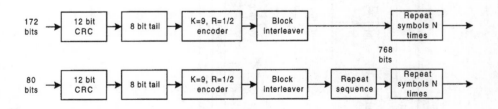

Figure 6.61: R-ACH coding parameters: (a) 9.6 kb/s, (b) 4.8 kb/s.

access attempts randomly, using a slotted ALOHA protocol. An access attempt is made up of access sub-attempts, which in turn consist of access probe sequences that have access probes. The access probe consists of an access preamble that is an integer multiple of 1.25 ms, where the duration of this preamble is known to an MS from the BS broadcast message. The R-PICH is part of the access preamble, and successive access probes, each consisting of an access preamble followed by an access message on the R-ACH, within an access probe sequence, are transmitted at increased power until either the BS receives the sequence and responds, or the maximum power level is reached without a BS response, which means the access attempt failed.

Having discussed the two reverse common control channels, namely the R-ACH and the R-PICH, we now consider the reverse dedicated channels. These are the reverse fundamental channel (R-FCH), the reverse supplemental channel (R-DCCH) and the reverse pilot channel (R-PICH). The R-PICH is identical to the R-PICH used with the R-ACH. When used as a dedicated channel, R-PICH is multiplexed with the power control (PC) bits in a frame consisting of three pilots followed by one PC bit. The duration of the frame is 1.25 ms. The PC bit is repeated $384N$ times, $N = 1, 3, 6, 9$ or 12, to give $384N$ chips in 1.25 ms, so the PC bit rate is 800 b/s transmitted at $307.2N$ kb/s. The three pilots and one PC bit every 1.25 ms constitutes a power control group (PCG) having $4 \times 384N$ chips transmitted at a rate of $1.2288N$ Mchips/s. We observe that each pilot has $384N$ chips, and that there are 16 PCGs in each 20 ms frame.

The R-FCH, R-SCH and R-DCCH are similar to the forward link counterparts, except that only Walsh codes of 2 or 4 are used. Table 6.17 provides the coding parameters for R-FCH for $N = 1$. For RS1, 5 ms frame, we have 48 bits per frame that are 1/4 FEC coded and then each symbol is repeated eight times to give 1536 bits in 5 ms, corresponding to a rate of 307.2 kb/s. Again, for RS1, 20 ms frame, 16 message bits per frame, the output rate is $30 \times 4 \times 8 \times (4/5) \times 8$ divided by 20 ms, namely 307.2 kb/s. This output rate of 307.2 kb/s is obtained for all RS1 and RS2 rates, and by replacing each output symbol by a four-chip Walsh code yields a chip rate of 1.2288 Mchips/s. Both R-FCH and R-DCCH use a four-chip Walsh code. R-DCCH operates at a 9.6 kb/s uncoded rate, and uses the same parameters as the R-FCH at this rate for either 5 ms or 20 ms frames; see rows one and two in Table 6.17, respectively.

R-SCH has a wide range of rates and coding parameters as discussed in connection with Table 6.16. However, only two-chip Walsh codes are used when a single R-SCH is employed, and a four-chip Walsh code when two R-SCH, are needed. This means that different sequence and symbol repetition rates are necessary for RS1 and RS2 rates in order that only two Walsh codes need be employed.

The low bit rate services only need one R-FCH, but higher rate and multimedia services will require additional dedicated channels. Figure 6.62 shows reverse dedicated channel

Table 6.17: R-FCH parameters for $N = 1$.

	Frame length	Bits per frame	CRC bits	Tail bits	FEC rate	Sequence repetition	Punct-uring	Symbol repetition	Rate out (kb/s)
	5	24	16	8	1/4			×8	307.2
	20	172	12	8	1/4			×8	307.2
RS 1	20	80	8	8	1/4	×2		×8	307.2
	20	40	6	8	1/4	×4	1/9	×8	307.2
	20	16	6	8	1/4	×8	1/5	×8	307.2
	20	267+1	12	8	1/4	×2	1/3	×4	307.2
RS 2	20	125+1	10	8	1/4	×4	1/3	×4	307.2
	20	55+1	8	8	1/4	×8	1/3	×4	307.2
	20	21+1	6	8	1/4	×16	1/3	×4	307.2

spreading and modulation at the MS transmitter for a service with a high bit rate require-ment. An R-FCH and an R-SCH1 are multiplied by their Walsh codes and gain factors before being added together to form one component of the baseband signal. The other base-band component is formed by multiplying the R-SCH2 and the R-DCCH by their Walsh codes and gain factors, and adding them and the pilot and power control bits together. The two baseband components are applied to a complex multiplier along with the products of the user's long code with the inphase and quadrature pseudo noise sequences, PN_I and PN_Q, re-spectively. Note that the two baseband components and the quadrature spreading sequences are applied to both the inphase and quadrature arms of the modulator.

6.4.2.1   Power control

This is an essential feature in order to decrease interference and thereby provide an accept-able frame erasure rate. The power control operates on both the forward and reverse links. The open-loop power control is where the transmit power of both the BS and the MS is determined by the level of their received signals, and is used to compensate for the path loss and slow fading in the radio channel. To combat fast fading and errors in the open-loop power control, a fast closed-loop power control is deployed.

The reverse pilot channel (R-PICH) is multiplexed with power control (PC) bits to inform the BS of the quality of the forward link. The R-PICH consists of power control groups (PCGs). A PCG lasts 1.25 ms and consists of three pilot codes followed by one power control (PC) bit in the last quarter of the PCG. Consequently, the PC sub-channel operates at a data rate of 800 b/s. There are 16 PCGs in one 20 ms frame. We observe that each pilot sequence has $384N$ chips in 1.25 ms, i.e. $307.2N$ chips/s, and the PC bit is sent as $384N$ chip code. So in one PCG there are $4 \times 384N$ chips giving a chip rate of $1.2288N$ chips/s.

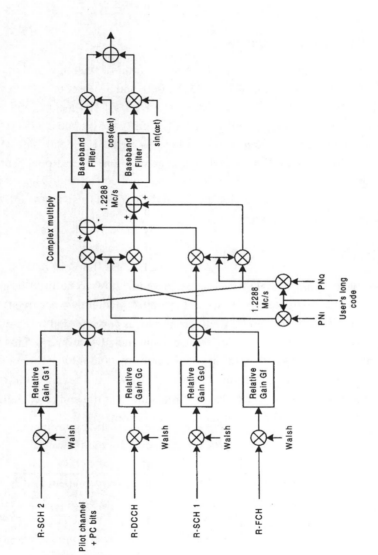

Figure 6.62: Reverse dedicated channel spreading and modulation at the MS transmitter.

The step size in the transmit power of the MS is normally 1 dB, but smaller step sizes of 0.5 dB and 0.25 dB may be implemented. A power control outer loop may be used to adjust the closed-loop power control thresholds at the BS in order to maintain the desired frame error rate.

6.4.2.2   cdma2000 FDD performance

The data relating to cdma2000 are very sparse at the time of writing. Simulation results for cdma2000 were presented to the ITU for IMT-2000, and a subset of these results is displayed in Table 6.18 [36]. The speech service was at 8 kb/s; LCD means long constrained data bearer services and the numbers following LCD signify the bit rate in kb/s. UDD stands for unconstrained delay data bearer services while LDD means low delay data. Antenna diversity was only used on the up-link, and all antennae were omnidirectional. Three carriers were available at each BS; hence 2 Mb/s services were not considered. The antenna diversity allowed the E_b/I_0 to be decreased for the desired QoS, and this significantly improved the up-link spectral efficiency compared with that of the down-link.

6.4.3   cdma2000 TDD

Like UTRA, cdma2000 has a TDD mode to enable cdma2000 to be used in an unpaired spectrum. The TDD burst structure is shown in Figure 6.63, where we observe that both the 5 ms and the 20 ms frame sizes can be used alternately as forward and reverse link bursts and are of 1.25 ms duration. Thus there are two bursts in each direction in a 5 ms frame, and eight forward and reverse link bursts in the 20 ms frames. These bursts may be assigned as required to support asymmetric traffic. To avoid forward and reverse link bursts from overlapping as the MS travels, a guard period is introduced at the end of each burst. Since the guard period is 52.08 μs, the maximum cell radius is some 7 km.

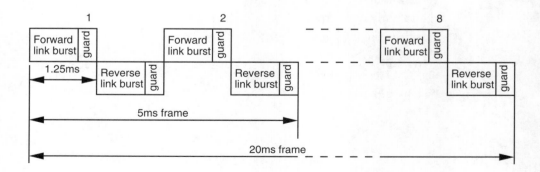

Figure 6.63: The burst structure for cdma2000 TDD.

Table 6.18: Summary of cdma2000 results

Environment	Service	Source bit rate (kb/s)	E_b/I_0 (dB) UL/DL	Spectral efficiency Erlangs per MHz per cell UL/DL
Vehicular	Speech	9.6	2.5/5.5	31/38
Outdoor to Indoor/Pedestrian	Speech	9.6	5.0/7.0	33/26
Vehicular	LCD	76.8	1.1/8.1	246/186
	LDD	76.8	1.5/7.6	231/158
	UDD	76.8	no data	210/142
	LCD	153.6	1.3/9.2	213/135
	LDD	153.6	2.0/8.1	197/107
	UDD	153.6	no data	187/70
Outdoor to Indoor/Pedestrian	LCD	76.8	1.6/6.1	291/119
	LDD	76.8	3.6/8.1	285/107
	UDD	76.8	no data	256/096
	LCD	460.8	1.8/7.0	172/116
	LDD	460.8	3.9/-	160/086
	UDD	460.8	no data	136/073

The modulation and spreading methods in the TDD mode are essentially the same as with cdma2000 FDD, but with some extra features. The guard periods following each burst may be varied by changing the puncturing level of the FEC codes. This is done at the I/Q mapping stage, in the same way as the power control bits are inserted into the coded data. The discontinuous data sequence is Walsh coded, and a burst generator compresses the data blocks to fit within the 1.25 ms bursts. PN spreading and modulation ensues as in the FDD mode. Since only one carrier is used for both forward and reverse transmissions the open-loop power control is more accurate compared with the FDD operation, and as a consequence the closed-loop power control may not be necessary in the TDD mode.

The forward and reverse link physical channels described for FDD apply for the TDD operation. The penalty of having a guard period in each 1.25 ms burst is a reduction in the power of the FEC code.

Bibliography

[1] System structure document, RACE 1043 (RACE MOBILE), Dec. 1991.

[2] de Boer H., RACE mobile communications, *IEE Electron. Commun. Eng. J.*, **5**(3), June 1993, 157–158.

[3] Chia S., The universal mobile telecommunication system, *IEEE Commun. Mag.*, **30**(12), Dec. 1992, 54–62.

[4] van den Broek ●●●, A.N. Brydon, J.M. Cullen, S. Kukkonen, A. Lensink, P.C. Mason and A. Tuoriniemi, Functional models of UMTS and integration into future networks, *IEE Electron. Commun. Eng. J.*, **5**(3), June 1993, 165–172.

[5] Steele R., The evolution of personal communications, *IEEE Pers. Commun.*, **1**(2), Second Quarter 1994, 6–11.

[6] Steele R., The therapy of Percy Comms: A dialogue on PCS issues, *Int. J. Wireless Commun.*, **2**(3), 1995, 123–132.

[7] Viterbi A.J., *CDMA, Principles of Spread Spectrum Communications*, Addison-Wesley, 1995.

[8] Prasad R., *CDMA for Wireless Personal Communications*, Artech House, 1996.

[9] Kohno R., R. Meidan and L.B. Milstein, Spread spectrum access methods for wireless communications, *IEEE Commun. Mag.*, Jan. 1995, 58–67.

[10] Gilhousen K.S., I.M. Jacobs, R. Padovani, A.J. Viterbi, L.A. Weaver and C.E. Wheatley, On the capacity of a cellular CDMA system, *IEEE Trans. Veh. Technol.*, **40**(2), 1991, 303–312.

[11] Padovani R., Reverse link performance of IS-95 based cellular systems, *IEEE Pers. Commun. Mag.*, **1**, Third Quarter 1994, 28–34.

[12] ETSI TS 101 038 Version 5.0.1 (1997-04), *Digital Cellular Telecommunicatiions System (Phase 2+); High Speed Circuit Switched Data (HSCSD)-Stage 2 (GSM 03.34).*

[13] ETSI TS 03 64 Version 5.1.0 (1997-11), *Digital Cellular Telecommunications System (Phase 2+); General Packet Radio Service (GPRS); Overall Description of the GPRS Radio Interface; Stage 2 (GSM 03.64, Version 5.1.0).*

[14] Furuskär A., S. Mazur, F. Müller and H. Olofsson, EDGE: Enhanced data rates for GSM and TDMA/136 evolution, *IEEE Pers. Commun.*, **6**(3), June 1999, 56–66.

[15] Webb W.T. and R.D. Shenton, Pan-European railway communications: Where PMR and cellular meet, *Electron. Commun. Eng. J.*, Aug. 1994, 195–202.

[16] Zvonar Z., P. Jung and K. Kammerlander, *GSM, Evolution Towards 3rd Generation Systems*, Kluwer Academic Press, 1999.

[17] GSM 05.02, *Digital Cellular Telecommunications System (Phase 2+); Multiplexing and Multiple Access on the Radio Path (GSM 05.02, Version 6.2.0, Release 1997).*

[18] GSM 03.64, *Digital Cellular Telecommunications System (Phase 2+); General Packet Radio Service (GPRS); Overall Description of the GPRS Radio Interface; Stage2 (GSM 03.64, Version 7.1.0, Release 1998).*

[19] GSM 03.60, *Digital Cellular Telecommunications System (Phase 2+); General Packet Radio Service (GPRS); Service Description; Stage2 (GSM 03.60, Version 7.1.1, Release 1998).*

[20] Balachandran K., R. Ejzak, S. Nanda, S. Vitebskiy and S. Seth, GPRS-136: High-rate packet data service for North America TDMA digital cellular systems, *IEEE Pers. Commun.*, **6**(3), June 1999, 34–47.

[21] EDGE Feasibility study, Work Item 184, *Improved Data Rates Through Optimised Modulation, Version 1.0*, 15–19 Dec. 1997.

[22] WWW.3GPP.ORG

[23] 3GPP Technical Specification Group (TSG) *RAN WG4 UTRA (BS) FDD; Radio Transmission and Reception, 3GPP, TS 25.104, Version 3.0.0.*

[24] 3GPP Technical Specification Group Radio Access Network; Spreading and Modulation (FDD), 3GPP, TS 25.213, Version 3.0.0.

[25] 3GPP Technical Specification Group Radio Access Network; Physical Channels and Mapping of Transport Channels (FDD), 3GPP, TS 25.211, Version 3.0.0.

[26] Verdu S., *Multiuser Detection*, Cambridge University Press, 1998.

[27] Ojanperä T., *Multirate Multiuser Detectors for Wideband CDMA*, Gummerus OY, Helsinki, Finland, 1999.

[28] Duel-Hallen A., J. Holtzman and Z. Zvonar, Multiuser detection for CDMA systems, *IEEE Pers. Commun.*, **2**(4) 1995, 46–58.

[29] Duel-Hallen A., A family of multiuser decision-feedback detectors for asynchronous CDMA channels, *IEEE Trans. Commun.*, **43**(2/3/4), 1995, 421–433.

[30] Optimum quadriphase primary scrambling codes for the uplink of UTRA/FDD, Tdoc SMG2 UMTS-L1, April 1998.

[31] UTRA Evaluation document (submitted to the ITU in September 1998), see ITU Web site www.itu.int.

[32] Oon T.-B., R. Steele and Y. Li, Performance of an adaptive successive serial-parallel cancellation scheme in flat Rayleigh fading channels, *Proc. IEEE VTC'97*, Phoenix, USA, May 1997, 193–197.

[33] Mowbray R., R. Pringle and P. Grant, Increased CDMA system capacity through adaptive channel interference regeneration and cancellation, *IEE Proc.-I*, **139**(5), 1992, 515–524.

[34] TIA/EIA Interim standard-95, 'Mobile station – base station compatibility standard for dual-mode wideband spread spectrum cellular systems, July 1993.

[35] Goodman D.J., *Wireless Personal Communication Systems*, Addison-Wesley, 1997.

[36] TIA Technical sub-committee TR-45.4, The cdma2000 ITU-RTT candidate submission, Document TR45.5.4/98.06.15.04, 2, June 1998.

Index